W9-BFA-150

SIMPLIFIED DESIGN OF STEEL STRUCTURES

Other titles in the Parker–Ambrose Series of Simplified Design Guides

Harry Parker and James Ambrose
Simplified Engineering for Architects and Builders, 7th Edition

James Ambrose
Study Guide for Simplified Engineering for Architects and Builders, 7th Edition

Harry Parker and James Ambrose
Simplified Design of Reinforced Concrete, 5th Edition

Harry Parker and James Ambrose
Simplified Design of Structural Wood, 4th Edition

Harry Parker and John W. MacGuire
Simplified Site Engineering for Architects and Builders

Harry Parker and James Ambrose
Simplified Mechanics and Strength of Materials, 4th Edition

James Ambrose
Simplified Design of Building Foundations, 2nd Edition

James Ambrose and Dimitry Vergun
Simplified Building Design for Wind and Earthquake Forces, 2nd Edition

SIMPLIFIED DESIGN OF STEEL STRUCTURES

Sixth Edition

THE LATE HARRY PARKER, M.S.
Formerly Professor of Architectural Construction
University of Pennsylvania

prepared by

JAMES AMBROSE, M.S.
Professor of Architecture
University of Southern California

A WILEY-INTERSCIENCE PUBLICATION

JOHN WILEY & SONS, INC.

New York · Chichester · Brisbane · Toronto · Singapore

Library of Congress Cataloging in Publication Data:

Parker, Harry, 1887–

Simplified design of steel structures / Harry Parker. — 6th ed. / prepared by James Ambrose.

p. cm.

"A Wiley-Interscience publication."

Rev. ed. of: Simplified design of structural steel. 5th ed. c1983.

1. Building, Iron and steel. 2. Steel, Structural. I. Ambrose, James E. II. Parker, Harry, 1887– Simplified design of structural steel. III. Title.

TA684.P33 1990

624.1'821—dc20 89-22631

ISBN 0-471-50539-0 CIP

Printed in the United States of America

10 9 8 7 6 5 4 3 2 1

CONTENTS

PREFACE TO THE SIXTH EDITION

The publication of yet another edition of Professor Parker's book testifies to its continued acceptance. While considerable revision of the work has occurred through the several editions following the first edition—which appeared almost 45 years ago—the general style and avowed purpose of the work remain essentially the same. The views and intensions expressed by Professor Parker in his preface to the first edition, which follows, are retained in the development of the work. This remains so despite the fact that this is the third edition prepared since Professor Parker's death.

The work presented in this edition retains the general body of content as contained in previous editions. Some materials considered to be less applicable to current construction practices have been deleted or reduced in scope. However, considerable new material has been added in this edition. This includes work relating to columns with bending, rigid frames, braced (trussed) frames, and composite construction. A major addition is the material in the new Chapter 12, consisting of discussions and illustrations of the design of structural systems for three example buildings.

In order to retain the compact size of the book, while incorporating considerable new material, most of the material presented in previous editions regarding basic mechanics and elementary analysis of beams has been omitted. It is assumed that readers will have been exposed to these basic introductory materials in other work. However, for the purpose of review, an abbreviated summary of basic concepts is presented in Appendix E.

New Appendix F contains study materials to assist readers in measuring their general comprehension of the topics developed in the individual chapters. While exercise problems are provided at the ends of book sections that contain computational work, to permit the reader to attempt similar work, the materials in Appendix F will permit better examination of basic concepts and meanings. Materials in the appendix consist of lists of significant words and terms and general questions for each of the book chapters. As all the study materials in the book are intended to provide assistance to the reader, complete answers to all exercise problems and general questions are provided.

As indicated in the title, the scope of this work is limited and does not purport to fully cover the topic of the design of steel structures. Still, as the diversity and complexity of the topic continues to grow, it becomes—if anything—increasingly necessary to reduce the most ordinary problems to simple terms, to permit the less fully trained to develop an understanding of basic issues and practical procedures for simple design problems.

I am once again grateful to the International Conference of Building Officials, the American Institute of Steel Construction, the Steel Deck Institute, and the Steel Joist Institute for their permissions to use materials from their publications. I am also highly indebted to the editors and production staff at John Wiley & Sons for their competent and consistent support. Finally, as most of my writing work is done in my home, I depend as always on the patient endurance and general encouragement of my family.

JAMES AMBROSE

Westlake Village, California
January 1990

PREFACE TO THE FIRST EDITION

"Simplified Design of Structural Steel" is the fourth of a series of elementary books dealing with the design of structural members used in the construction of buildings. The first volume, "Simplified Engineering for Architects and Builders," discussed rather briefly the design of structural members of timber, steel, and reinforced concrete. The primary objective of the first volume was to present the basic principles of design for younger men having no preliminary training in engineering. The book being elementary in character and the scope limited, many important subjects were necessarily omitted.

The present volume treats of the design of the most common structural steel members that occur in building construction. The solution of many structural problems is difficult and involved but it is surprising, on investigation, how readily many of the seemingly difficult problems may be solved. The author has endeavored to show how the application of the basic principles of mechanics simplifies the problems and leads directly to a solution. Using tables and formulas blindly is a dangerous procedure; they can only be used with safety when there is a clear understanding of the underlying principles upon which the tables or formulas are based. This book deals principally in the practical application of engineering principles and formulas and in the design of structural members. The derivations of the most commonly used formulas are given in order that the reader may comprehend fully why certain formulas are appropriate in the solution of specific problems.

In preparing material for this book the author has assumed that the reader is unfamiliar with the subject. Consequently the discussions advance by easy stages, beginning with problems relating to simple direct stresses and continuing to the more involved examples. Most of the fundamental principles of mechanics are reviewed and, in general, the only preparation needed is a knowledge of arithmetic and high school algebra.

The text has been arranged for use in the classroom as well as for home study. The tables essential to structural engineering have been included so that no additional reference books are required.

In addition to discussions and explanations of design procedure, it has been found that the solution of practical examples adds greatly to the value of a book of this character. Consequently, a great portion of the text consists of the solution of illustrative examples. The examples are followed by problems to be solved by the student.

The author is deeply grateful to the American Institute of Steel Construction for its kindness in granting permission to reproduce tables and data from its manual "Steel Construction." In general, the American Institute of Steel Construction specifications have been followed in the preparation of this book. Thanks are extended also to the American Welding Society for permission to reproduce data from its "Code for Arc and Gas Welding in Building Construction," and to The Lincoln Electric Company of Cleveland, Ohio, publishers of "Procedure Handbook of Arc Welding Design and Practice." This company has graciously permitted the reproduction of the data and design procedure from its excellent and comprehensive volume. The cooperation of the societies and company mentioned above is greatly appreciated;

without such cooperation a book of this character would not be possible.

The author proposes no new methods of design nor short cuts of questionable value. Instead, he has endeavored to present concise and clear explanations of the present-day design methods with the hope that the reader may obtain a foundation of sound principles of structural engineering.

HARRY PARKER

High Hollow, Southampton, Pa.
March 1945

SIMPLIFIED DESIGN OF STEEL STRUCTURES

1

INTRODUCTION

1.1 USE OF STEEL FOR BUILDING STRUCTURES

This book deals in general with the common uses of steel for the structures of buildings. If fully considered, this involves a considerable range of usage, as steel is used in one form or another in structures made from all the common materials: wood, concrete, masonry—and, of course, steel. Erection of frames of wood requires the use of nails, screws, bolts, anchorage devices, and various metal connectors—all ordinarily made of steel. Modern concrete and masonry construction is typically achieved with major use of steel reinforcement as well as various anchorage and attachment devices of steel. However, the topic of steel structures as developed here is limited essentially to situations in which steel is used as the major material for the primary structure. Exceptions to this are the use of steel spanning systems in combination with vertical bearing structures of concrete or masonry and the use of composite spanning elements of steel and concrete.

A major use of steel is that in which a primary frame is erected, consisting of linear elements of structural steel. The term "structural steel" is usually applied to structures in which the major elements are those produced by the process of hot-rolling, resulting in a linear element of constant cross section. Rolled sections—as produced in the United States—are ordinarily formed into shapes that conform with the standards established by the American Institute of Steel Construction and documented in its publication: *Manual of Steel Construction* (Ref. 1), referred to commonly, and hereinafter in this book, as the AISC Manual.

Another category of steel structures is one that utilizes elements of a size and weight generally one notch below the rolled products, although use may also be made of the lightest of the rolled shapes. Other than rolled shapes, the elements commonly used are those produced by cold-forming (stamping, folding, cold-rolling, etc.) of relatively thin sheets of steel. Roof and floor decking and some wall paneling is produced in this manner, and the products are often described as being of *formed steel.* Light framed elements, such as open web joists, may also use cold-formed elements for some parts.

A third category of usage for steel is that generally described by the term *miscellaneous metals*. This includes elements of the construction that are not parts of the primary structural system, but serve some secondary structural function, such as framing for curtain walls, suspended ceilings, door frames, and so on. As with the light steel structure, this construction may use the smaller rolled shapes or elements of cold-formed sheet steel.

Much of the material presented in this book deals with structural steel elements and systems—consisting of hot-rolled shapes. However, as mentioned in the Preface, the title of this edition has been changed to reflect the broader treatment of the subject, including major usage of cold-formed sheet steel products.

1.2 SOURCES OF DESIGN INFORMATION

The most widely used source of information for the design of steel structures is the AISC Manual. The Manual contains the *Specifi-*

cation for the Design, Fabrication, and Erection of Structural Steel for Buildings. This specification, in its several successive editions, has been generally adopted for reference by code-enforcing agencies in the United States. When referring to this document we call it the AISC Specification.

In addition to the specification and an accompanying extensive commentary, the AISC Manual contains essential data that pertain to the standard structural steel products used for the construction of building structures. There is also considerable general information and numerous design aids to assist the working structural designer. Although we have reprinted or adapted all of the information from the AISC Manual that is required for the work in this book, it is recommended that the serious reader acquire a personal copy of this highly useful reference.

In 1986 the AISC published the first edition of a new manual with the subtitle *Load and Resistance Factor Design.* This manual was intended to facilitate the implementation of the new AISC specification, titled Load and Resistance Factor Design Specification for Structural Steel Buildings, based on reliability theory. Also included in this manual was an expansion of the properties tables for W sections to include the so-called "jumbo shapes," consisting of additional, heavier sections in the 18-, 21-, 24-, 27-, 33-, and 36-in. groups. Another addition to the W shapes was a group of nominal 40-in. deep shapes.

In 1989 the AISC published the ninth edition of the AISC Manual with the subtitle *Allowable Stress Design,* consisting essentially of an update of the 8th edition, first published in 1980. The new 9th edition also contains the jumbo W shapes and, in addition, a new group of nominal 44-in. deep W shapes.

While the 1986 and 1989 editions of the AISC Manual represent the latest design standards and industry data, the work presented in this book generally involves data and specification requirements that remain the same as in the 1980 (8th) edition. To be accurate in the references for this book, therefore, we have used the 8th edition as the source for all the work shown here. Any readers who use this book as a guide for actual design work should ascertain the specific reference used for steel design by the administrative building code with jurisdiction for the work.

It should be noted that the jumbo and deeper W shapes are generally not available from producers in the United States.

Although the AISC is the principal service organizations in the area of steel construction, many other industrial and professional organizations provide material for the designer. The American Society for Testing and Materials (ASTM) establishes widely used standard specifications for types of steel, for welding and connector materials, and for various production and fabrication processes. Standard grades of steel and other materials are commonly referred to by short versions of their ASTM designation codes. Other organizations of note are the Steel Deck Institute (SDI), the Steel Joist Institute (SJI), and the American Iron and Steel Institute (AISI). Some of the materials distributed by these organizations are listed in the references at the back of this book.

Readers who wish to pursue topics in the area of design of steel structures beyond the level of development presented in this book are advised to seek the latest edition of one of the books used as a text on the topic in schools of civil engineering. For a general reference an excellent book is *Steel Structures: Design and Behavior* by Salmon and Johnson (Ref. 3).

1.3 UNITS OF MEASUREMENT

At the time of preparation of this edition, the building industry in the United States is still in a state of confused transition from the use of English units (feet, pounds, etc.) to the new metric-based system referred to as the SI units (for Système International). Although a complete phase-over to SI units seems inevitable, at the time of this writing, construction-materials and products suppliers in the United States are still resisting it. Consequently, the AISC Manual and most building codes and other widely used references are still in the old units. (The old system is now more appropriately called the U.S. system because England no longer uses it!) Although it results in some degree of clumsiness in the work, we have chosen to give the data and computations in this book in both units as much as is practicable. The technique is generally to perform the work in U.S. units and immediately

TABLE 1.1 Units of Measurement: U.S. System

Name of Unit	Abbreviation	Use
Length		
Foot	ft	large dimensions, building plans, beam spans
Inch	in.	small dimensions, size of member cross sections
Area		
Square feet	ft^2	large areas
Square inches	$in.^2$	small areas, properties of cross sections
Volume		
Cubic feet	ft^3	large volumes, quantities of materials
Cubic inches	$in.^3$	small volumes
Force, mass		
Pound	lb	specific weight, force, load
Kip	k	1000 pounds
Pounds per foot	lb/ft	linear load (as on a beam)
Kips per foot	k/ft	linear load (as on a beam)
Pounds per square foot	lb/ft^2, psf	distributed load on a surface
Kips per square foot	k/ft^2, ksf	distributed load on a surface
Pounds per cubic foot	lb/ft^3, pcf	relative density, weight
Moment		
Foot-pounds	ft-lb	rotational or bending moment
Inch-pounds	in.-lb	rotational or bending moment
Kip-feet	k-ft	rotational or bending moment
Kip-inches	k-in.	rotational or bending moment
Stress		
Pounds per square foot	lb/ft^2, psf	soil pressure
Pounds per square inch	$lb/in.^2$, psi	stresses in structures
Kips per square foot	k/ft^2, ksf	soil pressure
Kips per square inch	$k/in.^2$, ksi	stresses in structures
Temperature		
Degree Fahrenheit	°F	temperature

follow it with the equivalent work in SI units enclosed in brackets [thus] for separation and identity.

Table 1.1 lists the standard units of measurement in the U.S. system with the abbreviations used in this work and a description of the type of the use in structural work. In similar form, Table 1.2 gives the corresponding units in the SI system. The conver-

TABLE 1.2 Units of Measurement: SI System

Name of Unit	Abbreviation	Use
Length		
Meter	m	large dimensions, building plans, beam spans
Millimeter	mm	small dimensions, size of member cross sections
Area		
Square meters	m^2	large areas
Square millimeters	mm^2	small areas, properties of cross sections
Volume		
Cubic meters	m^3	large volumes
Cubic millimeters	mm^3	small volumes
Mass		
Kilogram	kg	mass of materials (equivalent to weight in U.S. system)
Kilograms per cubic meter	kg/m^3	density
Force (load on structures)		
Newton	N	force or load
Kilonewton	kN	1000 newtons
Stress		
Pascal	Pa	stress or pressure (1 pascal = 1 N/m^2)
Kilopascal	kPa	1000 pascals
Megapascal	MPa	1,000,000 pascals
Gigapascal	GPa	1,000,000,000 pascals
Temperature		
Degree Celsius	°C	temperature

TABLE 1.3 Factors for Conversion of Units

To Convert from U.S. Units to SI Units Multiply by	U.S. Unit	SI Unit	To Convert from SI Units to U.S. Units Multiply by
25.4	in.	mm	0.03937
0.3048	ft	m	3.281
645.2	in.^{2}	mm^2	1.550×10^{-3}
16.39×10^3	in.^{3}	mm^3	61.02×10^{-6}
416.2×10^3	in.^{4}	mm^4	2.403×10^{-6}
0.09290	ft^2	m^2	10.76
0.02832	ft^3	m^3	35.31
0.4536	lb (mass)	kg	2.205
4.448	lb (force)	N	0.2248
4.448	kip (force)	kN	0.2248
1.356	ft-lb (moment)	N-m	0.7376
1.356	kip-ft (moment)	kN-m	0.7376
1.488	lb/ft (mass)	kg/m	0.6720
14.59	lb/ft (load)	N/m	0.06853
14.59	kips/ft (load)	kN/m	0.06853
6.895	psi (stress)	kPa	0.1450
6.895	ksi (stress)	MPa	0.1450
0.04788	psf (load or pressure)	kPa	20.93
47.88	ksf (load or pressure)	kPa	0.02093
$0.566 \times (°F - 32)$	°F	°C	$(1.8 \times °C) + 32$

sion units used in shifting from one system to the other are given in Table 1.3.

For some of the work in this book, the units of measurement are not significant. What is required in such cases is simply to find a numerical answer. The visualization of the problem, the manipulation of the mathematical processes for the solution, and the quantification of the answer are not related to the specific units—only to their relative values. In such situations we have occasionally chosen not to present the work in dual units, to provide a less confusing illustration for the reader. Although this procedure may be allowed for the learning exercises in this book, the structural designer is generally advised to develop the habit of always indicating the units for any numerical answers in structural computations.

1.4 SYMBOLS

The following "shorthand" symbols are frequently used:

Symbol	Reading
$>$	is greater than
$<$	is less than
$\geqslant$	equal to or greater than
$\leqslant$	equal to or less than
$6'$	six feet
$6''$	six inches
Σ	the sum of
ΔL	change in L

1.5 NOMENCLATURE

The standard symbols (called the general nomenclature) used in steel design work are those established in the AISC Specification. Although an attempt is being made to standardize symbols for structural work, some specialized nomenclature remains for each of the areas of design: soils, wood, steel, concrete, and masonry. The following is a list of symbols used in this book; it is an abridged version of a more extensive list in the AISC Manual:

A Cross-sectional area (sq in.)
Gross area of an axially loaded compression member (sq in.)

A_e Effective net area of an axially loaded tension member (sq. in.)

A_n Net area of an axially loaded tension member (sq in.)

C_c Column slenderness ratio separating elastic and inelastic buckling

E Modulus of elasticity of steel (29,000 ksi)

F_a Axial compressive stress permitted in a prismatic member in the absence of bending moment (ksi)

F_b Bending stress permitted in a prismatic member in the absence of axial force (ksi)

F'_e Euler stress for a prismatic member divided by factor of safety (ksi)

F_p Allowable bearing stress (ksi)

F_t Allowable axial tensile stress (ksi)

F_u Specified minimum tensile strength of the type of steel or fastener being used (ksi)

F_v Allowable shear stress (ksi)

F_y Specified minimum yield stress of the type of steel being used (ksi). As used in this Manual, "yield stress' denotes either the specified minimum yield point (for those steels that have a yield point) or specified minimum yield strength (for those steels that have no yield point)

F'_y The theoretical maximum yield stress (ksi) based on the width/thickness ratio of one-half the unstiffened compression flange, beyond which a particular shape is not "compact."

$$= \left[\frac{65}{b_f/2t_f}\right]^2$$

F'''_y The theoretical maximum yield stress (ksi) based on the depth/thickness ratio of the web below which a particular shape may be considered "compact" for any condition of combined bending and axial stresses.

$$= \left[\frac{257}{d/t_w}\right]^2$$

I Moment of inertia of a section (in.^{4})

I_x Moment of inertia of a section about the X–X axis (in.^{4})

I_y Moment of inertia of a section about the Y–Y axis (in.^{4})

J Torsional constant of a cross section (in.^{4})

K Effective length factor for a prismatic member

L	Span length (ft)
	Length of connection angles (in.)
L_c	Maximum unbraced length of the compression flange at which the allowable bending stress may be taken at $0.66F_y$ or as determined by AISC Specification formulas (ft)
	Unsupported length of a column section (ft)
L_u	Maximum unbraced length of the compression flange at which the allowable bending stress may be taken at $0.6F_y$ (ft)
M	Moment (k-ft)
M_D	Moment produced by dead load
M_L	Moment produced by live load
M_p	Plastic moment (k-ft)
M_R	Beam resisting moment (k-ft)
N	Length of base plate (in.)
	Length of bearing of applied load (in.)
N_e	Length at end bearing to develop maximum web shear (in.)
P	Applied load (kips)
	Force transmitted by a fastener (kips)
P_e	Euler buckling load (kips)
Q_s	Axial stress reduction factor where width/thickness ratio of unstiffened elements exceeds limiting value
R	Maximum end reaction for $3\frac{1}{2}$ in. of bearing (kips)
	Reaction or concentrated load applied to beam or girder (kips)
	Radius (in.)
R_i	Increase in reaction (R) in kips for each additional inch of bearing
S	Elastic section modules (in.^{3})
S_x	Elastic section modulus about the $X–X$ (major) axis (in.^{3})
S_y	Elastic section modulus about the $Y–Y$ (minor) axis (in.^{3})
V	Maximum permissible web shear (kips)
	Statical shear on beam (kips)
Z	Plastic section modulus (in.^{3})

Z_x Plastic section modulus with respect to the major (X–X) axis (in.^{3})

Z_y Plastic section modulus with respect to the minor (Y–Y) axis (in.^{3})

b_f Flange width of rolled beam or plate girder (in.)

d Depth of column, beam, or girder (in.)
Nominal diameter of a fastener (in.)

e_o Distance from outside face of web to the shear center of a channel section (in.)

f_a Computed axial stress (ksi)

f_b Computed bending stress (ksi)

f'_c Specified compression strength of concrete at 28 days (ksi)

f_p Actual bearing pressure on support (ksi)

f_t Computed tensile stress (ksi)

f_v Computed shear stress (ksi)

g Transverse spacing located fastener gage lines (in.)

k Distance from outer face of flange to web toe of fillet of rolled shape or equivalent distance on welded section (in.)

l For beams, distance between cross sections braced against twist or lateral displacement of the compression flange (in.)
For columns, actual unbraced length of member (in.)
Length of weld (in.)

m Cantilever dimension of base plate (in.)

n Number of fasteners in one vertical row
Cantilever dimension of base plate (in.)

r Governing radius of gyration (in.)

r_x Radius of gyration with respect to the X–X axis (in.)

r_y Radius of gyration with respect to the Y–Y axis (in.)

s Longitudinal center-to-center spacing (pitch) of any two consecutive holes (in.)

t Girder, beam, or column web thickness (in.)
Thickness of a connected part (in.)
Wall thickness of a tubular member (in.)
Angle thickness (in.)

t_f Flange thickness (in.)

t_w Web thickness (in.)
x Subscript relating symbol to strong axis bending
y Subscript relating symbol to weak axis bending
Δ Beam deflection (in.).

1.6 DESIGN PROCESS AND METHODS

The computational work presented in this book is mostly simple and can be performed easily with a pocket calculator. Structural computations can for the most part be rounded off. Accuracy beyond the third place is seldom significant, and many results presented in the work here have been so rounded off. In lengthy computations, however, it is advisable to carry one or more places of accuracy beyond that desired for the final answer. In the most part, the work shown here was performed on an eight-digit pocket calculator.

In most professional design firms computations done for final design work are performed with computer-aided procedures. Many standard programs are available for routine work and many of the necessary data are accessible from computer-retrievable sources. In most cases this does not essentially change the theory or the basic relationships or the required types of investigation, but obviously affects the methodology for arriving at design decisions.

Use of computer-aided methods permits quicker accomplishment of tedious or complex investigations and easier design involving many interactive requirements. Design standards and building codes are developing requirements and procedures that pretty much imply use of computer-aided methods for practical design utilization. These procedures are built into standard computer programs, so that users are not necessarily aware of the difficulty or laborious nature of the computational work.

The value of computer-aided methods increases with the level of complexity or sheer length of the computational work. For the most part, the work shown here is hardly worth doing with a computer, and in many cases in practice is not—at least for preliminary design work. Much use is made of tabular and graphic aids, and the shortest route to a reliable answer is usually visualized as desirable. However, our purpose here is basically instructional, and the hand operation of the full computational process allows for more involvement in the problems and their operations. We therefore limit the work to the situations that permit relatively simple hand computation and give deliberately slow, step-by-step illustration for most work. This snail's pace of work would not be tolerated in a professional design office, of course.

2

STEEL: MATERIALS AND PRODUCTS

As a material, steel is subject to considerable variation in its basic properties and is used to form a wide range of products. The work in this book deals with the ordinary usage of commercially produced products for building structures: rolled structural shapes, fabricated structural shapes, steel connectors, cold-formed elements, and prefabricated trusses.

2.1 PROPERTIES OF STEEL

The strength, hardness, corrosion resistance, and some other properties of steel can be varied through a considerable range by changes in the production processes. Literally hundreds of different steels are produced, although only a few standard products are used for the majority of the elements of building structures. Working and forming processes, such as rolling, drawing, machining, and forging may also alter some properties. Certain properties, such as density (unit weight), stiffness (modulus of elasticity), thermal expansion, and fire resistance tend to remain constant for all steels.

Basic structural properties, such as strength, stiffness, ductility, and brittleness, can be interpreted from load tests. Figure 2.1 displays typical forms of curves that are obtained by plotting stress and strain values from such tests. An important characteristic of many structural steels is the plastic deformation (yield) phenomenon. This is demonstrated by curve 1 in Fig. 2.1. For steels with this character there are two different stress values of significance: the yield limit and the ultimate failure limit.

Generally, the higher the yield limit, the less the degree of ductility. Curve 1 in Fig. 2.1 is representative of ordinary structural steel (ASTM A36) and curve 2 indicates the typical effect as the yield strength is raised a significant amount. Eventually, the significance of the yield phenomenon becomes negligible when the yield strength approaches as much as three times the yield of ordinary steel.

Some of the highest-strength steels are produced only in thin sheet or drawn wire forms. Bridge strand is made from wire with strength as high as 300,000 psi. At this level yield is almost nonexistent and the wires approach the brittleness of glass rods.

For economical use of the expensive material, steel structures are generally composed of elements with quite relatively thin parts. This results in many situations in which the limiting strength of elements in bending, compression, and shear is determined by buckling, rather than the stress limits of the material. Since buckling is a function of stiffness (modulus of elasticity) of the material, and since this property remains the same for all steels, there is limited opportunity to make effective use of higher-strength steels in many situations. The grade of steel most commonly used is to some extent one that has the optimal effective strength for most typical tasks.

For various applications, other properties will be significant. Hardness effects the ease with which cutting, drilling, planing, and other working can be done. For welded connections there is a property of weldability of the base material that must be considered. Resistance to rusting is normally low, but can be enhanced by various materials added to the steel, producing various types

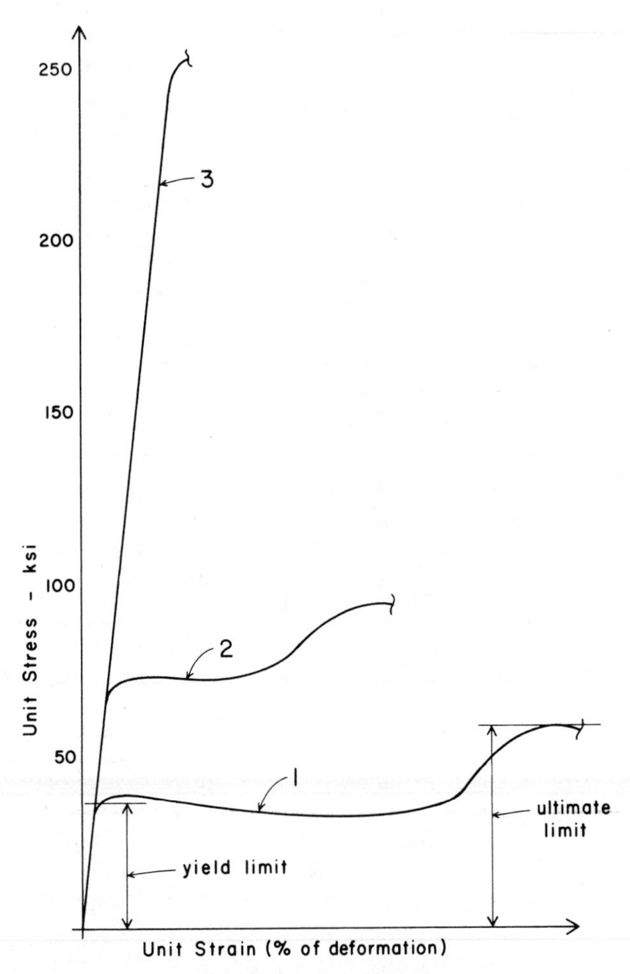

FIGURE 2.1 Form of stress-strain response of structural steel: (1) ordinary structural grade, A36; (2) higher grades used for rolled shapes; (3) high-strength strand.

of special steels, such as stainless steel and the so-called "rusting steel" that rusts at a very slow rate.

Since many structural elements are produced as some manufacturer's product line, choices of materials are often mostly out of the hands of individual building designers. The proper steel for the task—on the basis of many properties—is determined as part of the product design.

Steel that meets the requirements of the American Society for Testing and Materials (abbreviation: ASTM) Specification A36 is the grade of structural steel commonly used to produce rolled steel elements for building construction. It must have an ultimate tensile strength of 58 to 80 ksi and a minimum yield point of 36 ksi. It may be used for bolted, riveted, or welded fabrication. The current edition of the AISC Manual (Ref. 1) lists the following steels as those available for the production of rolled structural shapes, plates, and bars:

Structural Steel, ASTM A36

Structural Steel with 42,000 psi Minimum Yield Point, ASTM A529

High-Strength Low-Alloy Structural Manganese Vanadium Steel, ASTM A441

High-Strength Low-Alloy Columbium-Vanadium Steel of Structural Quality, ASTM A572

High-Strength Low-Alloy Structural Steel, ASTM A242

High-Strength Low-Alloy Structural Steel with 50,000 Minimum Yield Point to 4-in. Thick, ASTM A588

Quenched and Tempered Alloy Steel, A514 (for plates only)

Table 2.1 lists the critical stress properties of these steels and the various rolled products that are produced from them. Table 2.2 gives the groupings of rolled shapes referred to in Table 2.1. Of primary concern is the minimum yield stress, designated F_y, on which most allowable design stresses are based. The other limiting stress is the ultimate tensile stress, designated F_u (see Fig. 2.1). A few design stresses, mostly those relating to connection design, are based on the ultimate tensile stress. For some

TABLE 2.1 ASTM Structural Steel Grades for Rolled Products[a]

Steel Type	ASTM Designation	F_y Minimum Yield Stress (ksi)	F_u Tensile Stress[c] (ksi)	Shapes[b] Group per ASTM A6					Plates and Bars										
				1[d]	2	3	4	5	To 1/2″ Incl.	Over 1/2″ to 3/4″ Incl.	Over 3/4″ to 1 1/4″ Incl.	Over 1 1/4″ to 1 1/2″ Incl.	Over 1 1/2″ to 2″ Incl.	Over 2″ to 2 1/2″ Incl.	Over 2 1/2″ to 4″ Incl.	Over 4″ to 5″ Incl.	Over 5″ to 6″ Incl.	Over 6″ to 8″ Incl.	Over 8″
Carbon	A36	32	58-80																■
Carbon	A36	36	58-80[e]	■	■	■	■	■	■	■	■	■	■	■	■	■	■	■	
Carbon	A529	42	60-85	■					■										
High-Strength Low-Alloy	A441	40	60													■	■	■	
High-Strength Low-Alloy	A441	42	63				■	■					■	■	■				
High-Strength Low-Alloy	A441	46	67			■					■	■							
High-Strength Low-Alloy	A441	50	70	■	■				■	■									
High-Strength Low-Alloy	A572—Grade 42	42	60	■	■	■	■	■		■	■	■	■	■	■	■	■		
High-Strength Low-Alloy	A572—Grade 50	50	65	■	■	■	■	■	■	■	■	■	■						
High-Strength Low-Alloy	A572—Grade 60	60	75	■	■				■	■	■								
High-Strength Low-Alloy	A572—Grade 65	65	80	■					■	■	■								
Corrosion-Resistant High-Strength Low-Alloy	A242	42	63				■	■					■	■	■				
Corrosion-Resistant High-Strength Low-Alloy	A242	46	67			■					■	■							
Corrosion-Resistant High-Strength Low-Alloy	A242	50	70	■	■				■	■									
Corrosion-Resistant High-Strength Low-Alloy	A588	42	63														■	■	
Corrosion-Resistant High-Strength Low-Alloy	A588	46	67													■			
Corrosion-Resistant High-Strength Low-Alloy	A588	50	70	■	■	■	■	■	■	■	■	■	■	■	■				
Quenched & Tempered Alloy	A514[f]	90	100-130												■	■	■		
Quenched & Tempered Alloy	A514[f]	100	110-130						■	■	■	■	■	■					

[a] Shaded portion in table indicates availability of products for each grade.
[b] See Table 2.2.
[c] Minimum unless a range is shown.
[d] Includes bar-size shapes.
[e] For shapes over 426 lb/ft minimum of 58 ksi only applies.
[f] Plates only.

TABLE 2.2 Structural Shape Size Groupings for Tensile Property Classification

Structural Shape	Group 1	Group 2	Group 3	Group 4	Group 5
W Shapes	W 24x55, 62 W 21x44 to 57 incl W 18x35 to 71 incl W 16x26 to 57 incl W 14x22 to 53 incl W 12x14 to 58 incl W 10x12 to 45 incl W 8x10 to 48 incl W 6x9 to 25 incl W 5x16, 19 W 4x13	W 36x135 to 210 incl W 33x118 to 152 incl W 30x99 to 211 incl W 27x84 to 178 incl W 24x68 to 162 incl W 21x62 to 147 incl W 18x76 to 119 incl W 16x67 to 100 incl W 14x61 to 132 incl W 12x65 to 106 incl W 10x49 to 112 incl W 8x58, 67	W 36x230 to 300 incl W 33x201 to 241 incl W 14x145 to 211 incl W 12x120 to 190 incl	W 14x233 to 550 incl W 12x210 to 336 incl	W 14x605 to 730 incl
M Shapes	to 20 lb/ft incl				
S Shapes	to 35 lb/ft incl	over 35 lb/ft			
HP Shapes		to 102 lb/ft incl	over 102 lb/ft		
American Standard Channels (C)	to 20.7 lb/ft incl	over 20.7 lb/ft			
Miscellaneous Channels (MC)	to 28.5 lb/ft incl	over 28.5 lb/ft			
Angles (L), Structural & Bar-Size	to ½ in. incl	over ½ to ¾ in. incl	over ¾ in.		

Notes: Structural tees from W, M and S shapes fall in the same group as the structural shape from which they are cut.

Group 4 and Group 5 shapes are generally contemplated for application as compression members. When used in other applications or when subject to welding or thermal cutting, the material specification should be reviewed to determine if it adequately covers the properties and quality appropriate for the particular application. Where warranted, the use of killed steel or special metallurgical requirements should be considered.

grades the ultimate stress is given as a range rather than a single value, in which case it is advisable to use the lower value for design unless a higher value can be verified by a specific supplier for a particular rolled product.

Prior to 1963 a steel designated ASTM A7 was the basic product for structural purposes. It had a yield point of 33 ksi and was used primarily for riveted fabrication. With the increasing demand for bolted and welded construction A7 steel became less useful, and in a short time A36 steel was the material of choice for the majority of structural products. When a slightly higher strength is desired A529 steel is sometimes used, although its availability is limited to lighter-weight rolled shapes and plates and bars up to 0.5 in. thick.

These steels have yield points higher than those of the carbon steels used for the same products; thus the design allowable stresses will be higher and increased strength is possible for elements of the same size. A441 is a high-strength steel intended for use primarily in welded construction. A572 is suitable for bolting, riveting, or welding. A514 is used exclusively for plates in built-up sections for large structures: girders, piers, arches, and so on.

Because of their chemical composition, these steels exhibit higher degrees of resistance to atmospheric corrosion. For A242 this resistance is four to six times that of structural carbon steel; for A588 the resistance is about eight times that of A441. Several brands of A588 steel are available; each represents the proprietary product of a different manufacturer. These steels are used in the bare (uncoated) condition for exposed construction. Exposure to normal atmosphere causes formation of an oxide on the surface which adheres tightly and protects the steel from further oxidation.

Allowable Stresses for Structural Steel

For structural steel the AISC Specification expresses the allowable unit stresses in terms of some percent of the yield stress F_y or the ultimate stress F_u. Selected allowable unit stresses used in design are listed in Table 2.3. Specific values are given for ASTM A36 steel, with values of 36 ksi for F_y and 58 ksi for F_u. This is not

TABLE 2.3 Allowable Unit Stresses for Structural Steel: ASTM A36[a]

Type of Stress and Conditions	See Discussion in This Book in:	Stress Designation	AISC Specification	Allowable Stress ksi	Allowable Stress MPa
Tension					
1. On the gross (unreduced) area	Sec. 10.3	F_t	$0.60\ F_y$	22	150
2. On the effective net area, except at pinholes	Sec. 10.3		$0.50\ F_u$	29	125
3. Threaded rods on net area at thread			$0.33\ F_u$	19	80
Compression Shear	Chapter 6	F_a	See discussion		
1. Except at reduced sections	Sec. 5.3	F_v	$0.40\ F_y$	14.5	100
2. At reduced sections	Sec. 10.4		$0.30\ F_u$	17.4	120
Bending					
1. Tension and compression on extreme fibers of compact members braced laterally, symmetrical about and loaded in the plane of their minor axis	Sec. 5.2	F_b	$0.66\ F_y$	24	165
2. Tension and compression on extreme fibers of other rolled shapes braced laterally			$0.60\ F_y$	22	150
3. Tension and compression on extreme fibers of solid round and square bars, on solid rectangular sections bent on their weak axis, on qualified doubly symmetrical I and H shapes bent about their minor axis			$0.75\ F_y$	27	188
Bearing					
1. On contact area of milled surfaces		F_p	$0.90\ F_y$	32.4	225
2. On projected area of bolts and rivets in shear connections	Chapter 10		$1.50\ F_u$	87	600

[a] $F_y = 36$ ksi; assume that $F_u = 58$ ksi: some table values rounded off as permitted in the AISC Manual (Ref. 1). For SI units $F_y = 250$ MPa, $F_u = 400$ MPa.

a complete list, but it generally includes the stresses used in the examples in this book. Reference is made to the more complete descriptions in the AISC Specification which is included in the AISC Manual. There are in many cases a number of qualifying conditions, some of which are discussed in other portions of this book. Table 2.3 gives the location of some of these discussions.

Steel used for other purposes than the production of rolled products generally conforms to standards developed for the specific product. This is especially true for steel connectors, wire, cast and forged elements, and very high strength steels produced in sheet, bar, and rod form for fabricated products. The properties and design stresses for some of these product applications are discussed in other sections of this book. Standards used typically conform to those established by industry-wide organizations, such as the Steel Joist Institute (SJI) and the Steel Deck Institute (SDI). In some cases, larger fabricated products make use of ordinary rolled products, produced from A36 steel or other grades of steel from which hot-rolled products can be obtained.

2.2 TYPES OF STEEL PRODUCTS

Steel itself is formless, coming basically in the form of a molten material or a softened lump. The structural products produced derive their basic forms from the general potentialities and limitations of the industrial processes of forming and fabricating. Standard raw stock elements—deriving from the various production processes—are the following:

1. *Rolled Shapes.* These are formed by squeezing the heat-softened steel repeatedly through a set of rollers that shape it into a linear element with a constant cross section. Simple forms of round rods and flat bars, strips, plates, and sheets are formed, as well as more complex shapes of I, H, T, L, U, C, and Z. Special shapes, such as rails or sheet piling, can also be formed in this manner.
2. *Wire.* This is formed by pulling (called drawing) the steel through a small opening.
3. *Extrusion.* This is similar to drawing, although the sections produced are other than simple round shapes. This process is not much used for steel products of the sizes used for buildings.
4. *Casting.* This is done by pouring the molten steel into a form (mold). This is limited to objects of a three-dimen-

sional form, and is also not common for building construction elements.

5. *Forging*. This consists of pounding the softened steel into a mold until it takes the shape of the mold. This is preferred to casting because of the effects of the working on the properties of the finished material.

The raw stock steel elements produced by the basic forming processes may be reworked by various means, such as the following:

1. *Cutting*. Shearing, sawing, punching, or flame cutting can be used to trim and shape specific forms.
2. *Machining*. This may consist of drilling, planing, grinding, routing, or turning on a lathe.
3. *Bending*. Sheets, plates, or linear elements may be bent if made from steel with a ductile character.
4. *Stamping*. This is similar to forging; in this case sheet steel is punched into a mold that forms it into some three-dimensional shape, such as hemisphere.
5. *Rerolling*. This consists of reworking a linear element into a curved form (arched) or of forming a sheet or flat strip into a formed cross section.

Finally, raw stock or reformed elements can be assembled by various means into objects of multiple parts, such as a manufactured truss or a prefabricated wall panel. Basic means of assemblage include the following:

1. *Fitting*. Threaded parts may be screwed together or various interlocking techniques may be used, such as the tongue-and-groove joint or the bayonet twist lock.
2. *Friction*. Clamping, wedging, or squeezing with high-tensile bolts may be used to resist the sliding of parts in surface contact.
3. *Pinning*. Overlapping flat elements may have matching holes through which a pin-type device (bolt, rivet, or actual

pin) is placed to prevent slipping of the parts at the contact face.

4. *Nailing, Screwing.* Thin elements—mostly with some preformed holes—may be attached by nails or screws.
5. *Welding.* Gas or electric arc welding may be used to produce a bonded connection, achieved partly by melting the contacting elements together at the contact point.
6. *Adhesive Bonding.* This usually consists of some form of chemical bonding that results in some fusion of the materials of the connected parts.

We are dealing here with industrial processes which at any given time relate to the state of development of the technology, the availability of facilities, the existence of the necessary craft, and competition with other materials and products.

2.3 ROLLED STRUCTURAL SHAPES

The products of the steel rolling mills used as beams, columns, and other structural members are known as *sections* or *shapes* and their designations are related to the profiles of their cross sections. American standard I-beams (Fig. 2.2*a*) were the first beam sections rolled in the United States and are currently produced in sizes of 3 to 24 in. in depth. The wide-flange shapes (Fig. 2.2*b*) are a modification of the I cross section and are characterized by parallel flange surfaces as contrasted with the tapered inside flange surfaces of standard I-beams; they are available in depths of 4 to 44 in. In addition to the standard I and wide-flange sections, the structural steel shapes most commonly used in building construction are channels, angles, tees, plates, and bars. The tables in Appendix A list the dimensions, weights, and various properties of some of these shapes. Complete tables of structural shapes are given in the AISC Manual (Ref. 1).

Wide-Flange Shapes

In general, wide-flange shapes have greater flange widths and relatively thinner webs than standard I-beams; and, as noted

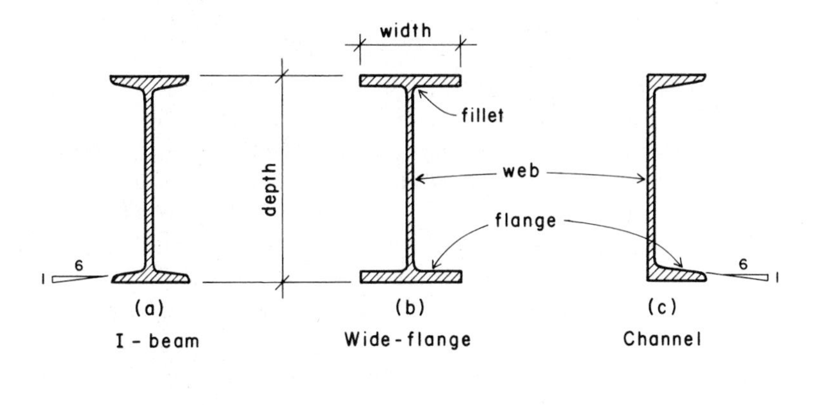

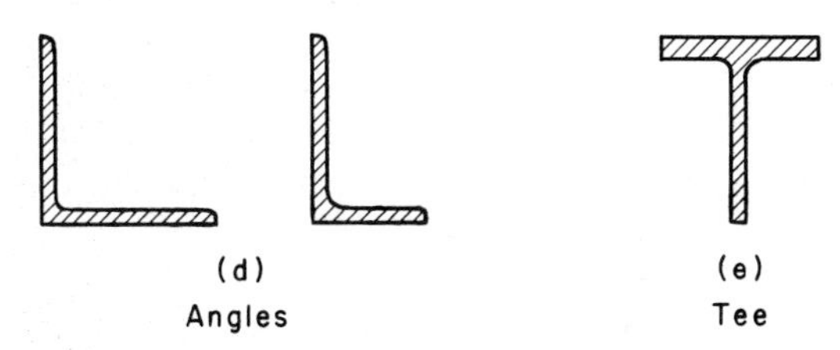

FIGURE 2.2 Rolled structural shapes.

above, the inner faces of the flanges are parallel to the outer faces. These sections are identified by the alphabetical symbol W, followed by the *nominal* depth in inches and the weight in pounds per linear foot. Thus the designation W 12 × 26 indicates a wide-flange shape of nominal-12-in. depth, weighing 26 lb per linear foot.

The actual depths of wide-flange shapes vary within the nominal depth groupings. By reference to Appendix Table A.1, it is found that a W 12 × 26 has an actual depth of 12.22 in., whereas the depth of a W 12 × 35 is 12.50 in. This is a result of the rolling process during manufacture in which the cross-sectional areas of wide-flange shapes are increased by spreading the rolls both vertically and horizontally. The additional material is thereby added to the cross section by increasing flange and web thickness as well as flange width (Fig. 2.2*b*). The resulting higher percentage of material in the flanges makes wide-flange shapes more efficient structurally than standard I-beams. A wide variety of weights is available within each nominal depth group.

In addition to shapes with profiles similar to the W 12 × 26, which has a flange width of 6.490 in., many wide-flange shapes are rolled with flange widths approximately equal to their depths. The resulting H configurations of these cross sections are much more suitable for use as columns than the I profiles. By reference to Appendix Table A.1 it is found that the following shapes, among others, fall into this category: W 14 × 90, W 12 × 65, W 10 × 60, and W 8 × 40. It is recommended that the reader compare these shapes with others listed in their respective nominal depth groups in order to become familiar with the variety of geometrical relationships.

Standard I-Beams

American standard I-beams are identified by the alphabetical symbol S, the designation S 12 × 35 indicating a standard shape 12 in. deep weighing 35 lb per linear foot. Unlike wide-flange sections, standard I-beams in a given depth group have uniform depths, and shapes of greater cross-sectional area are made by spreading the rolls in one direction only. Thus the depth remains constant, whereas the width of flange and thickness of web are increased.

All standard I-beams have a slope on the inside faces of the flanges of $16\frac{2}{3}\%$, or 1 in 6. In general, standard I-beams are not so efficient structurally as wide-flange sections and consequently are not so widely used. Also, the variety available is not nearly so large as that for wide-flange shapes. Characteristics that may favor the use of American standard I-beams in any particular situation are constant depth, narrow flanges, and thicker webs.

Standard Channels

The profile of an American standard channel is shown in Fig. 2.2*c*. These shapes are identified by the alphabetical symbol C. The designation C 10 × 20 indicates a standard channel 10 in.

deep and weighing 20 lb per linear foot. Appendix Table A.2 shows that this section has an area of 5.88 in.^{2}, a flange width of 2.739 in., and a web thickness of 0.379 in. Like the standard I-beams, the depth of a particular group remains constant and the cross-sectional area is increased by spreading the rolls to increase flange width and web thickness. Because of their tendency to buckle when used independently as beams or columns, channels require lateral support or bracing. They are generally used as elements of built-up sections such as columns and lintels. However, the absence of a flange on one side makes channels particularly suitable for framing around floor openings.

Angles

Structural angles are rolled sections in the shape of the letter L. Appendix Table A.3 gives dimensions, weights, and other properties of equal and unequal leg angles. Both legs of an angle have the same thickness.

Angles are designated by the alphabetical symbol L, followed by the dimensions of the legs and their thickness. Thus the designation L $4 \times 4 \times \frac{1}{2}$ indicates an equal leg angle with 4-in. legs, $\frac{1}{2}$ in. thick. By reference to Appendix Table A.3 it is found that this section weighs 12.8 lb per linear foot and has an area of 3.75 in.^{2}. Similarly, the designation L $5 \times 3\frac{1}{2} \times \frac{1}{2}$ indicates an unequal leg angle with one 5-in. and one $3\frac{1}{2}$-in. leg, both $\frac{1}{2}$ in. thick. Appendix Table A.3 shows that this angle weighs 13.6 lb per linear foot and has an area of 4 in.^{2}. To change the weight and area of an angle of a given leg length the thickness of each leg is increased the same amount. Thus if the leg thickness of the L5 $\times 3\frac{1}{2} \times \frac{1}{2}$ is increased to $\frac{5}{8}$ in., Appendix Table A.3 shows that the resulting L $5 \times 3\frac{1}{2} \times \frac{5}{8}$ has a weight of 16.8 lb per linear foot and an area of 4.92 in.^{2}. It should be noted that this method of spreading the rolls changes the leg lengths slightly.

Single angles are often used as lintels and pairs of angles, as members of light steel trusses. Angles were formerly used as elements of built-up sections such as plate girders and heavy columns, but the advent of the heavier wide-flange shapes has largely eliminated their usefulness for this purpose. Short lengths of angles are common connecting members for beams and columns.

Structural Tees

A structural tee is made by splitting the web of a wide-flange shape (Fig. 2.2*e*) or a standard I-beam. The cut, normally made along the center of the web, produces tees with a stem depth equal to half the depth of the original section. Structural tees cut from wide-flange shapes are identified by the symbol WT; those cut from standard I shapes, by ST. The designation WT 6 × 53 indicates a structural tee with a 6-in. depth and a weight of 53 lb per linear foot. This shape is produced by splitting a W 12 × 106 shape. Similarly, ST 9 × 35 designates a structural tee 9 in. deep, weighing 35 lb per linear foot and cut from a S 18 × 70. Tables of properties of these shapes appear in Appendix Table A.4. Structural tees are used for the chord members of welded steel trusses and for the flanges in certain types of plate girder.

Plates and Bars

Plates and bars are made in many different sizes and are available in all the structural steel specifications listed in Table 2.1.

Flat steel for structural use is generally classified as follows:

Bars. 6 in. or less in width, 0.203 in. and more in thickness.
6 to 8 in. in width, 0.230 in. and more in thickness.

Plates. More than 8 in. in width, 0.230 in. and more in thickness.
More than 48 in. in width, 0.180 in. and more in thickness.

Bars are available in varying widths and in virtually any required thickness and length. The usual practice is to specify bars in increments of $\frac{1}{4}$ in. for widths and $\frac{1}{8}$ in. in thickness.

For plates the preferred increments for width and thickness are the following:

Widths. Vary by even inches, although smaller increments are obtainable.

Thickness. $\frac{1}{32}$-in. increments up to $\frac{1}{2}$ in.
$\frac{1}{16}$-in. increments of more than $\frac{1}{2}$ to 2 in.
$\frac{1}{8}$-in. increments of more than 2 to 6 in.
$\frac{1}{4}$-in. increments of more than 6 in.

The standard dimensional sequence when describing steel plate is

$$\text{thickness} \times \text{width} \times \text{length}$$

All dimensions are given in inches, fractions of an inch, or decimals of an inch.

Column base plates and beam bearing plates may be obtained in the widths and thicknesses noted. For the design of column base plates and beam bearing plates, see Secs. 5.11 and 6.12, respectively.

Designations for Structural Steel Elements

As noted earlier, wide-flange shapes are identified by the symbol W, American standard beam shapes, by S. It was also pointed out that W shapes have essentially parallel flange surfaces, whereas S

TABLE 2.4 Standard Designations for Structural Steel Elements

Type of Element	Designation
Wide-flange shapes	W 12 × 27
American standard beams	S 12 × 35
Miscellaneous shapes	M 8 × 18.5
American standard channels	C 10 × 20
Miscellaneous channels	MC 12 × 45
Angles	
Equal legs	L 4 × 4 × $\frac{1}{2}$
Unequal legs	L 5 × 3$\frac{1}{2}$ × $\frac{1}{2}$
Structural tees	
Cut from wide flange shapes	WT 6 × 53
Cut from American standard beams	ST 9 × 35
Cut from miscellaneous shapes	MT 4 × 9.25
Plate	PL $\frac{1}{2}$ × 12
Structural tubing: square	TS 4 × 4 × 0.375
Pipe	Pipe 4 std.

shapes have a slope of approximately $16\frac{2}{3}\%$ on the inner flange faces. A third designation, M shapes, covers miscellaneous shapes that cannot be classified as W or S: these shapes have various slopes on their inner flange surfaces and many of them are of only limited availability. Similarly, some rolled channels cannot be classified as C shapes. These are designated by the symbol MC.

Table 2.4 lists the standard designations used for rolled shapes, formed rectangular tubing, and round pipe.

2.4 COLD-FORMED STEEL PRODUCTS

Many structural elements are formed from sheet steel. Elements formed by the rolling process must be heat-softened, whereas those produced from sheet steel are ordinarily made without heating the steel; thus the common description for these elements is *cold-formed*. Because they are typically formed from very thin sheets, they are also referred to as *light-gage* steel products.

Figure 2.3 illustrates the cross sections of some of these products. Large corrugated or fluted sheets are in wide use for all paneling and for structural decks for roofs and floors. Use of these elements for floor decking is discussed in Chapter 5. These products are made by a number of manufacturers and information

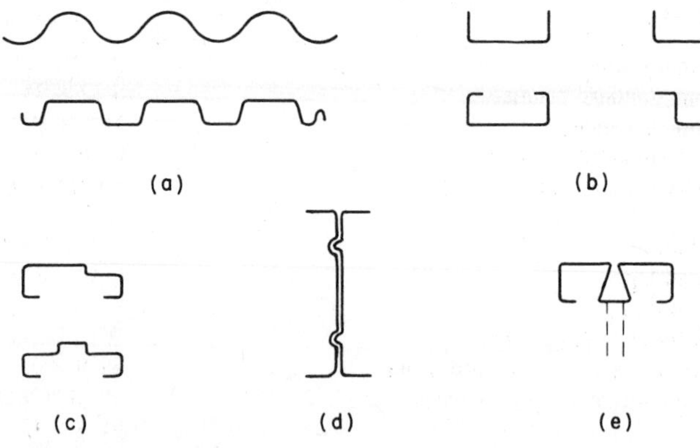

FIGURE 2.3 Cold-formed structural shapes.

regarding their structural properties may be obtained directly from the manufacturer. General information on structural decks may also be obtained from the Steel Deck Institute (see Refs. 9 and 10).

Cold-formed shapes range from the simple L, C, U, etc., to the special forms produced for various construction systems. Structures for some buildings may be almost entirely comprised of cold-formed products. Several manufacturers produce patented kits of these components for the formation of predesigned, packaged building structures. The design of cold-formed elements is described in the *Cold-Formed Steel Design Manual,* published by the American Iron and Steel Institute.

2.5 FABRICATED STRUCTURAL COMPONENTS

A number of special products are formed of both hot-rolled and cold-formed elements for use as structural members in buildings. Open-web steel joists consist of prefabricated, light steel trusses. For short spans and light loads a common design is that shown in Fig. 2-4*a* in which the web consists of a single, continuous bent steel rod and the chords of steel rods or cold-formed elements. For larger spans or heavier loads the forms more closely resemble those for ordinary light steel trusses; single angles, double angles, and structural tees constitute the truss members. Open-web joints for floor framing are discussed in Sec. 5.12.

Another type of fabricated joist is shown in Fig. 2.4*b*. This member is formed from standard rolled shapes by cutting the web in a zigzag fashion. The resulting product has a greatly reduced weight-to-depth ratio when compared with the lightest of the rolled shapes.

Other fabricated steel products range from those used to produce whole building systems to individual elements for construction of windows, doors, curtain wall systems, and interior partition walls. Many components and systems are produced as priority items by a single manufacturer, although some are developed under controls of industry-wide standards, such as those published by the Steel Joist Institute and Steel Deck Institute.

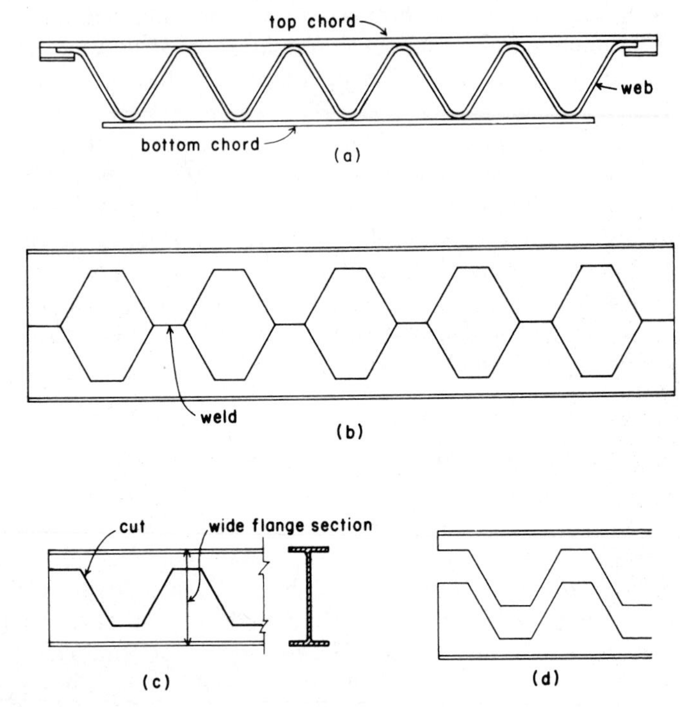

FIGURE 2.4 Fabricated structural joists: (a) open-web steel joist; (b) castelated steel joist; (c) and (d) cutting the parts for a castelated joist from a W shape.

Some structural products in this category are discussed and illustrated in the development of the building system design examples in Chapter 12.

2.6 CONNECTION METHODS

Connection of structural steel members that consist of rolled elements is typically achieved by direct welding or by steel rivets or bolts. Riveting for building structures has become generally obsolete in favor of high-strength bolts. The design of bolted connections and simple welded connections are discussed in Chapter 10. In general, welding is preferred for shop fabrication and bolting for field connections.

Thin elements of cold-formed steel may be attached by welding, bolting, or by sheet metal screws. Thin deck and wall paneling elements are sometimes attached to one another by simple interlocking at their abutting edges; the interlocked parts are sometimes folded or crimped to give further security to the connection.

Adhesives or sealants may be used to seal joints or to bond thin sheet materials in laminated fabrications. Some elements used in connections may be attached to connected parts by adhesion to facilitate the work of fabrication and erection, but adhesive connection is not used for major structural joints.

A major structural design problem is that of the connections of columns and beams in heavy frames for multistory buildings. For rigid frame action to resist lateral loads, these connections are achieved with vary large welds, generally developed to transfer the full strength of the connected members. Design of these connections is beyond the scope of this book, although lighter framing connections of various form are discussed in Chapter 10.

2.7 TABULATED DATA FOR STEEL PRODUCTS

Information in general regarding steel products used for building structures must be obtained from steel industry publications. The AISC is the primary source of design information regarding structural rolled products, which are the principal elements used for major structural components: columns, beams, large trusses, and so on. Several other industry-wide organizations also publish documents that provide information about particular products, such as manufactured trusses (open-web joists), cold-formed sections, and formed sheet steel decks. Many of these organizations and their publications are described in the appropriate chapters and sections of this part.

Individual manufacturers of steel products usually conform to some industry-wide standards in the design and fabrication of their particular products. Still, there is often some room for variation of products, so that it is advised that the manufacturers' own publications be used for specific data and details of the actual products. As in other similar situations in building design, the designer should strive to design and specify components of the building construction so that only those controls that are critical are predetermined; leaving flexibility in the choice of a particular manufactured product.

Some of the tabulated data presented in this book has been reproduced or abstracted from industry publications. In many cases the data presented here are abbreviated and limited to uses pertinent to the work displayed in the text example computations and necessary for the exercise problems. The reference sources cited should be consulted for more complete information, the more so since change is the name of the game, due to growth of the technology, advances in research, and modifications of codes and industry standards.

3

BEHAVIORAL AND USAGE CONSIDERATIONS

Steel is a relatively expensive, industrialized product. Use of steel for structures must generally be made with very careful consideration of the limitations of the material, attention to high efficiency in the volume of material used, and design with clear understanding of the practical aspects of production and erection of steel products. In this chapter we present discussions of a number of aspects regarding intelligent use of steel for structural applications.

3.1 STRESS AND STRAIN

Steel is one of the strongest materials used for building structures, but has limitations for various forms of stress development. Unlike wood, stress response tends to be nondirectional; and unlike concrete or masonry, stress resistance is high for all the basic stresses: tension, compression, and shear. Some specific stress limits for steel are the following:

1. Stress beyond the yield point will produce permanent deformations, which may be tolerable in the small dimensions of a joint, but create major problems within the general form of structural elements. Even though ultimate strength may be high, the much lower yield stress must be used for a limit of acceptable behavior in most situations.
2. Ordinary steel is formed by molten casting, resulting in a crystalline structure of the material. Certain forms of stress failure may be precipitated by fracture along crystalline fault lines; especially those related to dynamic, repetitive force actions. This is more of a problem in machinery, but dynamically loaded structures may need consideration for this effect.
3. Various actions, such as cold-forming, machining, or welding, may change the character of the material, resulting in hardening, loss of ductility, or locked-in stresses in the material. The processes of fabrication and erection must be carefully studied to be sure these do not produce undesirable conditions to complicate stress behavior under service load conditions.

In some cases the anticipated stress-strain responses may cause certain actions to occur that affect the overall structural resistance of a steel structure. An example of this is the formation of plastic hinges in rigid frames or in frames with eccentric bracing. The adjusted behavior of the structure in load response that occurs when a plastic hinge yields is a major element in visualization and computation of the structure's response. Some considerations for this action are discussed in Chapters 7 and 13.

Because of its strength, steel tends to carry a disproportionately high share of the load when sharing loads with other materials, such as wood or concrete. This is a major factor in design of composite structural elements, such as flitched beams of steel and wood and composite deck systems of steel and concrete (see Sec. 8.3). It is also a major consideration in design of reinforced concrete and masonry.

Although stress resistance is subject to variation, strain resistance as measured by the direct stress modulus of elasticity is not. This makes for a shifting relationship when stress capability is raised to produce higher grades of steel. While load resistance as measured by stress capacity may be increased, resistance to deformations is not. Thus deflection or buckling—both affected by the stiffness of the material—may become relatively more critical for structures made with higher grades of steel.

3.2 RUST

Exposed to air and moisture, most steels will rust at the surface of the steel mass. Rusting will generally continue at some rate until the entire steel mass is eventually rusted away. Response to this problem may involve one or more of the following actions.

1. Do nothing, if there is essentially no exposure, as when the steel element is encased in poured concrete or other encasing construction.
2. Paint the steel surface with rust-inhibiting material.
3. Coat the surface with nonrusting metal, such as zinc or aluminum.
4. Use a steel that contains ingredients in the basic material that prevent or retard the rusting action (see the discussion in Sec. 2.1 on corrosion-resistant steels).

Rusting is generally of greater concern when exposure conditions are more severe. It is also of greatest concern for the thinner elements, especially those formed of thin sheet steel, such as formed roof decks.

In some cases it may be necessary to leave the steel in an essentially bare condition, such as when field (on-site) welding is to be done or when steel items are to be encased in concrete. These are standard practices in building construction, but can be difficult to deal with when appearance is important due to the final exposed condition of the structure.

When structures are exposed to conditions likely to cause serious rusting, most designers tend to avoid use of excessively thin parts where possible. This somewhat reduces susceptibility of the structure to failure by loss of material in member cross sections, in the event that rust prevention methods are less than totally successful over the life of the structure.

Deterioration of steel can also be caused by exposure to various corrosive chemicals, such as acid rain, seacoast salt air, or air heavily polluted with various industrial wastes. Special protection or simply avoiding exposure as much as possible may be required for such conditions. Where not just appearance, but actual structural safety is at risk, these matters require serious attention by the structural designer.

3.3 FIRE

As with all materials, the stress and strain response of steel varies with its temperature. The rapid loss of strength (and stiffness, which may be more important when buckling is possible) at high temperatures, coupled with rapid heat gain due to the high conductivity of the material and the common use of thin parts, makes steel structures highly susceptible to fire. On the other hand, the material is noncombustible and less critical for some considerations in comparison to constructions with thin elements of wood.

The chief strategy for improving fire safety with steel structures is to prevent the fire (and the heat buildup) from getting to the steel by providing some coating or encasement with fire-resistant, insulative materials. Ordinary means for this include use of concrete, masonry, plaster, mineral fiber, or gypsum plasterboard elements. The general problem and some specific design situations are presented in Chapter 12.

Concrete is often used with steel framing as a fill on top of formed steel deck or as a structural concrete slab bearing directly on steel beams. In some situations the concrete may also be used to encase steel columns or beams, although building code acceptance of other means for achieving necessary fire ratings have largely eliminated this practice. One easy form of this construction occurs with the steel beam-plus-concrete slab construction shown in Fig. 3.1. A problem for concern in design of such a

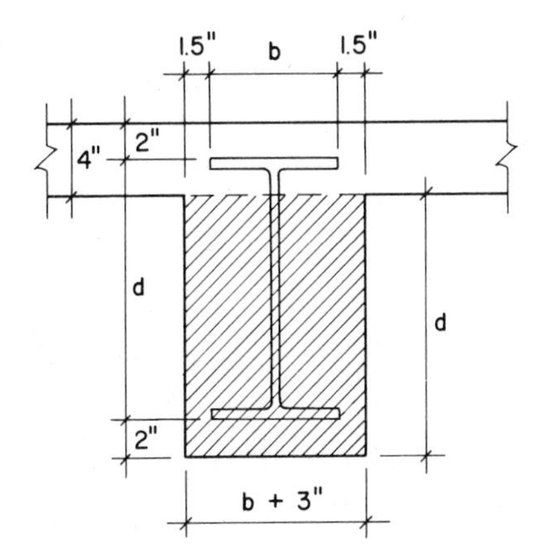

FIGURE 3.1 Cast-in-place concrete fireproofing for a steel beam.

system is the considerable added weight of the construction due to the concrete encasement.

3.4 COST

Steel is relatively expensive, on a volume basis. The real dollar cost of concern, however, is the final *installed cost,* that is, the total cost for the erected structure. Economy concerns begin with attempts to use the least volume of the material, but this is applicable only within the design of a single type of item. Rolled structural shapes do not cost the same per pound as fabricated open web joists. Furthermore, each item must be transported to the site and erected, using various auxiliary devices to complete the structure, such as connecting elements for structural components and bridging for joists.

Cost concerns for structures as a whole, and for the total building construction, are discussed in Chapter 12. In other parts of the book, when discussing design of single structural components, the usual approach is to generally seek to use the lightest-weight (least volume of material) elements that will satisfy the design criteria.

4

HORIZONTAL-SPAN FRAMING SYSTEMS

Elements of steel may be used to provide a variety of horizontal spanning floor or roof structures. The two primary spanning systems treated in this book are the rolled steel beam and the light, prefabricated truss. In this chapter we deal with some of the general issues involved in development of spanning systems. Design of beams and decks is treated in Chapter 5 and design of trusses is treated in Chapter 9. The special cases of rigid frames and bents are discussed in Chapter 8.

4.1 DECK–BEAM–GIRDER SYSTEMS

A framing system extensively used for buildings with large roof or floor areas is that in which columns are arranged in orderly rows for the support of a rectangular grid of steel beams or trusses. The actual roof or floor surface is then generated by a solid deck of wood, steel, or concrete, which spans in multiple, continuous spans over a parallel set of supports. Planning for such a system must begin with consideration of the general architectural design of the building, but should also respond to logical considerations for the development of the structure.

Consider the system shown in the partial framing plan in Fig. 4.1. In developing the layout for the system and choosing its components, considerations such as the following must be made:

1. *Deck Span.* The type of deck as well as its specific variation (thickness of plywood, gage of steel sheet, etc.) will relate to the deck span.
2. *Joist Spacing.* This determines the deck span and the magnitude of load on the joist. The type of joist selected may limit the spacing, based on the joist capacity. The type and spacing of joists must be coordinated with the selection of the deck.
3. *Beam Span.* For systems with some plan regularity, the joist spacing should be some full-number division of the beam span.
4. *Column Spacing.* The spacing of the columns determines the spans for the beams and joists, and is thus related to the planning modules for all the other components.

For a system such as that shown in Fig. 4.1, the basic planning begins with the location of the system supports, usually columns

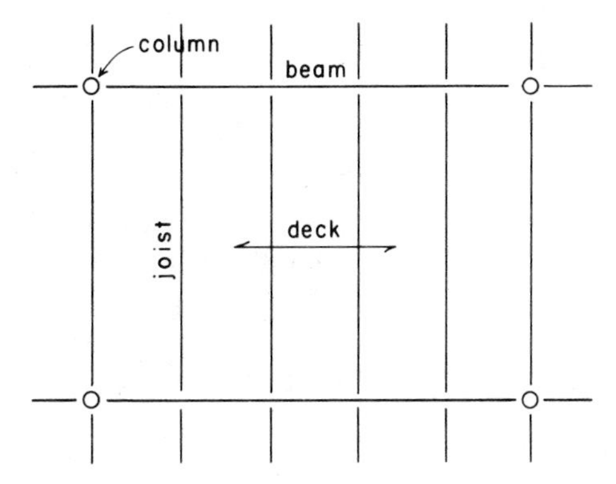

FIGURE 4.1 Common steel framing layout.

or bearing walls. The character of the spanning system is closely related to the magnitude of the spans it must achieve. Decks are mostly quite short in span, requiring relatively close spacing of the elements that provide their direct support. Joists and beams may be small or large, depending mostly on their spans. The larger they are, the less likely they will be very closely spaced. Thus very long span systems may have several levels of components before ending with the elements that directly support the deck.

Concerns for the design of individual components of the deck–beam–girder system are discussed in Chapter 5. Some aspects of planning for such a system are presented in Sec. 4.3. General discussion of design in the context of whole building system development is presented in the building system design examples in Chapter 12.

4.2 TRUSS SYSTEMS

Figure 4.2*a* shows a plan and elevation of a system that uses trusses for the major span. If the trusses are very large and the purlin spans quite long, the purlins may have to be quite widely spaced. A constraint on the purlin locations is usually that they coincide with the joints in the top of the truss, so as to avoid high shear and bending in the truss top chord. In the latter case, it may be advisable to use joists between the purlins to provide support for the deck. On the other hand, if the truss spacing is a modest distance, it may be possible to use a long-span deck with no purlins. The basic nature of the system can thus be seen to change with different positioning of the system supports. In any event, the truss span and panel module, the column spacing, the purlin span and spacing, the joist span and spacing, and the deck span are interrelated and the selection of the components is a highly interactive exercise.

General use of steel trusses is discussed in Chapter 9 and some illustrations of usage are presented in the examples in Chapter 12. Use of prefabricated, manufactured open-web steel joists and joist girders is discussed in Sec. 5.12.

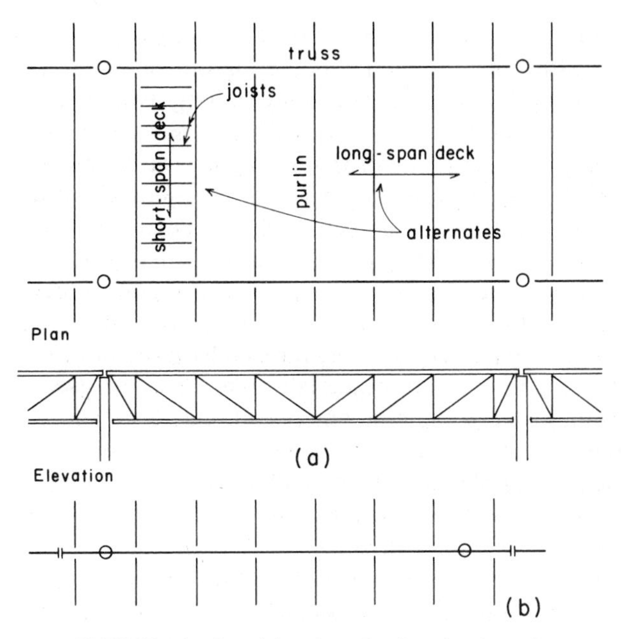

FIGURE 4.2 Considerations for framing layouts.

4.3 PLANNING CONSIDERATIONS FOR FRAMING SYSTEMS

For systems with multiple elements, some consideration must be given to the various intersections and connections of the components. For the framing plan shown in Fig. 4.1 there is a five-member intersection at the column, involving the column, the two beams, and the two joists (plus an upper column, if the building is multistory). Depending on the materials and forms of the members, the forms of connections, and the types of force transfer at the joint, this may be a routine matter of construction or a real mess. Some relief of the traffic congestion may be achieved by the plan layout shown in Fig. 4.2*a*, in which the module of the joist spacing is offset at the columns, leaving only the column and beam connections. A further reduction possible is that shown in

Fig. 4.2*b*, where the beam is made continuous through the column, with the beam splice occurring off the column. In the plan in Fig. 4.2*b* the connections are all only two-member relationships: column to beam, beam to beam, and beam to joist.

Bridging, blocking, and cross-bracing for trusses must also be planned with care. These members may interfere with ducting or other building elements as well as create connection problems similar to those just discussed. Use of these extra required elements for multiple purposes should be considered. Blocking required for plywood nailing may also function as edge nailing for ceiling panels and lateral bracing for slender joists or rafters. The cross-bracing required to brace tall trusses may be used to support ceilings, ducts, building equipment, catwalks, and so on.

In the end, structural planning must be carefully coordinated with the general planning of the building and its various subsystems. Real optimization of the structure may need to yield to other, more pragmatic concerns.

Additional concerns for structural planning are addressed in Sec. 11.3 and in the design examples in Chapter 12.

5

BEAMS, JOISTS, AND DECKS

In this chapter we deal with the various considerations involved in the design of the basic elements of the framing systems described in Chapter 4. Problems involved in the full development of building systems that utilize such systems are discussed in Chapters 11 and 12.

5.1 ROLLED SHAPES AS BEAMS

Various rolled shapes may serve beam functions, although the most commonly used shape for framing systems is the wide flange. Design of wide flange beams may involve any of the following considerations.

1. *Flexural Stress*. Bending stresses generated by moment effects are a primary consideration for beams. For steel beams, stress magnitude may be limited by basic allowable stress values as given in Table 2.1, or by reduced values due

to various special situations, such as noncompact sections, buckling, and so on. For the wide-flange section of A36 steel, the basic limiting flexural stress is two-thirds of F_y, or 24 ksi. The single property of the beam cross section that is most predictive of the moment capacity is the section modulus, S. Thus

$$\text{resisting moment} = (S)(F_b)$$

2. *Shear Stress*. Whereas shear is quite critical in wood and concrete beams, it is less often a problem with steel beams. Special problems may occur when beams carry exceptionally heavy concentrated loads, but otherwise shear is seldom a limiting condition. The diagonal compression effect or direct vertical compression at supports or concentrated loads may cause buckling of thin beam webs (see the discussion of web crippling in Sec. 5.10).
3. *Deflection*. Vertical deflection of beams must be controlled for various reasons. A critical factor is that of the depth of the beam with respect to the span length. Deflections must sometimes be computed, but mostly they can be expected not to be critical if particular span-to-depth ratios are maintained.
4. *Buckling*. In general, beams that are not adequately braced may be subject to lateral or rotational buckling. The more slender the beam cross section (deep and narrow profile), the greater its susceptibility to these effects. Design control may be achieved through use of reduced design stresses, but the most effective solution is usually to provide adequate bracing.
5. *Connections and Supports*. Framed structures contain many joints and the details of the connections and supports must be developed for proper construction methods as well as the transfer of the necessary structural forces through the joints.

In particular situations there may be many other concerns for complete functioning of the beam as a member of the framing system as well as a basic element of the whole building construction.

There are almost 300 wide-flange shapes and the selection of the proper shape for a particular beam may involve many factors, including all of those just mentioned. In general—as in most of structural design—it is desirable to use the shape with the least cross section, representing the lowest basic cost for the material. We refer to this choice as the *least-weight* member, and in general will prefer it unless special circumstances prevail.

In the following sections we discuss several individual considerations in the design of wide-flange beams.

5.2 DESIGN OF BEAMS FOR BENDING

Design for bending usually involves the determination of the maximum bending moment that the beam must resist and the use of the basic flexural formula ($f = M/S$) to determine the minimum section modulus required. Since weight is determined by area, not section modulus, the beam chosen may have more section modulus than required and still be the most economical choice in some cases. The following example illustrates the basic procedure.

Example. Design a simply supported beam to carry a superimposed load of 2 kips per ft [29.2 kN/m] over a span of 24 ft [7.3 m]. (The term *superimposed load* is used to denote any load other than the weight of a structural member itself.) The allowable bending stress is 24 ksi [165 MPa].

Solution. The bending moment due to the superimposed load is

$$M = \frac{wL^2}{8} = \frac{2 \times (24)^2}{8} = 144 \text{ kip-ft} \quad [195 \text{ kN-m}]$$

The required section modulus for this moment is

$$S = \frac{M}{F_b} = \frac{144 \times 12}{24} = 72.0 \text{ in.}^3 \quad [1182 \times 10^3 \text{ mm}^3]$$

Scanning Appendix Table A.1, we find a W 16 × 45 with a section modulus of 72.7 in.^{3} [1192 × 10^3 mm^3]. This value, however, is so close to that required that almost no margin is provided for the effect of the beam weight. Further scanning of the table reveals a W 16 × 50 with an S of 81.0 in.^{3} [1328 × 10^3 mm^3] and a W 18 × 46 with an S of 78.8 in.^{3} [1291 × 10^3 mm^3]. In the absence of any known restriction on the beam depth we try the lighter section. The bending moment at the center of the span with this beam is

$$M = \frac{wL^2}{8} = \frac{46 \times (24)^2}{8} = 3312 \text{ ft-lb} \quad \text{or} \quad 3.3 \text{ kip-ft} \;\; [4.46 \text{ kN-m}]$$

Thus the total bending moment at midspan is

$$M = 144 + 3.3 = 147.3 \text{ kip-ft} \;\; [199.5 \text{ kN-m}]$$

The section modulus required for this moment is

$$S = \frac{M}{F_b} = \frac{147.3 \times 12}{24} = 73.7 \text{ in.}^3 \;\; [1209 \times 10^3 \text{ mm}^3]$$

Because this required value is less than that of the W 18 × 46, this section is acceptable.

Use of Section Modulus Tables

Selection of rolled shapes on the basis of required section modulus may be achieved by the use of the tables in the AISC Manual (Ref. 1) in which beam shapes are listed in descending order of their section modulus values. Material from these tables is presented in Appendix Table B.1. Note that certain shapes have their designations listed in boldface type. These are sections that have an especially efficient bending moment resistance, indicated by the fact that there are other sections of greater weight but the same or smaller section modulus. Thus for a savings of material cost these *least-weight* sections offer an advantage. Consideration of other beam design factors, however, may sometimes make this a less important concern.

Data are also supplied in Appendix Table B.1 for the consideration of lateral support for beams. For consideration of lateral support the values are given for the two limiting lengths L_c and L_u. If a calculation has been made by assuming the maximum allowable stress of 24 ksi [165 MPa], the required section modulus obtained will be proper only for beams in which the lateral unsupported length is equal to or less than L_c.

A second method of using Appendix Table B.1 for beams of A36 steel omits the calculation of a required section modulus and refers directly to the listed values for the maximum bending resistance of the sections, given as M_R in the tables. Although the condition of the noncompact section may be noted, in this case the M_R values have taken the reduced values for bending stress into account.

Example. Rework the problem in the preceding example in this section by using Appendix Table B.1.

Solution. As before, we determine that the bending moment due to the superimposed loading is 144 kip-ft [195 kN-m]. Noting that some additional M_R capacity will be required because of the beam's own weight, we scan the tables for shapes with an M_R of slightly more than 144 kip-ft [195 kN-m]. Thus we find

Shape	M_R (kip-ft)	M_R (kN-m)
W 21 × 44	163	221
W 16 × 50	162	220
W 18 × 46	158	214
W 12 × 58	156	212
W 14 × 53	156	212

Although the W 21 × 44 is the least-weight section, other design considerations, such as restricted depth, may make any of the other shapes the appropriate choice.

It should be noted that not all of the available W shapes listed in Appendix Table A.1 are included in Appendix Table B.1. Specifically excluded are the shapes that are approximately square (depth equal to flange width) and are ordinarily used for columns rather than beams.

Structural Design Methods

Two different methods are used in the design of steel beams for bending stresses. The first, called *allowable stress design,* applies to this chapter and to the major part of the book. It is based on the idea of using F_b, an allowable extreme fiber stress, as a certain fraction of the yield stress (Sec. 2.1). These allowable stresses fall below the elastic limit of the material, and we speak of them as conforming to the elastic behavior of the material. This approach to the design of steel members has been standard practice for many years.

The second method, known as *plastic design,* is a more recent development and is based on the idea of computing an ultimate load and using a portion of the reserve strength, after initial yield stress has been reached, as part of the factor of safety (Sec. 13.7). A brief explanation of plastic design theory is given in Chapter 3.

The following problems involve design for bending stress only. Use A36 steel with an allowable bending stress of 24 ksi [165 MPa]. Assume that least-weight members are desired for each case.

Problem 5.2.A. Design for flexure a simple beam 14 ft [4.3 m] in length and having a total uniformly distributed load of 19.8 kips [88 kN].

Problem 5.2.B. Design for flexure a beam having a span of 16 ft [4.9 m] with a concentrated load of 12.4 kips [55 kN] at the center of the span.

Problem 5.2.C. A beam 15 ft [4.6 m] in length has three concentrated loads of 4 kips, 5 kips, and 6 kips at 4 ft, 10 ft, and 12 ft [17.8 kN, 22.2 kN, and 26.7 kN at 1.2 m, 3 m, and 3.6 m], respectively, from the left-hand support. Design the beam for flexure.

Problem 5.2.D. A beam 30 ft [9 m] long has concentrated loads of 9 kips [40 kN] each at the third points and also a total uniformly distributed load of 30 kips [133 kN]. Design the beam for flexure.

Problem 5.2.E. Design for flexure a beam 12 ft [3.6 m] in length, having a uniformly distributed load of 2 kips/ft [29 kN/m] and a concentrated load of 8.4 kips [37.4 kN] a distance of 5 ft [1.5 m] from one support.

Problem 5.2.F. A beam of 19 ft [5.8 m] in length has concentrated loads of 6 kips [26.7 kN] and 9 kips [40 kN] at 5 ft [1.5 m] and 13 ft [4 m], respectively, from the left-hand support. In addition, there is a uniformly distributed load of 1.2 kip-ft [17.5 kN/m] beginning 5 ft [1.5 m] from the left support and continuing to the right support. Design the beam for flexure.

Problem 5.2.G. A steel beam 16 ft [4.9 m] long has two uniformly distributed loads, one of 200 lb/ft [2.92 kN/m] extending 10 ft [3 m] from the left support and the other of 100 lb/ft [1.46 kN/m] extending over the remainder of the beam. In addition, there is a concentrated load of 8 kips [35.6 kN] at 10 ft [3 m] from the left support. Design the beam for flexure.

Problem 5.2.H. Design for flexure a simple beam 12 ft [3.7 m] in length, having two concentrated loads of 12 kips [53.4 kN] each, one 4 ft [1.2 m] from the left end and the other 4 ft [1.2 m] from the right end.

Problem 5.2.I. A cantilever beam 8 ft [2.4 m] long has a uniformly distributed load of 1600 lb/ft [23.3 kN/m]. Design the beam for flexure.

Problem 5.2.J. A cantilever beam 6 ft [1.8 m] long has a concentrated load of 12.3 kips [54.7 kN] at its unsupported end. Design the beam for flexure.

5.3 SHEAR EFFECTS

Shear stress in a steel beam is seldom a factor in determining its size. It is customary to determine first the size of the beam to resist bending stresses. Having done this, the beam is then investigated for shear, which means that we compute the actual maximum unit shear stress to see that it does not exceed the allowable stress. The AISC Specification gives F_v, the allowable shear stress in beam webs, as $0.40F_y$ on the gross section of the web, which is computed as the product of the web thickness and the overall beam depth. For A36 steel $F_v = 0.40 \times 36$ or 14.4 ksi. This value is rounded off at F_v = 14.5 ksi [100 MPa], as shown in Table 2.1.

Shearing stresses in beams are not distributed uniformly over the cross section but are zero at the extreme fibers, with the maximum value occurring at the neutral surface. Consequently, the material in the flanges of wide-flange sections, I-beams, and channels has little influence on shearing resistance, and the working formula for determining shearing stress is taken as

$$f_v = \frac{V}{A_w}$$

where f_v = unit shearing stress,
V = maximum vertical shear,
A_w = gross area of the web (actual depth of section times the web thickness, or $d \times t_w$).

The shearing stresses in beams are seldom excessive. If, however, the beam has a relatively short span with a large load placed near one of the supports, the bending moment is relatively small and the shearing stress becomes comparatively high. This situation is demonstrated in the following example:

Example. A simple beam of A36 steel is 6 ft [1.83 m] long and has a concentrated load of 36 kips [160 kN] applied 1 ft [0.3 m] from the left end. It is found that a W 10 × 17 is large enough to support this load with respect to bending stresses (required S = 15 in.^{3} [246×10^3 mm^3]). Investigate the beam for shear, neglecting the weight of the beam.

Solution. The two reactions are computed as follows:

$$\text{left reaction} = \frac{36 \times 5}{6} = 30 \text{ kips} \;\; [133.8 \text{ kN}]$$

$$\text{right reaction} = \frac{36 \times 1}{6} = 6 \text{ kips} \;\; [26.2 \text{ kN}]$$

The maximum vertical shear is thus 30 kips [133.8 kN].

From Appendix Table A.1 we find that d = 10.11 in. [256.8 mm] and t_w = 0.240 in. [6.1 mm] for the W 10 × 17. Then

$$A_w = d \times t_w = 10.11 \times 0.240 = 2.43 \text{ in.}^2 \; [1566 \text{ mm}^2]$$

and

$$f_v = \frac{V}{A_w} = \frac{30}{2.43} = 12.35 \text{ ksi} \; [85 \text{ MPa}]$$

Because this is less than the allowable value of 14.5 ksi [100 MPa], the W 10 × 17 is acceptable.

Problems 5.3.A, B, C. Compute the maximum permissible values for web shear for the following beams of A36 steel: (a) W 24 × 84; (b) W 12 × 40; (c) W 10 × 19.

5.4 DEFLECTION OF BEAMS

The deformation that accompanies the bending of a beam is called *deflection*. If the beam is in a horizontal position, it is the vertical distance moved from a horizontal line. The deflection of a beam may not be apparent visually but it is, nevertheless, always present. Figure 5.1 illustrates the deflection of a simple beam; we are principally concerned with its maximum value, which in this instance occurs at the center of the span.

A beam may be strong enough to withstand the bending stresses without failure, but the curvature may be so great that cracks will appear in suspended plaster ceilings, water will collect in low spots on roofs, and the general lack of stiffness will result

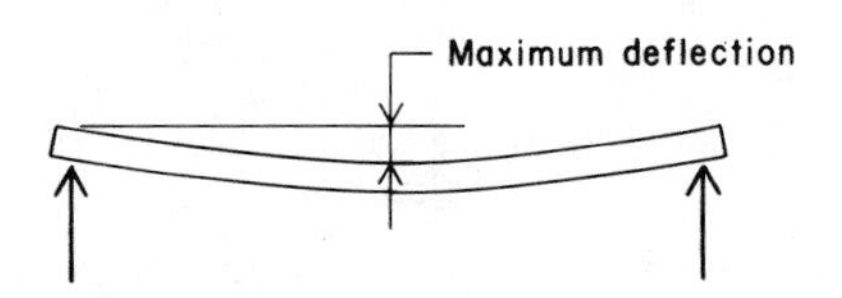

FIGURE 5.1 Deflection of a simple beam.

in an excessively springy floor. When designing floor construction for buildings in which machinery will be used, particular care should be given to the deflection of beams because excessive deflection may be the cause of inordinate vibrations or misalignments. Another fault that results from excessive deflection occurs where floor beams frame into girders. At these points there is a tendency for cracks to develop in the flooring directly over the girder. This is illustrated in Fig. 5.2.

If, in the design of a beam, the deflection is computed to be excessive, the remedy is to select a deeper beam—a beam with a greater moment of inertia. For a given span and loading the deflection of a beam varies directly as the fiber stress (bending stress) and inversely as the depth. For this reason it is preferable to select a beam that has the greatest practicable depth so that the deflection will be a minimum. When ample headroom is available a deeper beam is preferable to a shallow beam with the same section modulus.

In general, the procedure is to design a beam of ample dimensions to resist bending stresses (design for strength) and then to investigate the beam for deflection.

Allowable Deflection

There is common agreement that deflection of beams should be limited, but authorities differ with respect to the maximum degree of deflection to be permitted. Some codes require that deflection be limited to 1/360 of the span but say nothing of the kind of load that produces the deflection. The AISC Specification requires

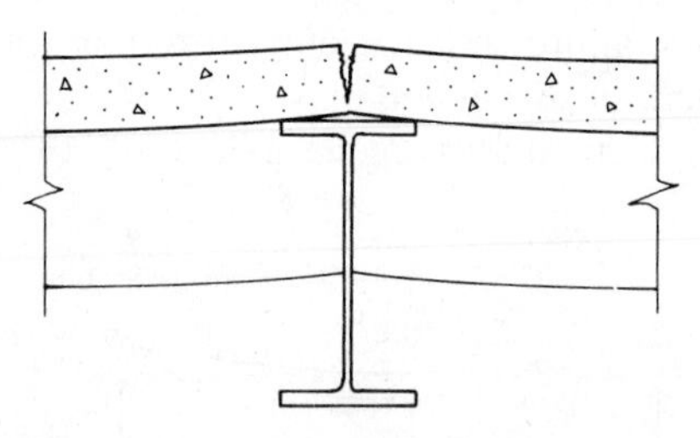

FIGURE 5.2 Effect of continuity of decks.

that beams and girders that support plastered ceilings have a maximum deflection of 1/360 of the span due to the live load. (See the discussion of loads in Chapter 11.) For actual design work the local building code should be consulted for any deflection limits.

Deflection for Uniformly Distributed Load

By referring to Appendix Fig. E.6 we find formulas that can be used to compute the maximum deflection for the several types of loading that occur most frequently. The Greek letter delta (Δ) represents the maximum deflection.

A number of terms make up these formulas and their solution may be a tedious task. There are, however, simplified methods of computing deflections. Consider the following example:

A designer is engaged in determining the size of steel floor beams; an allowable bending stress of 24 ksi has been chosen. The immediate problem is a simple beam with a span of 16 ft [4.88 m] and a uniformly distributed load of 34 kips [151 kN]. The computations show that a W 12 × 26 is large enough to resist the bending stresses and the next step is to compute the maximum deflection. To do this the designer uses the formula

$$\Delta = \frac{0.02483L^2}{d} \left[\frac{0.1719L^2}{d}\right]$$

where L = span length of the beam, in feet [meters],
d = depth of the beam, in inches [meters],
Δ = maximum deflection of the beam, in inches [millimeters].

Referring to Appendix Table A.1, we find that d for the section is 12.22 in. Then

$$\Delta = \frac{0.02483 \times (16)^2}{12.22} = 0.520 \text{ in. } [13.2 \text{ mm}]$$

Assume that the allowable deflection is 1/360 of the span, or

$$\Delta = \frac{16 \times 12}{360} = 0.533 \text{ in. } [13.56 \text{ mm}]$$

The actual deflection does not exceed the allowable.

Now let us see how the foregoing formula, identified later as Formula 5.4.3, is determined.

The maximum bending moment for this type of loading is $M = wl^2/8$. In Sec. E.2 we find

$$f = \frac{Mc}{I}$$

Consequently, by substituting the value of M we obtain

$$f = \frac{wl^2c}{8I}$$

The maximum deflection for a simple beam with a uniformly distributed load is given in Case 2, Appendix Fig. E.6:

$$\Delta = \frac{5}{384} \times \frac{Wl^3}{EI}$$

or, since $W = wl$,

$$\Delta = \frac{5}{384} \times \frac{wl^4}{EI} = \left(\frac{wl^2c}{8I}\right)\left(\frac{5l^2}{48Ec}\right) = (f)\left(\frac{5l^2}{48Ec}\right) = \frac{5fl^2}{48Ec}$$

(Formula 5.4.1)

This is a basic formula; it can be used for any material by substituting the appropriate values for f and E.

For sections symmetrical with respect to a horizontal axis through the centroid, such as rectangular and I-sections, $c = d/2$. Then by substituting this value of c in Formula 5.4.1 we have

$$\Delta = \frac{5}{48}\left(\frac{f}{E}\right)\left(\frac{2l^2}{d}\right) \qquad \text{(Formula 5.4.2)}$$

Now, let $f = 24$ ksi, $E = 29{,}000$ ksi, and convert the span to feet ($l = 12L$). Then

$$\Delta = \frac{5}{48}\left(\frac{24}{29{,}000}\right)\left\{\frac{2\,(12L)^2}{d}\right\}$$

$$= \frac{0.02483L^2}{d} \qquad \text{(Formula 5.4.3)}$$

In SI units, with $f = 165$ MPa and $E = 200$ GPa,

$$\Delta = \frac{5}{48}\left(\frac{165}{200{,}000}\right)\left(\frac{2L^2}{d}\right)$$

$$= \frac{0.0001719L^2}{d} \text{ (in meters)}$$

$$= \frac{0.1719L^2}{d} \text{ (in millimeters)} \qquad \text{(Formula 5.4.4)}$$

Remember that these equations apply only to simple beams with uniformly distributed loading and only when the stress is 24 ksi [165 MPa]. Because both stress and deflection are directly proportional to the magnitude of the load, the deflection that accompanies some other value of stress may be directly proportioned as shown in the examples.

Example 1. A simple beam has a span of 20 ft [6.10 m] and a uniformly distributed load of 39 kips [173.5 kN]. The section used for this load is a W 14 × 34 and the extreme fiber stress is 24 ksi [165 MPa]. Compute the deflection.

Solution. Formula 5.4.3 is appropriate without modification. Referring to Appendix Table A.1, we find $d = 13.98$ in. [0.355 m]. Then

$$\Delta = \frac{0.02483L^2}{d} = \frac{0.02483(20)^2}{13.98} = 0.710 \text{ in. [18.0 mm]}$$

If we had not known that $f = 24$ ksi, we might have used the formula for deflection given in Fig. E.6, Case 2. Thus

$$\Delta = \frac{5}{384} \times \frac{Wl^3}{EI}$$

Appendix Table A.1 shows that I for the W 14 × 34 is 340 in.^{4} [141.5 × 10^6 mm^4]. Then

$$\Delta = \frac{5}{384} \times \frac{39 \times (20 \times 12)^3}{29{,}000 \times 340} = 0.712 \text{ in. } [18.1 \text{ mm}]$$

Example 2. A W 12 × 26 is used as a simple beam on a span of 19 ft [5.79 m]. It supports a uniformly distributed load and the computed maximum bending stress is 20 ksi [138 MPa]. Find the deflection.

Solution. Referring to Appendix Table A.1, we find $d = 12.22$ in. [0.310 m]. To use Formula 5.4.3 we must adjust for the actual stress of 20 ksi [138 MPa], which is done by a simple proportion:

$$\Delta = \frac{20}{24} \times \frac{0.02483(19)^2}{12.22} = 0.611 \text{ in. } [15.5 \text{ mm}]$$

Deflections by Span Coefficients

Formula 5.4.3 may be used to derive a single coefficient for a given span, which—when divided by a beam depth—will yield the deflection for all beams of that depth, provided that they are stressed to exactly 24 ksi maximum bending stress. This application has been used to obtain the deflection factors for the beams in the load–span value table in Appendix Table D.1, as described in the next section. Formula 5.4.3 can also be used to plot the deflection of beams of a single depth for various spans, as shown in Fig. 5.3. Use of these graphs presents yet another method for finding deflections.

The reader is surely confused at this point as to which of the several methods to use for actual problem work. Indeed, in a

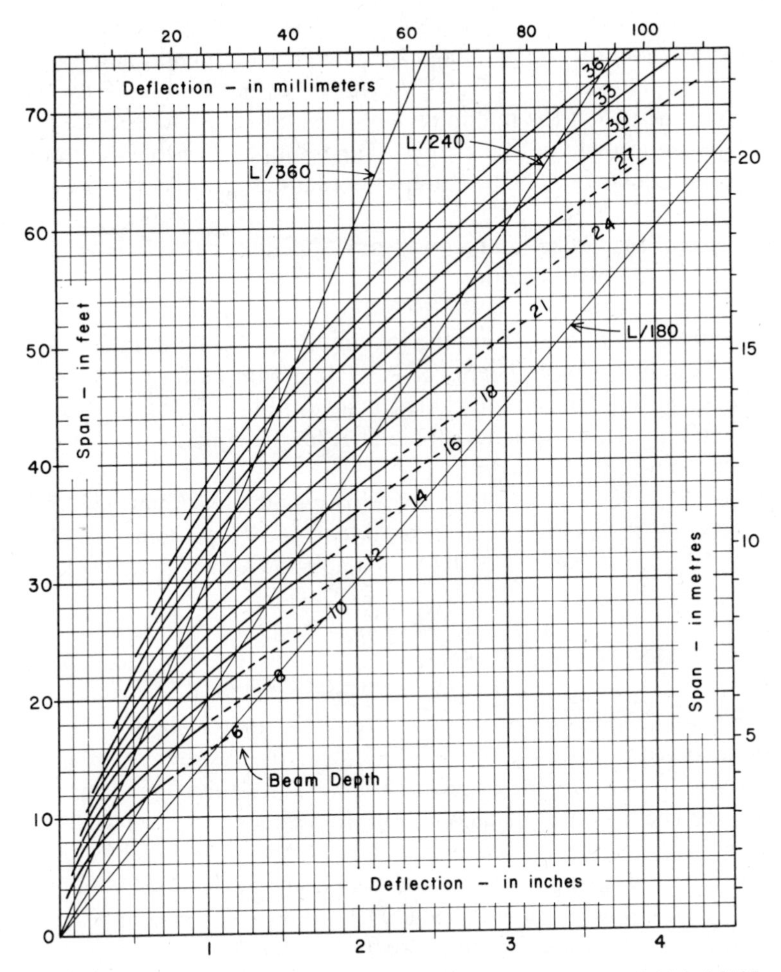

FIGURE 5.3 Deflection of steel beams with bending stress of 24 ksi [165 MPa].

specific situation, any of the methods described could be used, although one may prove to be the simplest in application. For hand computations, the method using deflection coefficients (factors in Appendix Table D.1) is surely quite simple.

The real value of the graphs in Fig. 5.3 is in the design process. Given a span for a beam and a specific deflection limitation (typically, a value such as $L/360$, $L/240$, or $L/180$) it may easily be determined what the minimum depth for a beam is to assure that its deflection will not be critical. Then if a beam of that depth or greater is chosen, computation of actual deflection is not necessary.

Use of the deflection factors from Appendix Table D.1 is illustrated in the next section.

Problems 5.4.A, B, C. Find the maximum deflection in inches for the following simple beams of A36 steel with uniformly distributed load. Find deflection values using: (a) Fig. 5.3; (b) the deflection factors from Appendix table D.1; (c) the equation for Case 2 in Appendix Fig. E.6.

(A) W 10 × 33, span = 18 ft, total load = 30 kips [5.5 m, 133 kN]
(B) W 16 × 36, span = 20 ft, total load = 50 kips [6 m, 222 kN]
(C) W 18 × 46, span = 24 ft, total load = 55 kips [7.3 m, 245 kN]

5.5 SAFE LOAD TABLES

The simple span beam loaded entirely with uniformly distributed load occurs so frequently in steel structural systems that it is useful to have a rapid design method for quick selection of shapes for a given load and span condition. In preceding articles we have demonstrated the use of the section modulus tables and the charts for laterally unsupported beams which are in the AISC Manual (Ref. 1). Use of Appendix Table D.1 allows an even simpler procedure when design conditions permit their use.

For the simple beam with uniformly distributed load we note the maximum bending moment to be $WL/8$. If we equate this to the resisting moment of the beam, expressed as $S \times F_b$, we obtain the following expression for the limiting load on the beam:

$$W = \frac{8SF_b}{L}$$

where W = total uniformly distributed load,
S = section modulus of the beam (S_x),
F_b = allowable bending stress,
L = beam span.

If we assume a doubly symmetrical shape (W, M, or S), a compact section, and a laterally unsupported length not greater than L_c, we may use the maximum allowable bending stress of $0.66F_y$. Then, for a given shape and span, the allowable value of W can be computed for a given grade of steel.

Appendix Table D.1 has been developed by this process, assuming A36 steel. Use of the tables requires only the determination of the span and total load if bending stress is the single concern. If the distance between points of lateral bracing exceeds the value of L_c for a given shape, the charts in Appendix C should be used instead of Appendix Table D.1. For a check the values of L_c are given in the table for each shape. The loads in the tables will not result in excessive shear stress on the beam webs if the full section is available for resistance. Shear stress becomes increasingly critical as the span becomes shorter and should be investigated as described in Chapter 10 if the web section is reduced to form end connections.

Deflection due to the loads in the table may be determined by use of the deflection factors given at the top of the table for each span. Deflections for loads less than those tabulated may be found by proportion, as demonstrated in previous examples. As the span increases, deflection becomes increasingly critical, and the point at which the load will cause a deflection greater than 1/360 of the span is noted in the table by a heavy vertical line.

The following examples illustrate the use of Appendix Table D.1 for some common design situations.

Example 1. A simple span beam of A36 steel is required to carry a total uniformly distributed load of 40 kips [178 kN] on a span of 30 ft [9.14 m]. Find (a) the lightest shape permitted and (b) the shallowest shape permitted.

Solution. From Appendix Table D.1 we find the following:

Shape	Allowable Load (kips)
W 21 × 44	43.5
W 18 × 46	42.0
W 16 × 50	43.2
W 14 × 53	41.5

Thus the lightest section is the W 21 × 44 and the shallowest is the W 14 × 53.

Example 2. A simple span beam of A36 steel is required to carry a total uniformly distributed load of 25 kips [111 kN] on a span of 24 ft [7.32 m] while sustaining a maximum deflection of 1/360 of the span. Find the lightest shape permitted.

Solution. In Appendix Table D.1 we find that the lightest shape that will carry this load is the W 16 × 26. For this beam the deflection will be

$$\Delta = \frac{25}{25.6} \times \frac{14.3}{16} = 0.873 \text{ in.}$$

which exceeds the allowable (24 × 12)/360 = 0.80 in.

The next heaviest beam in the table is a W 16 × 31, for which the deflection will be

$$\Delta = \frac{25}{31.5}\,\frac{14.3}{16} = 0.709 \text{ in.}$$

which is less than the limit; therefore, the W 16 × 31 is the lightest choice.

Problems 5.5.A, B, C, D, E, F. For each of the following conditions find (a) the lightest permitted shape; (b) the shallowest permitted shape of A36 steel.

	Span	Total Uniformly Distributed Load (kips)	Deflection Limited to 1/360 of the Span
A	16	10	No
B	20	30	No
C	36	40	No
D	18	16	Yes
E	32	20	Yes
F	42	50	Yes

5.6 EQUIVALENT TABULAR LOADS

The safe loads shown in Appendix Table D.1 are uniformly distributed loads on simple beams. By the use of coefficients we can convert other types of loading to equivalent uniform loads and thereby greatly extend the usefulness of the tables.

The maximum bending moments for typical loadings are shown in Appendix Fig. E.6. For a simple beam with a uniformly distributed load $M = WL/8$. For a simple beam with equal concentrated loads at the third points $M = PL/3$. These values are shown in Fig. E.6, Cases 2 and 3, respectively. By equating these values

$$\frac{WL}{8} = \frac{PL}{3} \quad \text{and} \quad W = 2.67 \times P$$

which shows that if the value of one of the concentrated loads (in Case 3) were multiplied by the coefficient 2.67 we would have an equivalent distributed load that would produce the same bending moment as the concentrated loads. The coefficients for finding equivalent uniform loads for other beams and loadings are given in Fig. E.6. Because of their use with safe load tables, equivalent uniform loads are usually called *equivalent tabular loads*, abbreviated *ETL*.

It is important to remember that an *ETL* does not include the weight of the beam, for which an estimated amount should be added. Beams found by this method should be investigated for shear and deflection; it is assumed that they are adequately supported laterally. Also, when recording the beam reactions they must be determined from the *actual* loading conditions without regard to the *ETL*.

5.7 LATERAL SUPPORT OF BEAMS

A beam may fail by sideways buckling of the top (compression) flange when lateral deflection is not prevented. The tendency to buckle increases as the compressive bending stress in the flange increases and as the unbraced length of the span increases. Consequently the full value of the allowable extreme fiber stress $F_b = 0.66F_y$ can be used only when the compression flange is adequately braced. The value of this required length includes the variable of the F_y of the beam steel.

When the compression flanges of compact beams (see Sec. 5.8) are supported laterally at intervals not greater than L_c the full allowable stress of $0.66F_y$ may be used. For lateral unsupported lengths greater than L_c, but not greater than L_u, the allowable bending stress is reduced to $0.60\ F_y$. (In certain instances the AISC Specification permits a proportionate reduction in the allowable stress between the two limits; however, in the examples in this book we assume that the drop occurs totally as L_c is exceeded.)

When the laterally unsupported length exceeds L_u the specifications provide a formula for the determination of the allowable stress, based on the specific value of the unsupported length. The design of beams based on these requirements is not a simple matter and the AISC Manual (Ref. 1) contains supplementary charts to aid the designer. These beam charts include "Allowable Moments in Beams" as a function of the laterally unsupported length and provide a workable approach to this otherwise rather cumbersome problem. Reproductions of the charts for beams of A36 steel appear in Appendix C of this book.

After determining the maximum bending moment in kip-ft and noting the longest unbraced length of the compression flange these two coordinates are located on the sides of the chart and are projected to their intersection. Any beam whose curve lies above and to the right of this intersection point satisfies the bending stress requirement. The nearest curve that is a solid (versus dashed) line represents the most economical, or least-weight, section in terms of beam weight, a relationship similar to that discussed in Sec. 5.2 with regard to selection from the section modulus table in Appendix B. It should be noted that selection from the charts incorporates the considerations of bending stress, compact sections, and lateral support; however, deflection, shear, and other factors may also have to be considered.

Example. A simple beam carries a total uniform load of 19.6 kips [87.2 kN], including its own weight, over a span of 20 ft [6.10 m]. It has no lateral support except at the ends of the span. Assuming A36 steel, select from the charts of Appendix C a beam that will meet bending strength requirements and not deflect more than 0.75 in. [19 mm] under the full load.

Solution. For this loading the maximum bending moment is

$$M = \frac{WL}{8} = \frac{19.6 \times 20}{8} = 49 \text{ kip-ft} \quad [66.5 \text{ kN-m}]$$

Entering the appropriate chart with the values for the moment and the unsupported length, we locate the critical intersection point. The nearest solid line curve above and to the right of this point is that for a W 8 × 31.

For consideration of the deflection limit we may now find the actual deflection for the W 8 × 31 for the given span and load. If this is excessive, we then return to the chart to read the other shapes whose curves are above and to the right of the critical intersection point and examine the corresponding deflections until we can make an acceptable choice. An alternative to this pick-and-try method is a separate calculation for the required moment of inertia of the beam with a transformed version of the formula for maximum deflection for the beam (Appendix Fig. E.6, Case 2). Thus

$$I = \frac{5}{384} \times \frac{Wl^3}{E\Delta} = \frac{5 \times 19.6 \times (20 \times 12)^3}{384 \times 29{,}000 \times 0.75} = 162 \text{ in.}^4$$

$$[0.06782 \text{ m}^4 \quad \text{or} \quad 67.82 \times 10^6 \text{ mm}^4]$$

For the W 8 × 31, from Appendix Table A.1, I = 110 in.[4] Because this is less than that required, we return to the chart and select the W 10 × 33 for which I = 170 in.[4] The W 10 × 33 is an acceptable selection for the criteria established.

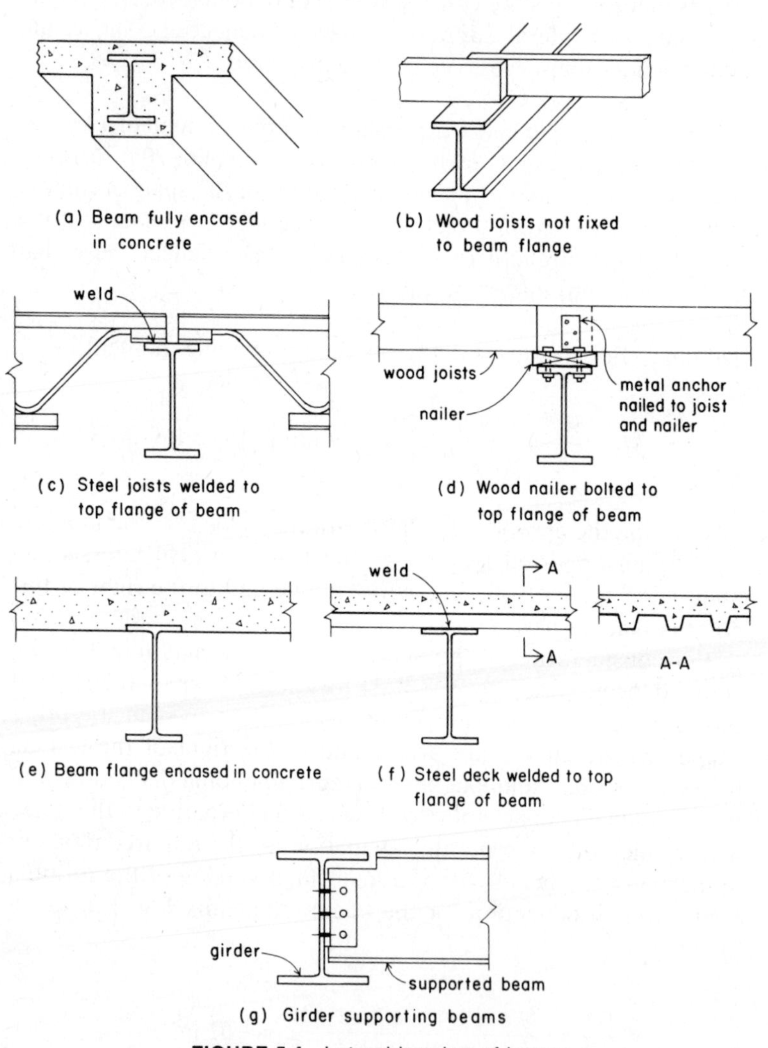

FIGURE 5.4 Lateral bracing of beams.

It is not always a simple matter to decide that a beam is laterally supported. In cases such as that shown in Fig. 5.4*a* lateral support is supplied to beams by the floor construction; it is evident from the figure that lateral deflection of the top flange is prevented by the concrete slab. On the other hand, the type of floor construction shown in Fig. 5.4*b*, where wood joists simply rest on the top flange of a steel beam, offers no resistance to sideways buckling. Floor systems of the types indicated in Fig. 5.4*c*–*f* usually furnish adequate lateral support of the top flange. However, metal or precast floor systems held in place by clips generally have insufficiently rigid connections to the flange to provide adequate lateral bracing. If a beam acts as a girder and supports other beams with connections like those shown in Fig. 5.4*g*, lateral bracing is provided at the connections and the laterally unsupported length of the top flange to be checked against L_c or L_u becomes the distance between the supported beams.

Problem 5.7.A. A W 21 × 83 is used as a simple beam to carry a total uniformly distributed load, including the beam weight, of 53 kips [236 kN] on a span of 36 ft [11 m]. Lateral support is provided only at the ends and at the midspan. Is the section an adequate choice for these conditions?

Problem 5.7.B. A simple beam is required for a span of 32 ft [9.75 m]. The load, including the beam weight, is uniformly distributed and consists of a total of 100 kips [445 kN]. If lateral support is provided only at the ends and midspan and the deflection under full load is limited to 0.6 in. [15.2 mm], find an adequate rolled shape for the beam.

5.8 COMPACT AND NONCOMPACT SECTIONS

To qualify for use of the maximum allowable bending stress of $0.66F_y$ a beam consisting of a rolled section must satisfy several qualifications, principal among which are the following:

The beam section must be symmetrical about its minor ($Y-Y$) axis and the plane of the loading must coincide with the plane of this axis.

The web and flanges of the section must have width-to-thickness ratios that qualify the section as *compact*.

The compression flange of the beam must be adequately braced against lateral buckling.

The criteria for determining whether a section is compact includes as a variable the F_y of the steel. It is therefore not possible to identify sections for this condition strictly on the basis of their geometric properties. The yield stress limit for qualification as a compact section is given as the value F'_y in Appendix Table A.1 for W shapes. When members are subjected to combined bending and compression the qualifications for consideration as compact get even tougher and the value in the tables given as F'''_y indicates the limit for this condition.

When sections do not qualify as compact the allowable bending stress must be reduced by using the formulas in Sec. 1.5.1.4 of the AISC Specification. In some cases these reductions have been incorporated into the design aids in the AISC Manual (Ref. 1) for A36 steel, as described in the example problems.

5.9 TORSIONAL EFFECTS

In various situations steel beams may be subjected to torsional twisting effects in addition to the primary conditions of shear and bending. These effects may occur when the beam is loaded in a plane that does not coincide with the shear center of the section. Even for the doubly symmetrical W, M, and S shapes loadings may produce torsion when applied off-center, as shown in Fig. 5.5.

A special torsional effect is that of the rotational effect known as torsional buckling. Beams that are weak on their minor axes are subject to this effect, which occurs at the points of support or at the location of concentrated loads (Fig. 5.6*a*).

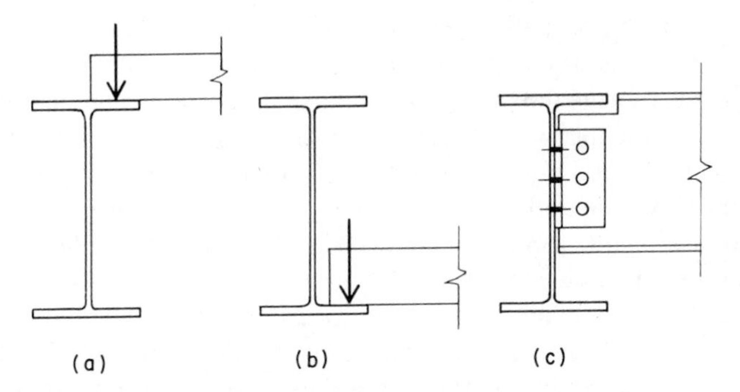

FIGURE 5.5 Torsion produced by off-center loading.

When potential torsional effects threaten they may be dealt with in one of two ways. In the first we simply compute the torsional moments and design the beam to resist them or we determine the torsional buckling and reduce the allowable stress accordingly. The analysis required, however, is complex and beyond the scope of this book. In the second method, which is usually the preferred, adequate bracing is provided for the beam

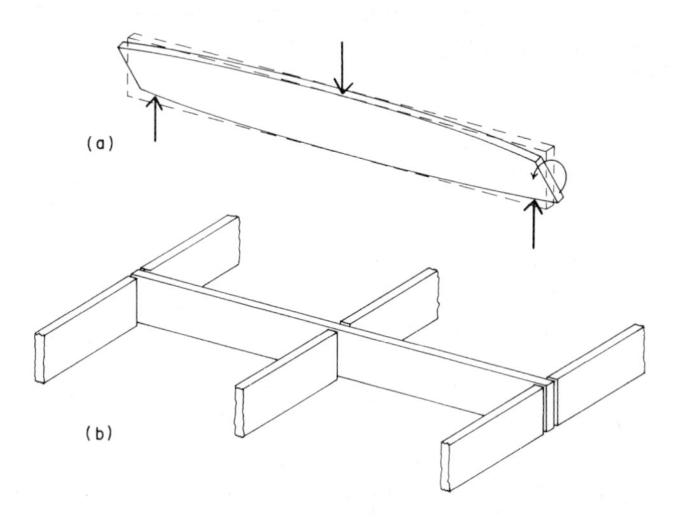

FIGURE 5.6 Torsional (rotational) buckling: (a) of an unbraced beam; (b) prevented by framing.

to prevent the potential torsional rotation, in which case the torsional effect is essentially avoided.

In many cases the ordinary details of construction will result in adequate bracing against torsion. In the detail shown in Fig. 5.5*c*, for example, although the supported beam applies a slightly off-center load, the connection between the beam and girder may be adequately stiff to prevent torsional twisting of the girder. Some judgment must be exercised, of course, with regard to the relative sizes of girder, beam, and the connection itself.

In Fig. 5.6*b* the torsional buckling of the beam shown in Fig. 5.6*a* is prevented by the framing that is attached at right angles to the beam.

5.10 CRIPPLING OF BEAM WEBS

An excessive end reaction on a beam or an excessive concentrated load at some point along the interior of the span may cause crippling or localized yielding of the beam web. The AISC Specification requires that end reactions or concentrated loads for beams without stiffeners or other web reinforcement shall not exceed the following (Fig. 5.7):

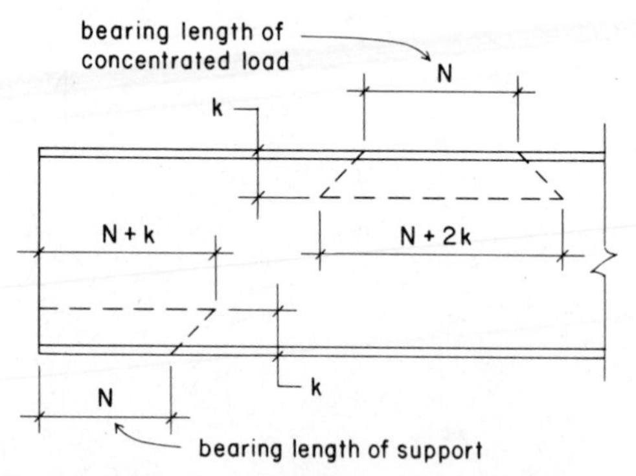

FIGURE 5.7 Determination of effective length for computation of web crippling.

$$\text{maximum end reaction} = 0.75F_y t(N + k)$$

$$\text{maximum interior load} = 0.75F_y t(N + 2k)$$

where t = thickness of beam *web*, in inches,
N = length of bearing or length of concentrated load (not less than k for end reactions), in inches,
k = distance from outer face of flange to web toe of fillet, in inches,
$0.75F_y$ = 27 ksi for A36 steel [186 MPa].

When these value are exceeded the webs of the beams should be reinforced with stiffeners, the length of bearing increased, or a beam with a thicker web selected.

Example 1. A W 21 × 57 beam of A36 steel has an end reaction that is developed in bearing over a length of N = 10 in. [254 mm]. Check the beam for web crippling if the reaction is 44 kips [196 kN].

Solution. In Appendix Table A.1 we find that k = 1.375 in. [35 mm] and the web thickness is 0.405 in. [10 mm]. To check for web crippling we find the maximum end reaction permitted and compare it with the actual value for the reaction. Thus

$$R = F_p \times t \times (N + k) = 27 \times 0.405 \times (10 + 1.375)$$

$$= 124 \text{ kips (the allowable reaction) [538 kN]}$$

Because this is greater than the actual reaction, the beam is not critical with regard to web crippling.

Example 2. A W 12 × 26 of A36 steel supports a column load of 70 kips [311 kN] at the center of the span. The bearing length of the column on the beam is 10 in. [254 mm]. Investigate the beam for web crippling under this concentrated load.

Solution. In Appendix Table A.1 we find that $k = 0.875$ in. [22 mm] and the web thickness is 0.230 in. [5.84 mm]. The allowable load that can be supported on the given bearing length is

$$P = F_p \times t \times (N + 2k) = 27 \times 0.230 \times \{10 + (2 \times 0.875)\}$$
$$= 73 \text{ kips } [324 \text{ kN}]$$

which exceeds the required load. Because the column load is less than this value, the beam web is safe from web crippling.

Problem 5.10.A. Compute the maximum allowable reaction with respect to web crippling for a W 14 × 30 of A36 steel with an 8 in. [203 mm] bearing-plate length.

Problem 5.10.B. A column load of 81 kips [360 kN] with a bearing-plate length of 11 in. [279 mm] is placed on top of the beam in Problem 5.10.A. Are web stiffeners required to prevent web crippling?

5.11 BEAM BEARING PLATES

Beams that are supported on walls or piers of masonry or concrete usually rest on steel bearing plates. The purpose of the plate is to provide an ample bearing area. The plate also helps to seat the beam at its proper elevation. Bearing plates provide a level surface for a support and, when properly placed, afford a uniform distribution of the beam reaction over the area of contact with the supporting material.

By reference to Fig. 5.8 the area of the bearing plate is $B \times N$. It is found by dividing the beam reaction by F_p, the allowable bearing value of the supporting material. Then

$$A = \frac{R}{F_p}$$

where $A = B \times N$, area of the plate, in sq in.,
R = reaction of beam, in pounds or kips,

F_p = allowable bearing pressure on the supporting material, in psi or ksi (see Table 5.1).

The thickness of the wall generally determines N, the dimension of the plate parallel to the length of the beam. If the load from the beam is unusually large, the dimension B may become excessive. For such a condition one or more shallow-depth I-beams, placed parallel to the wall length, may be used instead of a plate. The dimensions B and N are usually in even inches and a great variety of thicknesses is available.

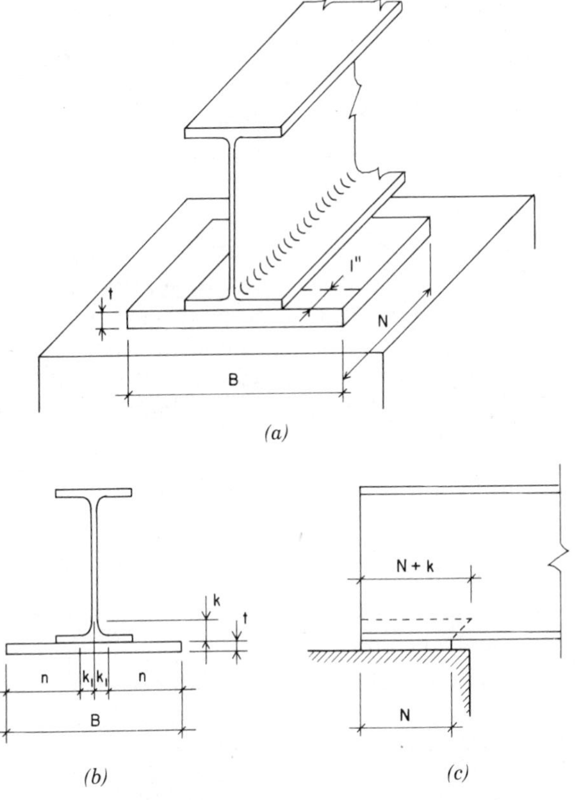

FIGURE 5.8 Reference dimensions for beam and bearing plates.

TABLE 5.1 Allowable Bearing Pressure on Masonry and Concrete

Type of Material and Conditions	Allowable Unit Stress in Bearing, F_p psi	kPa
Solid brick, reinforced, type S mortar		
f'_m = 1500 psi	170	1200
f'_m = 4500 psi	338	2300
Hollow unit masonry, unreinforced, type S mortar, f'_m = 1500 psi (on net area of masonry)	225	1500
Concrete[a]		
(1) Bearing on full area of support		
f'_c = 2000 psi	500	3500
f'_c = 3000 psi	750	5000
(2) Bearing on $\frac{1}{3}$ or less of support area		
f'_c = 2000 psi	750	5000
f'_c = 3000 psi	1125	7500

[a] Stresses for areas between these limits may be determined by direct proportion.

The thickness of the plate is determined by considering the projection n (Fig. 5.8*b*) as an inverted cantilever; the uniform bearing pressure on the bottom of the plate tends to curl it upward about the beam flange. The required thickness may be computed readily by the following formula, which does not involve direct computation of bending moment and section modulus:

$$t = \sqrt{\frac{3f_p n^2}{F_b}}$$

where t = thickness of plate, in inches,

f_p = *actual* bearing pressure of the plate on the masonry, in psi or ksi,

F_b = allowable bending stress in the plate (the AISC Specification gives the value of F_b as $0.75F_y$; for A36 steel $F_y = 36$ ksi; therefore, $F_b = 0.75 \times 36 = 27$ ksi),

$n = \dfrac{B}{2} - k_1$, in inches (see Fig. 5.8*b*),

k_1 = the distance from the center of the web to the toe of the fillet; values of k_1 for various beam sizes may be found in Appendix Table A.1.

The foregoing formula is derived by considering a strip of plate 1 in. wide (Fig. 5.8*a*) and t in. thick, with a projecting length of n inches, as a cantilever. Because the upward pressure on the steel strips is f_p, the bending moment at distance n from the edge of the plate is

$$M = f_p n \times \frac{n}{2} = \frac{f_p n^2}{2}$$

For this strip with rectangular cross section

$$\frac{I}{c} = \frac{bd^2}{6} \qquad \text{(Fig. E.24)}$$

and because $b = 1$ in. and $d = t$ in.

$$\frac{I}{c} = \frac{1 \times t^2}{6} = \frac{t^2}{6}$$

Then, from the beam formula,

$$\frac{M}{F_b} = \frac{I}{c}$$

Substituting the values of M and I/c determined above,

$$\frac{f_p n^2}{2} \times \frac{1}{F_b} = \frac{t^2}{6}$$

and

$$t^2 = \frac{6 f_p n^2}{2F_b} \quad \text{or} \quad t = \sqrt{\frac{3 f_p n^2}{F_b}}$$

When the dimensions of the bearing plate are determined the beam should be investigated for web crippling on the length ($N + k$) shown in Fig. 5.8*c*. This is explained in Sec. 5.10.

Example 1. A W 21 × 57 of A36 steel transfers an end reaction of 44 kips [196 kN] to a wall built of solid brick by means of a bearing plate of A36 steel. Assume type S mortar and a brick with f'_m = 1500 psi. The N dimension of the plate (see Fig. 5.8) is 10 in. [254 mm]. Design the bearing plate.

Solution. In Appendix Table A.1 we find that k_1 for the beam is 0.875 in. [22 mm]. From Table 5.1 the allowable bearing pressure F_p for this wall is 170 psi [1200 kPa]. The required area of the plate is then

$$A = \frac{R}{F_p} = \frac{44{,}000}{170} = 259 \text{ in.}^2 \quad [163 \times 10^3 \text{ mm}^2]$$

Then, because N = 10 in. [254 mm],

$$B = \frac{259}{10} = 25.9 \text{ in.}^2 \quad [643 \text{ mm}]$$

which is rounded off to 26 in. [650 mm].

With the true dimensions of the plate we now compute the true bearing pressure:

$$f_p = \frac{R}{A} = \frac{44{,}000}{10 \times 26} = 169 \text{ psi} \quad [1187 \text{ kPa}]$$

To find the thickness we first determine the value of n:

$$n = \frac{B}{2} - k_1 = \frac{26}{2} - 0.875 = 12.125 \text{ in.} \quad [303 \text{ mm}]$$

Then

$$t = \sqrt{\frac{3f_p n^2}{F_b}} = \sqrt{\frac{3 \times 169 \times (12.125)^2}{27{,}000}} = \sqrt{2.2760} = 1.66 \text{ in.}$$

$$\sqrt{1758} = [42 \text{ mm}]$$

The complete design for this problem would include a check of the web crippling in the beam. This has already been done as Example 1 in Sec. 5.10.

In the event that a comparatively light reaction or a high allowable bearing pressure reduces the bearing area required so that a beam may be supported without a bearing plate, the beam flange should be checked for bending induced by the bearing pressure to make certain that it does not exceed F_b. This may be accomplished by use of the formula

$$f_b = \frac{3f_p n^2}{t^2}$$

where f_b = actual bending stress in the beam flange,
f_p = actual bearing pressure of the beam flange on the supporting structure,
n = (flange width/2) − k_1,
t = thickness of the flange.

Example 2. A W 12 × 53 of A36 steel transfers a load of 16 kips [71 kN] to a brick wall laid up with type S mortar and brick with f'_m of 4500 psi. The beam has an 8-in. [203-mm] bearing length (dimension N) on the wall. If a bearing plate is not required, compute the maximum bending stress in the beam flange. Does the bending stress exceed 27,000 psi [186 MPa], the allowable in the flange when acting as a bearing plate?

Solution. In Appendix Table A.1 we find that the flange width is 9.995 in. [254 mm], the flange thickness, 0.575 in. [14.6 mm], and the dimension k_1, 0.8125 in. [20.6 mm]. From Table 5.1 we find that the allowable bearing pressure on the brick wall is 338 psi [2300 kPa].

The bearing area of the flange on the wall is

$$A = 8 \times 10 = 80 \text{ in.}^2 \quad [52 \times 10^3 \text{ mm}^2]$$

and because the reaction is 16 kips [71 kN] the actual bearing pressure on the wall is

$$f_p = \frac{R}{A} = \frac{16{,}000}{80} = 200 \text{ psi } [1377 \text{ kPa}]$$

Because this value is less than the allowable pressure, no bearing plate is required unless the bending stress in the beam flange exceeds the allowable. Investigating the beam flange for bending stress, we obtain

$$n = \frac{10}{2} - 0.8125 = 4.1875 \text{ in. } [106 \text{ mm}]$$

and

$$f_b = \frac{3f_p n^2}{t^2} = \frac{3 \times 200 \times (4.1875)^2}{(0.575)^2} = 31{,}822 \text{ psi } [219 \text{ MPa}]$$

which is greater than that allowable.

Although the bearing stress is not critical, in this case a bearing plate must be used unless the bearing length can be increased.

Problem 5.11.A. A W 14 × 30 with a reaction of 20 kips [89 kN] rests on a brick wall with brick of f'_m = 1500 psi and type S mortar. The beam has a bearing length of 8 in. [203 mm] parallel to the length of the beam. If the bearing plate is A36 steel, determine its dimensions.

Problem 5.11.B. A wall of brick with f'_m of 1500 psi and type S mortar supports a W 18 × 50 of A36 steel. The beam reaction is 25 kips [111 kN] and the bearing length N is 9 in. [229 mm] Design the beam bearing plate.

5.12 MANUFACTURED TRUSSES

Shop-fabricated parallel-chord trusses are produced in a wide range of sizes with various fabrication details by a number of manufacturers. Most producers comply with the regulations developed by the major industry organization in this area: the Steel

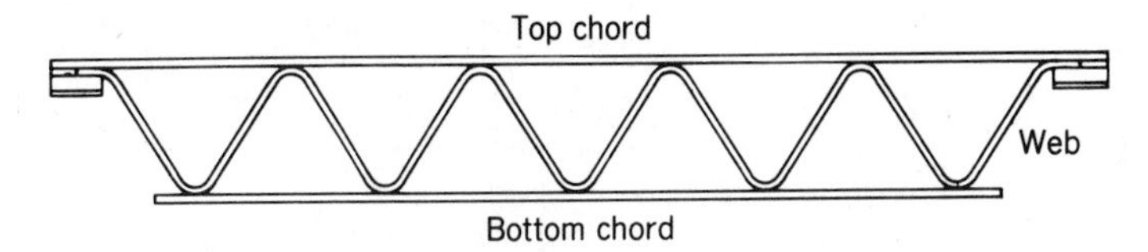

FIGURE 5.9 Short-span, open-web steel joist.

Joist Institute. Publications of the institute (called the SJI) are a chief source of design information (see Ref. 9), although the products of individual manufacturers vary some, so that more information may be provided by suppliers or the manufacturers themselves.

The smallest and lightest members produced, called open-web joists, are used for the direct support of roof and floor decks, sustaining essentially only uniformly distributed loads on their top chords. A popular form for these is that shown in Fig. 5.9, with chords of cold-formed sheet steel and webs of steel rods. Chords may also be double angles, with the rods sandwiched between the angles at joints.

Table 5.2 is adapted from the standard tables of the Steel Joist Institute. This table lists the range of joist sizes available in the basic K series. (*Note:* A few of the heavier sizes have been omit-

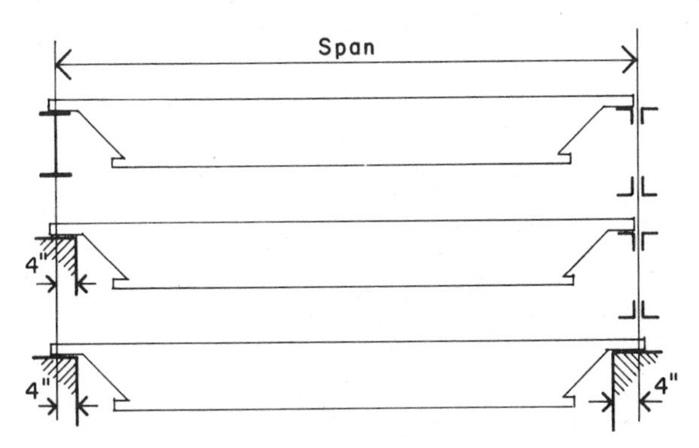

FIGURE 5.10 Definition of span for open-web steel joists, as given in Ref. 9. Reprinted by permission of the Steel Joist Institute.

TABLE 5.2 Allowable Loads for K-Series Open-Web Steel Joists[a]

Joist Designation	12K1	12K3	12K5	14K1	14K3	14K4	14K6	16K2	16K3	16K4	16K6	18K3	18K4	18K5	18K7	20K3	20K4	20K5	20K7
Weight (lb/ft)	5.0	5.7	7.1	5.2	6.0	6.7	7.7	5.5	6.3	7.0	8.1	6.6	7.2	7.7	9.0	6.7	7.6	8.2	9.3
Span (ft)																			
20	241 (142)	302 (177)	409 (230)	284 (197)	356 (246)	428 (287)	525 (347)	368 (297)	410 (330)	493 (386)	550 (426)	463 (423)	550 (490)	550 (490)	550 (490)	517 (517)	550 (550)	550 (550)	550 (550)
22	199 (106)	249 (132)	337 (172)	234 (147)	293 (184)	353 (215)	432 (259)	303 (222)	337 (247)	406 (289)	498 (351)	382 (316)	460 (370)	518 (414)	550 (438)	426 (393)	514 (461)	550 (490)	550 (490)
24	166 (81)	208 (101)	282 (132)	196 (113)	245 (141)	295 (165)	362 (199)	254 (170)	283 (189)	340 (221)	418 (269)	320 (242)	385 (284)	434 (318)	526 (382)	357 (302)	430 (353)	485 (396)	550 (448)
26				166 (88)	209 (110)	251 (129)	308 (156)	216 (133)	240 (148)	289 (173)	355 (211)	272 (190)	328 (222)	369 (249)	448 (299)	304 (236)	366 (277)	412 (310)	500 (373)
28				143 (70)	180 (88)	216 (103)	265 (124)	186 (106)	207 (118)	249 (138)	306 (168)	234 (151)	282 (177)	318 (199)	385 (239)	261 (189)	315 (221)	355 (248)	430 (298)
30								161 (86)	180 (96)	216 (112)	266 (137)	203 (123)	245 (144)	276 (161)	335 (194)	227 (153)	274 (179)	308 (201)	374 (242)
32								142 (71)	158 (79)	190 (92)	233 (112)	178 (101)	215 (118)	242 (132)	294 (159)	199 (126)	240 (147)	271 (165)	328 (199)
36												141 (70)	169 (82)	191 (92)	232 (111)	157 (88)	189 (103)	213 (115)	259 (139)
40																127 (64)	153 (75)	172 (84)	209 (101)

Joist Designation	22K4	22K5	22K6	22K9	24K4	24K5	24K6	24K9	26K5	26K6	26K9	28K6	28K7	28K8	28K10	30K7	30K8	30K9	30K12
Weight (lb/ft)	8.0	8.8	9.2	11.3	8.4	9.3	9.7	12.0	9.8	10.6	12.2	11.4	11.8	12.7	14.3	12.3	13.2	13.4	17.6
Span (ft)																			
28	348 (270)	392 (302)	427 (328)	550 (413)	381 (323)	429 (362)	467 (393)	550 (456)	466 (427)	508 (464)	550 (501)	548 (541)	550 (543)	550 (543)	550 (543)				
30	302 (219)	341 (245)	371 (266)	497 (349)	331 (262)	373 (293)	406 (319)	544 (419)	405 (346)	441 (377)	550 (459)	477 (439)	531 (486)	550 (500)	550 (500)	550 (543)	550 (543)	550 (543)	550 (543)
32	265 (180)	299 (201)	326 (219)	436 (287)	290 (215)	327 (241)	357 (262)	478 (344)	356 (285)	387 (309)	519 (407)	418 (361)	466 (400)	515 (438)	549 (463)	501 (461)	549 (500)	549 (500)	549 (500)
36	209 (126)	236 (141)	257 (153)	344 (201)	229 (150)	258 (169)	281 (183)	377 (241)	280 (199)	305 (216)	409 (284)	330 (252)	367 (280)	406 (306)	487 (366)	395 (323)	436 (353)	475 (383)	487 (392)
40	169 (91)	190 (102)	207 (111)	278 (146)	185 (109)	208 (122)	227 (133)	304 (175)	227 (145)	247 (157)	331 (207)	266 (183)	297 (203)	328 (222)	424 (284)	319 (234)	353 (256)	384 (278)	438 (315)
44	139 (68)	157 (76)	171 (83)	229 (109)	153 (82)	172 (92)	187 (100)	251 (131)	187 (108)	204 (118)	273 (155)	220 (137)	245 (152)	271 (167)	350 (212)	263 (176)	291 (192)	317 (208)	398 (258)
48					128 (63)	144 (70)	157 (77)	211 (101)	157 (83)	171 (90)	229 (119)	184 (105)	206 (117)	227 (128)	294 (163)	221 (135)	244 (148)	266 (160)	365 (216)
52									133 (65)	145 (71)	195 (93)	157 (83)	175 (92)	193 (100)	250 (128)	188 (106)	208 (116)	226 (126)	336 (184)
56												135 (66)	151 (73)	166 (80)	215 (102)	162 (84)	179 (92)	195 (100)	301 (153)
60																141 (69)	156 (75)	169 (81)	262 (124)

Source: Data adapted from more extensive tables in the *Standard Specifications, Load Tables, and Weight Tables for Steel Joists and Joist Girders*, 1988 ed. (Ref. 8), with permission of the publishers, the Steel Joist Institute. The Steel Joist Institute publishes both specifications and load tables; each of these contains standards that are to be used in conjunction with one another.

[a] Loads in pounds per foot of joist span; first entry represents the total joist capacity; entry in parentheses is the load that produces a maximum deflection of 1/360 of the span. See Fig. 5.10 for a definition of span.

ted to shorten the table.) Joists are identified by a three-unit designation. The first number indicates the overall depth of the joist, the letter tells the series, and the second number gives the class of size of the members used; the higher the number, the heavier the joist.

Table 5.2 can be used to select the proper joist for a determined load and span condition. There are usually two entries in the table for each span; the first number represents the total load capacity of the joist, and the number in parentheses identifies the load that will produce a deflection of 1/360 of the span. The following examples illustrate the use of the table data for some typical design situations.

Example 1. Open-web steel joists are to be used to support a roof with a unit live load of 20 psf and a unit dead load of 15 psf (not including the weight of the joists) on a span of 40 ft. Joists are spaced at 6 ft center to center. Select the lightest joist if deflection under live load is limited to 1/360 of the span.

Solution. We first determine the load per foot on the joist:

$$\begin{aligned} &\text{Live load:} && 6 \times 20 = 120 \text{ lb/ft} \\ &\text{Dead load:} && 6 \times 15 = \underline{\ 90} \text{ lb/ft} \\ &\text{Total load:} && = 210 \text{ lb/ft} \end{aligned}$$

We then scan the entries in Table 5.2 for the joists that will just carry these loads, noting that the joist weight must be deducted from the entry for total capacity. The possible choices for this example are summarized in Table 5.3. Although the joist weights are all very close, the 22K9 is the lightest choice.

TABLE 5.3 Possible Choices for the Roof Joist

	Load per Foot for the Indicated Joists			
Load Condition	22K9	24K6	26K5	28K6
Total capacity (from Table 5.2)	278	227	227	266
Joist weight (from Table 5.2)	11.3	9.7	9.8	11.4
Net usable capacity	266.7	217.3	217.2	254.6
Load for 1/360 deflection (from Table 5.2)	146	133	145	183

Example 2. Open-web steel joists are to be used for a floor with a unit live load of 75 psf and a unit dead load of 40 psf (not including the joists) on a span of 30 ft. Joists are 2 ft center to center, and deflection is limited to 1/360 of the span under live load only and to 1/240 of the span under total load. Determine the lightest joist possible and the joist with the least depth possible.

Solution. As in Example 1, we first find the loads on the joist:

$$\begin{aligned} &\text{Live load:} && 2 \times 75 = 150 \text{ lb/ft} \\ &\text{Dead load:} && 2 \times 40 = \underline{80} \text{ lb/ft} \\ &\text{Total load:} && = 230 \text{ lb/ft} \end{aligned}$$

To satisfy the deflection criteria, we must find a table entry in parentheses of 150 lb/ft (for live load only) or 240/360 × 230 = 153 lb/ft (for total load). The possible choices obtained from scanning Table 5.2 are summarized in Table 5.4, from which we observe:

The lightest joist is the 20K4.
The shallowest joist is the 18K5.

In real situations there may be compelling reasons for selection of a deeper joist, even though its load capacity may be redundant.

For heavier loads and longer spans, trusses are produced in series described as long span and deep long span, the latter achieving depths of 7 ft and spans approaching 150 ft. In some situations the particular loading and span may clearly indicate the choice of the series, as well as the specific size of member. In

TABLE 5.4 Possible Choices for the Floor Joist

	Load per Foot for the Joists Indicated		
Load Condition	18K5	20K4	22K4
Total capacity (from Table 5.2)	276	274	302
Joist weight (from Table 5.2)	7.7	7.6	8.0
Net usable capacity	268.3	266.4	294
Load for 1/360 deflection (from Table 5.2)	161	179	219

many cases, however, the separate series overlap in capabilities, making the choice dependent on the product costs.

Open-web, long-span, and deep long-span trusses are all essentially designed for the uniformly loaded condition. This load may be due only to a roof or floor deck on the top, or may include a ceiling attached to the bottom. For roofs, an often used potential is that for sloping the top chord to facilitate drainage, while maintaining the flat bottom for a ceiling. Relatively small concentrated loads may be tolerated, the more so if they are applied close to the truss joints.

Load tables for all the standard products, as well as specifications for lateral bracing, support details, and so on, are available from the SJI (see Ref. 9). Planar trusses, like very thin beams, need considerable lateral support. Attached roof or floor decks and even ceilings may provide the necessary support if the attachment and the stiffness of the bracing construction are adequate. For trusses where no ceiling exists, bridging or other forms of lateral bracing must be used.

Development of bracing must be carefully studied in many cases. This is a three-dimensional problem, involving not only the stability of the trusses, but the general development of the building construction. Elements used for bracing can often do service as supports for ducts, lighting, building equipment, catwalks, and so on.

For development of a complete truss system, a special type of prefabricated truss available is that described as a *joist girder*. This truss is specifically designed to carry the regularly spaced, concentrated loads consisting of the end supports of joists. The general form of joist girders is shown in Fig. 5.11. Also shown in Fig. 5.11 is the form of standard designation for a joist girder, which includes indications of the nominal girder depth, weight, spacing of members, and the unit load to be carried.

Predesigned joist girders may be selected from the catalogs of manufacturers in a manner similar to that for open-web joists. The procedure is usually as follows:

1. The designer first determines the desired spacing of joists, loads to be carried, and span of the girder. The joist spacing must be an even-number division of the girder span.

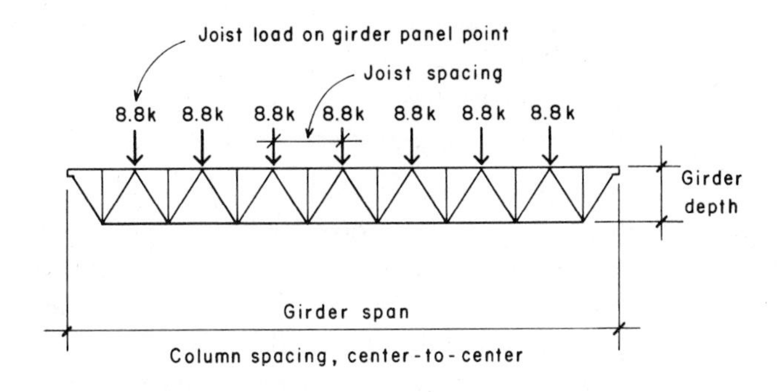

Standard Designation:

48G	8N	8.8K
Depth in inches	Number of joist spaces	Load on each panel point in kips

Specify: 48G8N8.8K

FIGURE 5.11 Considerations for joist girders, from Ref. 9.

2. The total design load for one supported joist is determined. This is the unit concentrated load on the girder.
3. With the girder span, joist spacing, and joist load, the weight and depth for the girder is found from design tables in the manufacturer's catalog. Specification of the girder is as shown in Fig. 5.11.

An illustration of the use of a complete truss system using open-web joists and joist girders is shown in the design example in Sec. 12.2.

Problem 5.12.A. Open-web steel joists are to be used for a roof with a live load of 25 psf and a dead load of 20 psf (not including joists) on a span of 48 ft. Joists are 4 ft center to center, and the deflection under live load is limited to $\frac{1}{360}$ of the span. Select the lightest joist possible.

Problem 5.12.B. Open-web steel joists are to be used for a roof with a live load of 30 psf and a dead load of 18 psf (not including joists) on a span of 44 ft. Joists are 5 ft center to center, and deflection under live load is limited to $\frac{1}{360}$ of the span. Select the lightest joist.

Problem 5.12.C. Open-web steel joists are to be used for a floor with a live load of 50 psf and a dead load of 45 psf (not including joists) on a span of 36 ft. Joists are 2 ft center to center, and deflection is limited to $\frac{1}{360}$ of the span under live load and to $\frac{1}{240}$ of the span under total load. Select (a) the lightest possible joist and (b) the shallowest possible joist.

Problem 5.12.D. Repeat Problem 5.12.C except that the live load is 100 psf, the dead load 35 psf, and the span 26 ft.

5.13 STEEL DECKS

Steel decks consist of formed sheet steel, produced in a variety of configurations, as shown in Fig. 5.12. The simplest is the corrugated sheet, shown in Fig. 5.12*a*. This may be used as the single, total surface for walls or roofs of utilitarian buildings (tin shacks?). For more serious buildings it is mostly used only in pairing with a light steel joist system where the joist spacing is quite close and a poured concrete fill is used.

A widely used product is that shown in three variations in Fig. 5.12*b–d*. When used for roof deck, where loads are light and a need exists for a flat top surface to support insulation and roofing, the deck unit shown in Fig. 5.12*b* is popular. For heavier floor loads and use with a structural-grade concrete fill, the units shown in Fig. 5.12*c* and *d* are used. These come in different sizes as measured by the overall height of the deck pleats; common sizes being 1.5, 3, and 4.5 in. The units shown in Fig. 5.12*b–d* are produced by many different manufacturers.

There are also steel deck units produced as priority items by individual manufacturers. Figure 5.12*e* and *f* shows two units of considerable depth, capable of achieving quite long spans. Use of these special units is somewhat a matter of regional marketing.

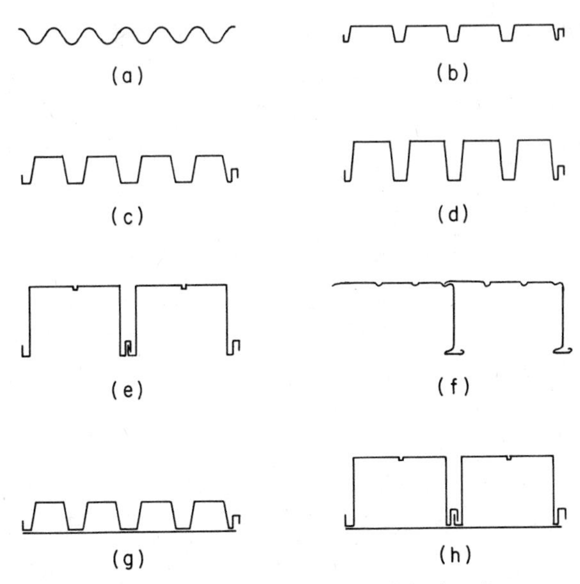

FIGURE 5.12 Formed sheet steel deck units.

Although less used now, with the advent of other wiring products and techniques, a possible use for the steel deck is as a conduit for power or communication wiring. This is accomplished by closing the deck cells with a flat sheet of steel, as shown in Fig. 5.12*g* and *h*. This provides for wiring in one direction in a wiring grid; the perpendicular wiring being achieved in conduits buried in the concrete fill on top of the deck.

Decks vary in form, including ones now shown in Fig. 5.12, and in the thickness (gage) of the sheet steel. Choice relates to form and to the load and span conditions for the deck. Units are typically available in single pieces of 30 or more ft length. For the shallower decks, this usually assures a multiple-span support condition. Required fire ratings for the construction and use of the deck for diaphragm action in resisting lateral forces must also be considered.

Roof decks are most often used with either a very low density poured concrete fill or preformed rigid foam plastic insulation units on top of the deck. The deck itself is the functioning struc-

TABLE 5.5 Load Capacity of Formed Steel Roof Deck

Deck[a] Type	Span Condition	Weight[b] (psf)	Total (Dead & Live) Safe Load[c] for Spans Indicated in ft-in.												
			4-0	4-6	5-0	5-6	6-0	6-6	7-0	7-6	8-0	8-6	9-0	9-6	10-0
NR22	Simple	1.6	73	58	47										
NR20		2.0	91	72	58	48	40								
NR18		2.7	121	95	77	64	54	46							
NR22	Two	1.6	80	63	51	42									
NR20		2.0	96	76	61	51	43								
NR18		2.7	124	98	79	66	55	47	41						
NR22	Three or More	1.6	100	79	64	53	44								
NR20		2.0	120	95	77	63	53	45							
NR18		2.7	155	123	99	82	69	59	51	44					
IR22	Simple	1.6	86	68	55	45									
IR20		2.0	106	84	68	56	47	40							
IR18		2.7	142	112	91	75	63	54	46	40					
IR22	Two	1.6	93	74	60	49	41								
IR20		2.0	112	88	71	59	50	42							
IR18		2.7	145	115	93	77	64	55	47	41					
IR22	Three or More	1.6	117	92	75	62	52	44							
IR20		2.0	140	110	89	74	62	53	46	40					
IR18		2.7	181	143	116	96	81	69	59	52	45	40			
WR22	Simple	1.6			(89)	(70)	(56)	(46)							
WR20		2.0			(112)	(87)	(69)	(57)	(47)	(40)					
WR18		2.7			(154)	(119)	(94)	(76)	(63)	(53)	(45)				
WR22	Two	1.6			98	81	68	58	50	43					
WR20		2.0			125	103	87	74	64	55	49	43			
WR18		2.7			165	137	115	98	84	73	65	57	51	46	41
WR22	Three or More	1.6			122	101	85	72	62	54	(46)	(40)			
WR20		2.0			156	129	108	92	80	(67)	(57)	(49)	(43)		
WR18		2.7			207	171	144	122	105	(91)	(76)	(65)	(57)	(50)	(44)

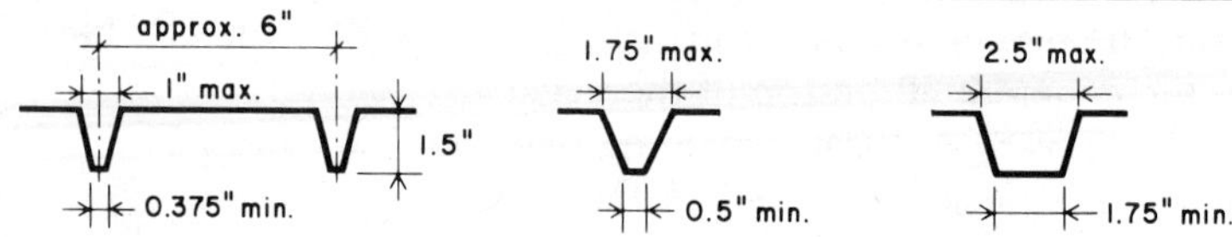

Source: Adapted from the *Steel Deck Institute Design Manual for Composite Decks, Form Decks, and Roof Decks,* 1981–1982 issue (Ref. 10), with permission of the publishers, the Steel Deck Institute. May not be reproduced without express permission of the Steel Deck Institute.

[a] Letters refer to rib type (see key). Numbers indicate gage (thickness) of steel.

[b] Approximate weight with paint finish; other finishes also available.

[c] Total safe allowable load in lb/sq ft. Loads in parentheses are governed by live load deflection not in excess of $\frac{1}{240}$ of the span, assuming a dead load of 10 psf.

ture in these cases. For floor decks, however, the concrete fill is usually of sufficient strength to perform structural tasks, so the steel unit + concrete fill combination is often designed as a composite structural element. See discussion in Sec. 8.3.

Table 5.5 presents data relating to the use of the type of deck unit shown in Fig. 5.12*b* for roof structures. These data are adapted from a publication distributed by an industry-wide organization referred to in the table notes. In general, data for these manufactured products is best obtained from the manufacturers of the products, especially for the design of the composite floor units.

In general, economy is obtained with the use of the thinnest sheet steel (highest gage number) that can be utilized for a particular application. Where lateral forces are critical, this choice may be determined on the basis of consideration of diaphragm actions.

Rusting is a critical problem with the very thin deck elements, so that some rust-inhibiting finish is commonly used, except for surfaces that will be covered with a concrete fill. The deck weights given in Table 5.5 are based on painted surfaces, which is the least expensive and most common finish.

Steel deck units are available from a large number of manufacturers. Specific information regarding the type of deck available, possible range of sizes, rated load-carrying capacities, and so on, should be obtained directly from those who supply these products to the region in which a proposed building is to be built. The information in Table 5.5 is provided by the national organization that provides standards to the deck manufacturers and indicates one widely used type of deck.

Problem 5.13.A,B,C. Using data from Table 5.5, select the lightest steel deck for the following:

A Simple span of 7 ft, total load of 45 psf

B Two-span condition, span of 8.5 ft, total load of 45 psf

C Three-span condition, span of 6 ft, total load of 50 psf

5.14 ALTERNATIVE DECKS WITH STEEL FRAMING

Figure 5.13 shows four possibilities for a floor deck used in conjunction with a framing system of rolled steel sections. When a wood deck is used, it is usually nailed to a series of wood joists or trusses, which are supported by the steel beams. However, in some cases the deck may be nailed to wood strips bolted to the top of the steel framing, as shown in the illustration.

A site-cast concrete slab may be used, with the slab forms placed on the underside of the top flanges of the steel beams,

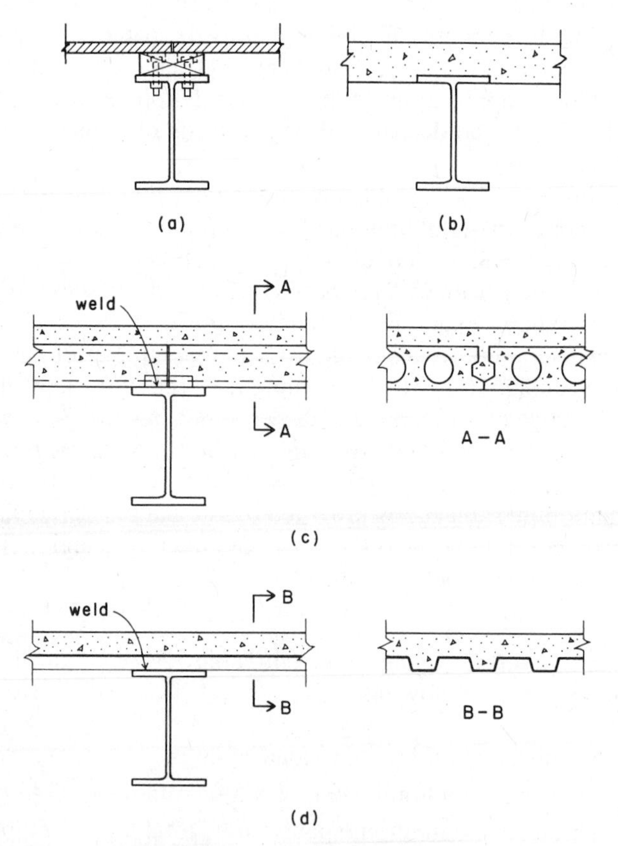

FIGURE 5.13 Typical floor decks.

producing the detail shown in Fig. 5.13*b*. It is common in this case to use steel devices welded to the top of the beams to develop composite action of the beams and the slab.

Concrete may also be used in the form of precast units that are welded to the beams using steel devices cast into the units. A concrete fill is ordinarily used with such units in order to develop a smooth surface for flooring (see Fig. 5.13*c*).

As described in Section 5.13, formed sheet steel units may be used with a concrete fill, the steel units being welded to the tops of the beams. It is also now common to use a concrete fill on top of a plywood deck. For both cases, the fill adds stiffness, mass to resist bouncing, enhanced acoustic properties, and some possibility for buried wiring.

Figure 5.14 shows three possibilities for a roof deck used in conjunction with a framing system of steel. A fourth possibility is that of the plywood deck shown in Fig. 5.13*a*. Many of the issues discussed for the floor decks also apply here. However, roof loads are usually lighter and there is less concern for bounciness due to people walking; thus some lighter systems are feasible here and not for floors.

Formed steel units such as those presented in Table 5.5 are normally used with a rigid insulation (as shown in Fig. 5.14*a*) or a poured concrete fill. If the concrete fill is of the very low density type, it does not have significant strength and the deck is the functioning spanning structure. However, a slightly stronger fill may be used in some cases with a light steel deck, the steel units functioning primarily only to form the concrete deck.

To facilitate roof drainage, concrete fill may be varied in thickness, allowing the deck units to remain flat. Rigid insulation units can also be obtained in modular packages with tapered pieces. This is only possible for a few inches of elevation change; the structure must be tilted for substantial slopes.

A special deck is that shown in Fig. 5.14*c*, consisting of a low-density concrete fill poured on top of a forming system of inverted steel tees and lay-in rigid panel units. The panel units can be used to form the finished ceiling surface where it is possible to leave the steel framing elements exposed.

There are many factors relating to regional concerns that make

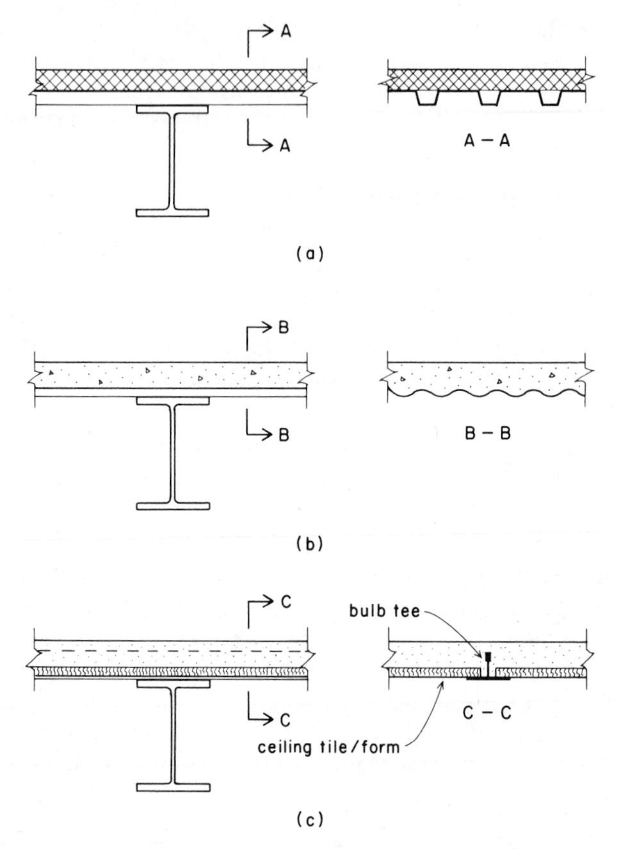

FIGURE 5.14 Typical roof decks.

certain forms of decking popular in different parts of the United States. In regions of high seismic risk, required diaphragm actions by both roof and floor decks present a major performance requirement. In regions of severe windstorms major uplift forces on roofs may affect both the choice of the structural deck and the general details of roof construction materials. Ready availability and general market competition of products is always a major consideration, and may favor certain products when choices are otherwise somewhat arbitrary; that is, there are several alternatives that may more or less equally satisfy design requirements.

In the end, it is the complete development of the building construction that is most important, and the selection of any individual component must be judged in the larger context.

6

STEEL COLUMNS

Steel columns in buildings vary from single-piece members (pipe, tube, wide-flange, etc.) to built-up assemblages of endless variety, including gigantic sections for large towers and high-rise buildings. The basic column function is the development of axial compression; thus the ordinary form of a column is a linear member of constant cross section. However, many columns must also sustain bending, shear, or torsion, and even in some cases tension. In this chapter we present a discussion of general column design issues, with an emphasis on the simple, single-piece column designed for axial compression.

6.1 INTRODUCTION

A column or strut is a compression member, the length of which is several times greater than its least lateral dimension. The term column denotes a relatively heavy vertical member, whereas the

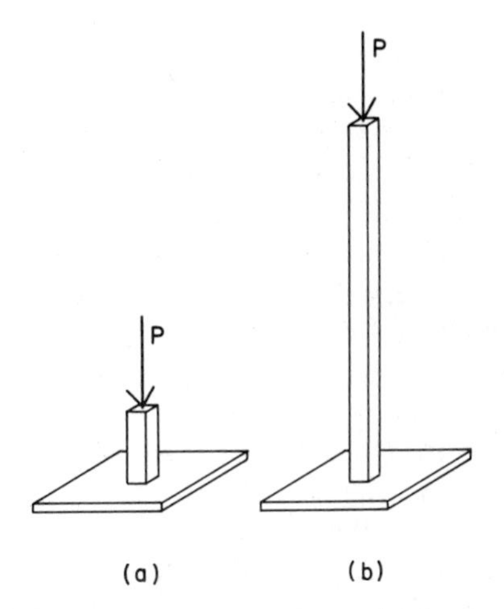

FIGURE 6.1 Effect of column slenderness.

lighter vertical and inclined members, such as braces and the compression members of roof trusses, are called struts.

The unit compressive stress in the short block shown in Fig. 6.1*a* could be expressed by the direct stress formula $f_a = P/A$; but this relationship becomes invalid as the ratio of the length of the compression member to its least width is increased. To pursue this further consider a small block of steel 1 in. by 1 in. in cross section and 1 or 2 in. high. If the allowable compressive stress is 20 ksi, the block will safely support a load P (Fig. 6.1*a* of 20 kips. If, however, we consider a bar of the same cross section with a length of, say, 30 to 40 in., we find that the value of P it will sustain is considerably less because of the tendency of this slenderer bar to buckle or bend (Fig. 6.1*b*). Therefore in columns the element of *slenderness* must be taken into account when determining allowable loads. A short column or block fails by crushing but long slender columns fail by stresses that result from bending.

6.2 COLUMN SECTIONS

The most common building columns are the round, cylindrical pipe, the rectangular tubular element, and the wide-flange shape (see Fig. 6.2). Pipes and tubes are useful for one-story buildings in which attachment to supported framing is achieved by setting the spanning members on top of the column. Framing that requires spanning members to be attached to the side of a column is more easily achieved with wide-flange members. The most commonly used columns for multistory buildings are the approximately square wide-flange shapes of nominal 10-, 12-, and 14-in. size. These are available in a wide range of flange and web thicknesses, up to the heaviest nominal 14-in. rolled sections.

Pipes and square tubes are ideally suited by the geometry of their cross sections for resistance of axial compression. The wide flange section has a strong axis (the X–X axis) and a weak axis (the Y–Y axis), although the close-to-square column shapes have the least difference in stiffness and radius of gyration on the two principal axes. The wide-flange section is well suited to the development of bending in combination with the axial force, if it is bent about its strong axis.

For various reasons, it is sometimes necessary to make a column section by assembling two or more individual steel elements. Figure 6.3 shows some commonly used assemblages which are used for special purposes. These are often required where a particular size or shape is not available from the inventory of stock rolled sections, where a particular structural task is required, or where details of the framing require a particular form. These are somewhat less used now, as the range of size and shape of stock sections has steadily increased. They are now mostly used for exceptionally large columns or for special framing problems. The customized fabrication of built-up sections is usually costly, so a single piece is typically favored if one is available.

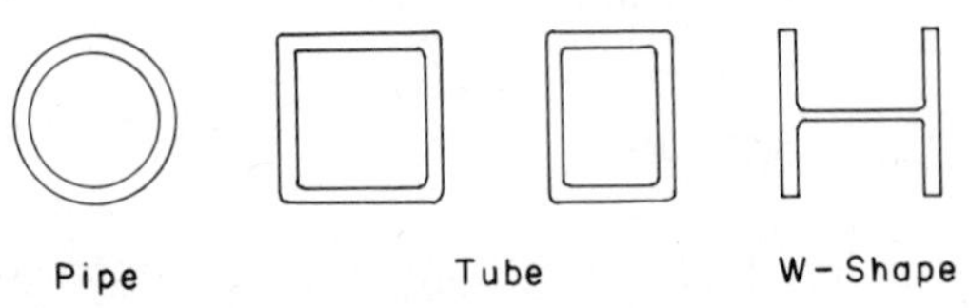

FIGURE 6.2 Ordinary column sections.

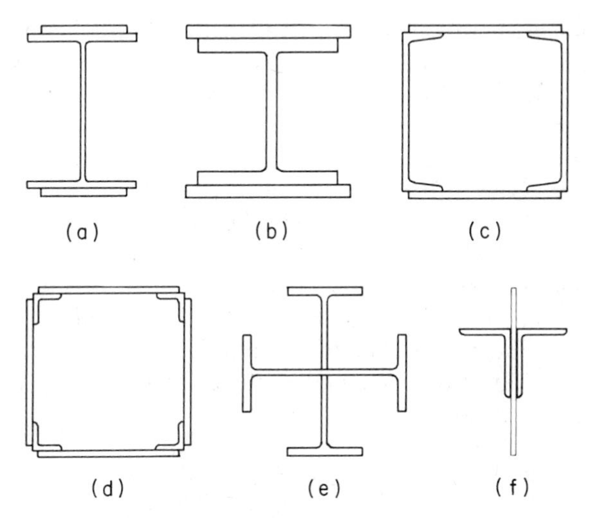

FIGURE 6.3 Built-up column sections.

One widely used built-up section is that of the double angle, shown in Fig. 6.3*f*. This occurs most often as a member of a truss or as a bracing member in a frame, the general stability of the paired members being much better than that of the single angle. This section is not much used as a building column.

6.3 SLENDERNESS RATIO

In the design of timber columns the term *slenderness ratio* is defined as the unbraced length divided by the dimension of the least side, both in inches. For structural shapes such as those shown in Fig. 6.3 the least lateral dimension is not an accurate criterion; and the radius of gyration r, which relates more precisely to the stiffness of columns in general, is used in steel column design. For rolled sections the value of the radius of gyration with respect to both major axes is given in the tables of properties for designing. For built-up sections it may be necessary to compute its value. The slenderness ratio of a steel column is then l/r, where l is the effective length of the column in inches and r is the *least* radius of gyration of the cross section, also in inches. The slenderness ratio for compression members should not exceed 200.

Example. A W 10 × 49 is used as a column whose effective length is 20 ft [6.10 m]. Compute the slenderness ratio.

Solution. Reference to Appendix Table A.1 reveals that the radii of gyration for this section are $r_X = 4.35$ in. and $r_Y = 2.54$ in. Therefore the *least* radius of gyration is 2.54 in.

Because the effective length of the column is 20 ft [6.10 m], the slenderness ratio is

$$\frac{L}{r} = \frac{20 \times 12}{2.54} = 9.45$$

It should be remembered that the tendency to bend due to buckling under the compression load is in a direction perpendicular to the axis about which the radius of gyration is least.

Effective Column Length

The AISC Specification requires that, in addition to the unbraced length of a column, the condition of the ends must be given consideration. The slenderness ratio is Kl/r, where K is a factor dependent on the restraint at the ends of a column and the means available to resist lateral motion. Figure 6.4 shows diagrammatically six idealized conditions in which joint rotation and joint translation are illustrated. The term K is the ratio of the effective column length to the actual unbraced length. For average conditions in building construction the value of K is taken as 1; therefore the slenderness ratio Kl/r becomes simply l/r (see Fig. 6.4*d*).

Problem 6.3.A. The effective length of a W 8 × 31 used as a column is 16 ft [4.88 m]. Compute the slenderness ratio.

Problem 6.3.B. What is the slenderness ratio of a column whose section is a W 12 × 65 with an effective length of 30 ft [9.14 m]?

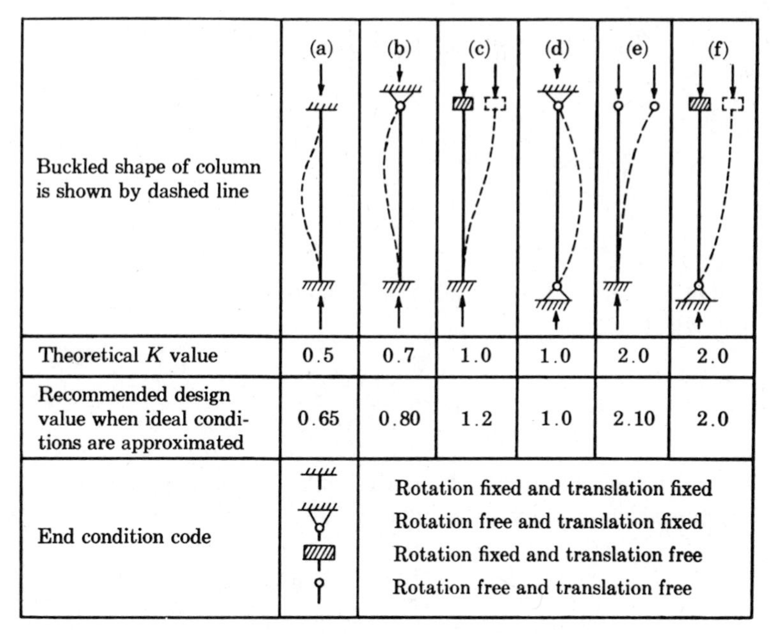

Buckled shape of column is shown by dashed line	(a)	(b)	(c)	(d)	(e)	(f)
Theoretical K value	0.5	0.7	1.0	1.0	2.0	2.0
Recommended design value when ideal conditions are approximated	0.65	0.80	1.2	1.0	2.10	2.0

End condition code	
	Rotation fixed and translation fixed
	Rotation free and translation fixed
	Rotation fixed and translation free
	Rotation free and translation free

FIGURE 6.4 Determination of effective column length. Reprinted from the *Manual of Steel Construction* (Ref. 1), with permission of the publishers, the American Institute of Steel Construction.

6.4 COLUMN FORMULAS

The AISC Specification gives the following requirements for use in the design of compression members. The allowable unit stresses shall not exceed the following values:

On the gross section of axially loaded compression members, when Kl/r, the largest effective slenderness ratio of any unbraced segment is less than C_c,

$$F_a = \frac{[1 - (Kl/r)^2/2C_c^2]F_y}{\text{FS}} \qquad \text{(Formula 6.5.1)}$$

where

$$\text{FS} = \text{factor of safety} = \frac{5}{3} + \frac{3(Kl/r)}{8C_c} - \frac{(Kl/r)^3}{8C_c^3}$$

and

$$C_c = \sqrt{\frac{2\pi^2 E}{F_y}}$$

On the gross section of axially loaded columns, when Kl/r exceeds C_c,

$$F_a = \frac{12\pi^2 E}{23(Kl/r)^2} \qquad \text{(Formula 6.5.2)}$$

On the gross section of axially loaded bracing and secondary members when l/r exceeds 120 (for this case, K is taken as unity),

$$F_{as} = \frac{F_a(\text{by Formula 6.5.1 or 6.5.2})}{1.6 - l/200r} \qquad \text{(Formula 6.5.3)}$$

In these formulas

F_a = axial compression stress permitted in the absence of bending stress
K = effective length factor (see Sec. 6.3)
l = actual unbraced length
r = governing radius of gyration (usually the least)
C_c = $\sqrt{2\pi^2 E/F_y}$; for A36 steel $C_c = 126.1$
F_y = minimum yield point of the steel being used (for A36 steel $F_y = 36{,}000$)
FS = factor of safety (see above)
E = modulus of elasticity of structural steel, 29,000 ksi
F_{as} = axial compressive stress permitted in the absence of bending stress for bracing and other secondary members

To determine the allowable axial load that a main column will support F_a, the allowable unit stress, is computed by Formula

TABLE 6.1 Allowable Unit Stresses for Columns of A36 Steel (ksi)[a]

Main and Secondary Members Kl/r not over 120						Main Members Kl/r 121 to 200				Secondary Members[b] l/r 121 to 200			
$\frac{Kl}{r}$	F_a (ksi)	$\frac{Kl}{r}$	F_a (ksi)	$\frac{Kl}{r}$	F_a (ksi)	$\frac{Kl}{r}$	F_a (ksi)	$\frac{Kl}{r}$	F_a (ksi)	$\frac{l}{r}$	F_{as} (ksi)	$\frac{l}{r}$	F_{as} (ksi)
1	21.56	41	19.11	81	15.24	121	10.14	161	5.76	121	10.19	161	7.25
2	21.52	42	19.03	82	15.13	122	9.99	162	5.69	122	10.09	162	7.20
3	21.48	43	18.95	83	15.02	123	9.85	163	5.62	123	10.00	163	7.16
4	21.44	44	18.86	84	14.90	124	9.70	164	5.55	124	9.90	164	7.12
5	21.39	45	18.78	85	14.79	125	9.55	165	5.49	125	9.80	165	7.08
6	21.35	46	18.70	86	14.67	126	9.41	166	5.42	126	9.70	166	7.04
7	21.30	47	18.61	87	14.56	127	9.26	167	5.35	127	9.59	167	7.00
8	21.25	48	18.53	88	14.44	128	9.11	168	5.29	128	9.49	168	6.96
9	21.21	49	18.44	89	14.32	129	8.97	169	5.23	129	9.40	169	6.93
10	21.16	50	18.35	90	14.20	130	8.84	170	5.17	130	9.30	170	6.89
11	21.10	51	18.26	91	14.09	131	8.70	171	5.11	131	9.21	171	6.85
12	21.05	52	18.17	92	13.97	132	8.57	172	5.05	132	9.12	172	6.82
13	21.00	53	18.08	93	13.84	133	8.44	173	4.99	133	9.03	173	6.79
14	20.95	54	17.99	94	13.72	134	8.32	174	4.93	134	8.94	174	6.76
15	20.89	55	17.90	95	13.60	135	8.19	175	4.88	135	8.86	175	6.73
16	20.83	56	17.81	96	13.48	136	8.07	176	4.82	136	8.78	176	6.70
17	20.78	57	17.71	97	13.35	137	7.96	177	4.77	137	8.70	177	6.67
18	20.72	58	17.62	98	13.23	138	7.84	178	4.71	138	8.62	178	6.64
19	20.66	59	17.53	99	13.10	139	7.73	179	4.66	139	8.54	179	6.61
20	20.60	60	17.43	100	12.98	140	7.62	180	4.61	140	8.47	180	6.58
21	20.54	61	17.33	101	12.85	141	7.51	181	4.56	141	8.39	181	6.56
22	20.48	62	17.24	102	12.72	142	7.41	182	4.51	142	8.32	182	6.53
23	20.41	63	17.14	103	12.59	143	7.30	183	4.46	143	8.25	183	6.51
24	20.35	64	17.04	104	12.47	144	7.20	184	4.41	144	8.18	184	6.49
25	20.28	65	16.94	105	12.33	145	7.10	185	4.36	145	8.12	185	6.46
26	20.22	66	16.84	106	12.20	146	7.01	186	4.32	146	8.05	186	6.44
27	20.15	67	16.74	107	12.07	147	6.91	187	4.27	147	7.99	187	6.42
28	20.08	68	16.64	108	11.94	148	6.82	188	4.23	148	7.93	188	6.40
29	20.01	69	16.53	109	11.81	149	6.73	189	4.18	149	7.87	189	6.38
30	19.94	70	16.43	110	11.67	150	6.64	190	4.14	150	7.81	190	6.36
31	19.87	71	16.33	111	11.54	151	6.55	191	4.09	151	7.75	191	6.35
32	19.80	72	16.22	112	11.40	152	6.46	192	4.05	152	7.69	192	6.33
33	19.73	73	16.12	113	11.26	153	6.38	193	4.01	153	7.64	193	6.31
34	19.65	74	16.01	114	11.13	154	6.30	194	3.97	154	7.59	194	6.30
35	19.58	75	15.90	115	10.99	155	6.22	195	3.93	155	7.53	195	6.28
36	19.50	76	15.79	116	10.85	156	6.14	196	3.89	156	7.48	196	6.27
37	19.42	77	15.69	117	10.71	157	6.06	197	3.85	157	7.43	197	6.26
38	19.35	78	15.58	118	10.57	158	5.98	198	3.81	158	7.39	198	6.24
39	19.27	79	15.47	119	10.43	159	5.91	199	3.77	159	7.34	199	6.23
40	19.19	80	15.36	120	10.28	160	5.83	200	3.73	160	7.29	200	6.22

[a] Reprinted from the *Manual of Steel Construction*, 8th ed, with permission of the publishers, the American Institute of Steel Construction.

[b] K is taken as 1.0 for secondary members.

16.5.1 or 16.5.2, and this stress is multiplied by the cross-sectional area of the column. If the column is a secondary member or is used for bracing, Formula 6.5.3 gives the allowable unit stress; these allowable unit stresses are somewhat greater than those permitted for main members. Table 6.1 gives allowable stresses computed in accordance with these formulas. It should be examined carefully because it will be of great assistance. Note particularly that this table is for use with A36 steel; tables based on other grades of steel are contained in the AISC Manual.

6.5 ALLOWABLE COLUMN LOADS

The allowable axial load that a steel column will support is found by multiplying the allowable unit stress by the cross-sectional area of the column. The value of Kl/r is first determined, and by referring to Table 6.1 we can establish the allowable unit stress.

Example 1. A W 12 × 65 is used as a column with an unbraced length of 16 ft [4.88 m]. Compute the allowable load.

Solution. Referring to Appendix Table A.1, we find that $A = 19.1$ in.^{2} [12,323 mm^2], $r_X = 5.28$ in. [134 mm], and $r_Y = 3.02$ in. [76.7 mm]. Because the column is unbraced with respect to both axes, the least radius of gyration is used to determine the slenderness ratio. Also, with no qualifying conditions given, $K = 1.0$. The slenderness ratio is then

$$\frac{KL}{r} = \frac{1 \times 16 \times 12}{3.02} = 63.6$$

In design work it is usually considered acceptable to round the slenderness ratio off in front of the decimal point because a typical lack of accuracy in the design data does not warrant greater precision. Therefore, we consider the slenderness ratio to be 64; the allowable stress given in Table 6.1 is $F_a = 17.04$ ksi [117.5 MPa]. The allowable load on the column is then

$$P = A \times F_a = 19.1 \times 17.04 = 325.5 \text{ kips } [1448 \text{ kN}]$$

Example 2. A built-up column of A36 steel consists of a W 14 × 311 core section with two 18 × 1 in. [457 × 25 mm] cover plates (Fig. 6.5). If K is 1 and the unbraced height is 20 ft [6.096 m], compute the axial load that this combined section can support.

Solution. The first step in this problem is to find the properties of the combined section. We need the total area, the moment of inertia about the Y–Y axis, and the value for r_y. From Appendix Table A.1, for the W 14 × 311, we find $A = 91.4$ in.^{2} [58,971 mm^2] and $I_y = 1610$ in.^{2} [670.1 × 10^6 mm^4]. The total area is thus

$$A = 91.4 + 2 \times (18 \times 1) = 91.4 + 36 = 127.4 \text{ in.}^2 \quad [1821 \text{ mm}^2]$$

Reference to Appendix Table E.20 shows that the moment of inertia of one plate about the Y–Y axis is

$$I = \frac{bd^3}{12} = \frac{1 \times (18)^3}{12} = 486 \text{ in.}^4 \quad [198.8 \times 10^6 \text{ mm}^4]$$

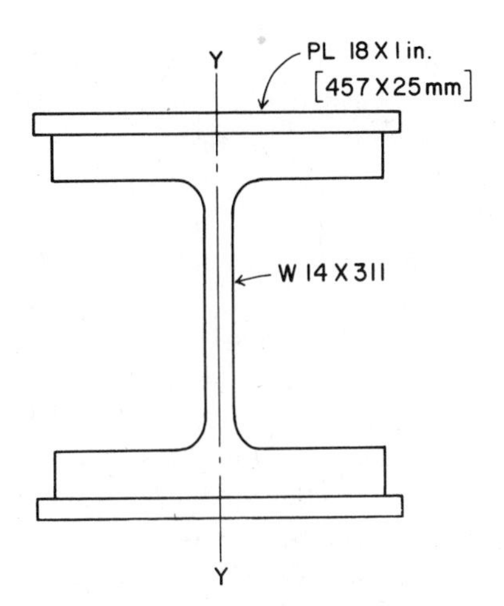

FIGURE 6.5 The built-up section.

and the total I for the section is thus

$$I = 1610 + (2 \times 486) = 1610 + 972$$
$$= 2582 \text{ in.}^4 \ [1067.7 \times 10^6 \text{ mm}^4]$$

Referring to Appendix Sec. E.8, we find that

$$r = \sqrt{\frac{I}{A}} = \sqrt{\frac{2582}{127.4}} = 4.50 \text{ in.} \ [114.2 \text{ mm}]$$

Then

$$\frac{KL}{r} = \frac{1 \times 20 \times 12}{4.50} = 53.33$$

Rounding the slenderness ratio off to 53, we find from Table 6.1 that $F_a = 18.08$ ksi [124.7 MPa]. The allowable load on the section is

$$P = A \times F_a = 127.4 \times 18.08 = 2303 \text{ kips} \ [10{,}203 \text{ kN}]$$

Conditions frequently exist to cause different modification of the basic column action on the two axes of a wide-flange column. In this event it may be necessary to investigate the conditions relating to the separate axes in order to determine the limiting condition. The following example illustrates such a case.

Example 3. Figure 6.6*a* shows an elevation of the steel framing at the location of an exterior wall. The column is laterally restrained but rotationally free at the top and bottom (end conditions as in case *d* in Fig. 6.4). With respect to the *x*-axis of the section, the column is laterally unbraced for its full height. However, the existence of horizontal framing in the wall plane provides lateral bracing at a point between the top and bottom with respect to the *y*-axis of the section. If the column is a W 12 × 58 of A36 steel, L_1 is 30 ft, and L_2 is 18 ft (see Fig. 6.6*b*), what is the allowable compression load?

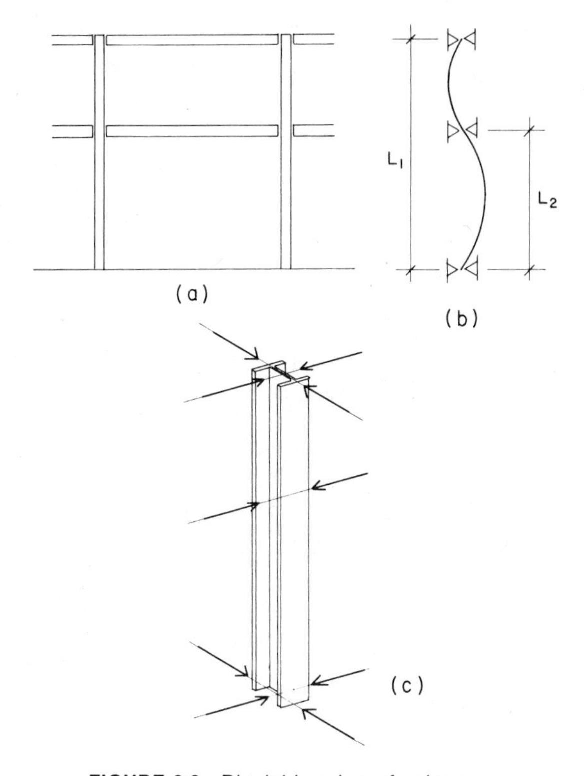

FIGURE 6.6 Biaxial bracing of columns.

Solution. With respect to the x-axis, the column functions as a pin-ended member for its full height (Fig. 6.4*d*). However, with respect to the y-axis, the form of buckling is as shown in Fig. 6.6*b*, and the laterally unsupported height is 18 ft [5.49 m]. For both conditions the K factor is 1, and the investigation is as follows (see Example 1 for data for the section):

$$x\text{-axis:} \quad \frac{KL}{r} = \frac{30(12)}{5.28} = 68.2$$

$$y\text{-axis:} \quad \frac{KL}{r} = \frac{18(12)}{2.51} = 86.1, \text{ say } 86$$

Despite the bracing, the column is still critical on its y-axis. From Table 6.1 we obtain $F_a = 14.67$ ksi [101 MPa] and the allowable load is thus

$$P = AF_a = 17.0(14.67) = 249.4 \text{ kips } [1108 \text{ kN}]$$

Note: For the following problems use A36 steel.

Problem 6.5.A. Determine the allowable axial load for a W 10 × 88 column with an unbraced height of 15 ft [4.57 m].

Problem 6.5.B. Determine the allowable axial load for a W 12 × 65 column with an unbraced height of 22 ft [6.71 m].

Problem 6.5.C. A built-up column section is composed of a W 14 × 342 with a 22 × 2 in. [559 × 50 mm] cover plate attached to each flange, as shown in Fig. 6.5. If the unbraced length is 18 ft [5.49 m], determine the allowable axial compression.

Problem 6.5.D. Determine the allowable axial compression load for the column in Problem 6.5.B, if the conditions are as shown in Fig. 6.6, with $L_1 = 15$ ft and $L_2 = 8$ ft.

6.6 DESIGN OF STEEL COLUMNS

In practice, the design of steel columns is accomplished largely by the use of safe load tables; if these tables are not available, design is carried out by the trial method. Data include the load and length of the column. The designer selects a trial cross section on the basis of his experience and judgment and, by means of a column formula, computes the allowable load that it can support. If this load is less than the actual load the column will be required to support, the trial section is too small and another section is tested in a similar manner.

Table 6.2 lists allowable loads on a number of column sections. It has been compiled from more extensive tables in the AISC Manual and the loads are computed in accordance with the formulas in Sec. 6.4. Note particularly that these allowable loads are for main members of A36 steel. Loads for Kl/r ratios between 120 and 200 are for main members. The significance of the bending factors, given at the extreme right of the table, is considered in Sec. 6.10.

To illustrate the use of Table 6.2, refer to Example 1 of Sec. 6.5. This problem asked for the allowable load that could be supported by a W 12 × 65 of A36 steel with an unbraced height of 16 ft. Referring to Table 6.2, we see at a glance that the allowable axial load is 326 kips, which agrees closely with the value found by computation.

Although the designer may select the proper column section by merely referring to the safe load tables, it is well to understand the application of the formulas by means of which the tables have been computed. To that end the *design procedure* is outlined below. When the design load and length have been ascertained, the following steps are taken:

Step 1: Assume a trial section and note from the table of properties the cross-sectional area and the least radius of gyration.

Step 2: Compute the slenderness ratio Kl/r, l being the unsupported length of the column in inches. For the value of K, see Sec. 6.3.

Step 3: Compute F_a, the allowable unit stress, by using a column formula or Table 6.1.

Step 4: Multiply F_a found in step 3 by the area of the column cross section. This gives the allowable load *on the trial column section*.

Step 5: Compare the allowable load found in step 4 with the design load. If the allowable load on the trial section is less than the design load (or if it is so much greater that it makes use of the section uneconomical), try another section and test it in the same manner. The reader should note that, except for assuming a trial section, these operations were carried out in Example 1 of Sec. 6.5.

Problem 6.6.A. Using Table 6.2, select a column section to support an axial load of 148 kips [658 kN] if the unbraced height is 12 ft [3.66 m]. A36 steel is to be used and K is assumed to be 1.

TABLE 6.2 Allowable Column Loads for Selected W and M Shapes[a]

	Effective length (KL) in feet										Bending factor	
Shape	8	9	10	11	12	14	16	18	20	22	B_x	B_y
M 4 × 13	48	42	35	29	24	18					0.727	2.228
W 4 × 13	52	46	39	33	28	20	16				0.701	2.016
W 5 × 16	74	69	64	58	52	40	31	24	20		0.550	1.560
M 5 × 18.9	85	78	71	64	56	42	32	25			0.576	1.768
W 5 × 19	88	82	76	70	63	48	37	29	24		0.543	1.526
W 6 × 9	33	28	23	19	16	12					0.482	2.414
W 6 × 12	44	38	31	26	22	16					0.486	2.367
W 6 × 16	62	54	46	38	32	23	18				0.465	2.155
W 6 × 15	75	71	67	62	58	48	38	30	24	20	0.456	1.424
M 6 × 20	98	92	87	81	74	61	47	37	30	25	0.453	1.510
W 6 × 20	100	95	90	85	79	67	54	42	34	28	0.438	1.331
W 6 × 25	126	120	114	107	100	85	69	54	44	36	0.440	1.308
W 8 × 24	124	118	113	107	101	88	74	59	48	39	0.339	1.258
W 8 × 28	144	138	132	125	118	103	87	69	56	46	0.340	1.244
W 8 × 31	170	165	160	154	149	137	124	110	95	80	0.332	0.985
W 8 × 35	191	186	180	174	168	155	141	125	109	91	0.330	0.972
W 8 × 40	218	212	205	199	192	127	160	143	124	104	0.330	0.959
W 8 × 48	263	256	249	241	233	215	196	176	154	131	0.326	0.940
W 8 × 58	320	312	303	293	283	263	240	216	190	162	0.329	0.934
W 8 × 67	370	360	350	339	328	304	279	251	221	190	0.326	0.921
W 10 × 33	179	173	167	161	155	142	127	112	95	78	0.277	1.055
W 10 × 39	213	206	200	193	186	170	154	136	116	97	0.273	1.018
W 10 × 45	247	240	232	224	216	199	180	160	138	115	0.271	1.000
W 10 × 49	279	273	268	262	256	242	228	213	197	180	0.264	0.770
W 10 × 54	306	300	294	288	281	267	251	235	217	199	0.263	0.767
W 10 × 60	341	335	328	321	313	297	280	262	243	222	0.264	0.765
W 10 × 68	388	381	373	365	357	339	320	299	278	255	0.264	0.758
W 10 × 77	439	431	422	413	404	384	362	339	315	289	0.263	0.751
W 10 × 88	504	495	485	475	464	442	417	392	364	335	0.263	0.744
W 10 × 100	573	562	551	540	428	503	476	446	416	383	0.263	0.735
W 10 × 112	642	631	619	606	593	565	535	503	469	433	0.261	0.726
W 12 × 40	217	210	203	196	188	172	154	135	114	94	0.227	1.073
W 12 × 45	243	235	228	220	211	193	173	152	129	106	0.227	1.065
W 12 × 50	271	263	254	246	236	216	195	171	146	121	0.227	1.058
W 12 × 53	301	295	288	282	275	260	244	227	209	189	0.221	0.813
W 12 × 58	329	322	315	308	301	285	268	249	230	209	0.218	0.794
W 12 × 65	378	373	367	361	354	341	326	311	294	277	0.217	0.656
W 12 × 72	418	412	406	399	392	377	361	344	326	308	0.217	0.651
W 12 × 79	460	453	446	439	431	415	398	379	360	339	0.217	0.648
W 12 × 87	508	501	493	485	477	459	440	420	398	376	0.217	0.645
W 12 × 96	560	552	544	535	526	506	486	464	440	416	0.215	0.635
W 12 × 106	620	611	602	593	583	561	539	514	489	462	0.215	0.633

TABLE 6.2 *(Continued)*

	Effective Length (*KL*) in Feet										Bending Factor	
Shape	8	10	12	14	16	18	20	22	24	26	B_x	B_y
W 12 × 120	702	692	660	636	611	584	555	525	493	460	0.217	0.630
W 12 × 136	795	772	747	721	693	662	630	597	561	524	0.215	0.621
W 12 × 152	891	866	839	810	778	745	710	673	633	592	0.214	0.614
W 12 × 170	998	970	940	908	873	837	798	757	714	668	0.213	0.608
W 12 × 190	1115	1084	1051	1016	978	937	894	849	802	752	0.212	0.600
W 12 × 210	1236	1202	1166	1127	1086	1042	995	946	894	840	0.212	0.594
W 12 × 230	1355	1319	1280	1238	1193	1145	1095	1041	985	927	0.211	0.589
W 12 × 252	1484	1445	1403	1358	1309	1258	1203	1146	1085	1022	0.210	0.583
W 12 × 279	1642	1600	1554	1505	1452	1396	1337	1275	1209	1141	0.208	0.573
W 12 × 305	1799	1753	1704	1651	1594	1534	1471	1404	1333	1260	0.206	0.564
W 12 × 336	1986	1937	1884	1827	1766	1701	1632	1560	1484	1404	0.205	0.558
W 14 × 43	230	215	199	181	161	140	117	96	81	69	0.201	1.115
W 14 × 48	258	242	224	204	182	159	133	110	93	79	0.201	1.102
W 14 × 53	286	268	248	226	202	177	149	123	104	88	0.201	1.091
W 14 × 61	345	330	314	297	278	258	237	214	190	165	0.194	0.833
W 14 × 68	385	369	351	332	311	289	266	241	214	186	0.194	0.826
W 14 × 74	421	403	384	363	341	317	292	265	236	206	0.195	0.820
W 14 × 82	465	446	425	402	377	351	323	293	261	227	0.196	0.823
W 14 × 90	536	524	511	497	482	466	449	432	413	394	0.185	0.531
W 14 × 99	589	575	561	546	529	512	494	475	454	433	0.185	0.527
W 14 × 109	647	633	618	601	583	564	544	523	501	478	0.185	0.523
W 14 × 120	714	699	682	663	644	623	601	578	554	528	0.186	0.523
W 14 × 132	786	768	750	730	708	686	662	637	610	583	0.186	0.521
W 14 × 145	869	851	832	812	790	767	743	718	691	663	0.184	0.489
W 14 × 159	950	931	911	889	865	840	814	786	758	727	0.184	0.485
W 14 × 176	1054	1034	1011	987	961	933	904	874	842	809	0.184	0.484
W 14 × 193	1157	1134	1110	1083	1055	1025	994	961	927	891	0.183	0.477
W 14 × 211	1263	1239	1212	1183	1153	1121	1087	1051	1014	975	0.183	0.477
W 14 × 233	1396	1370	1340	1309	1276	1241	1204	1165	1124	1081	0.183	0.472
W 14 × 257	1542	1513	1481	1447	1410	1372	1331	1289	1244	1198	0.182	0.470
W 14 × 283	1700	1668	1634	1597	1557	1515	1471	1425	1377	1326	0.181	0.465
W 14 × 311	1867	1832	1794	1754	1711	1666	1618	1568	1515	1460	0.181	0.459
W 14 × 342		2022	1985	1941	1894	1845	1793	1738	1681	1621	0.181	0.457
W 14 × 370		2181	2144	2097	2047	1995	1939	1881	1820	1756	0.180	0.452
W 14 × 398		2356	2304	2255	2202	2146	2087	2025	1961	1893	0.178	0.447
W 14 × 426		2515	2464	2411	2356	2296	2234	2169	2100	2029	0.177	0.442
W 14 × 455		2694	2644	2589	2430	2467	2401	2332	2260	2184	0.177	0.441
W 14 × 500		2952	2905	2845	2781	2714	2642	2568	2490	2409	0.175	0.434
W 14 × 550		3272	3206	3142	3073	3000	2923	2842	2758	2670	0.174	0.429
W 14 × 605		3591	3529	3459	3384	3306	3223	3136	3045	2951	0.171	0.421
W 14 × 665		3974	3892	3817	3737	3652	3563	3469	3372	3270	0.170	0.415
W 14 × 730		4355	4277	4196	4100	4019	3923	3823	3718	3609	0.168	0.408

Source: Adapted from data in the *Manual of Steel Construction*, 8th ed. (Ref. 1), with permission of the publishers, the American Institute of Steel Construction.

[a] Loads in kips for shapes of steel with yield stress of 36 ksi [250 MPa], based on buckling with respect to the *y*-axis.

Problem 6.6.B. Same data as in Problem 6.6.A except that the load is 258 kips [1148 kN] and the unbraced height is 16 ft [4.88 m].

Problem 6.6.C. Same data as in Problem 6.6.A except that the load is 355 kips [1579 kN] and the unbraced height is 20 ft [6.10 m].

6.7 STEEL PIPE COLUMNS

Round steel pipe columns are frequently installed in both steel and wood framed buildings. In routine work they are designed for simple axial load by the use of safe load tables.

Table 6.3 gives allowable axial loads for standard weight steel pipe columns with a yield point of 36 ksi [250 MPa]. The outside diameters at the head of the table are *nominal* dimensions that designate the pipe sizes. True outside diameters are slightly larger and can be found from Appendix Table A.6. The AISC Manual (Ref. 1) contains additional tables that list allowable loads for the two heavier weight groups of steel pipe: extra strong and double-extra strong.

Example. Using Table 6.3, select a steel pipe column to carry a load of 41 kips [182 kN] if the unbraced height is 12 ft [3.58 m]. Verify the value in the table by computing the allowable axial load.

Solution. Entering Table 6.3 with an effective length of 12 ft, we find that a load of 43 kips can be supported by a 4-in. column. From Appendix Table A.1 we find that this section has $A = 3.17$ in.^{2} [2045 mm^2] and $r = 1.51$ in. [38.35 mm]. The slenderness ratio, with $K = 1$, is

$$\frac{KL}{r} = \frac{1 \times 12 \times 12}{1.51} = 95.4, \text{ say } 95$$

TABLE 6.3 Allowable Column Loads, Standard Steel Pipe of A36 Steel (kips)

Nominal Dia.	12	10	8	6	5	4	$3\frac{1}{2}$	3
Wall Thickness	0.375	0.365	0.322	0.280	0.258	0.237	0.226	0.216
Weight per Foot	49.56	40.48	28.55	18.97	14.62	10.79	9.11	7.58
F_y	36 ksi							
Effective length in feet KL with respect to radius of gyration								
0	315	257	181	121	93	68	58	48
6	303	246	171	110	83	59	48	38
7	301	243	168	108	81	57	46	36
8	299	241	166	106	78	54	44	34
9	296	238	163	103	76	52	41	31
10	293	235	161	101	73	49	38	28
11	291	232	158	98	71	46	35	25
12	288	229	155	95	68	43	32	22
13	285	226	152	92	65	40	29	19
14	282	223	149	89	61	36	25	16
15	278	220	145	86	58	33	22	14
16	275	216	142	82	55	29	19	12
17	272	213	138	79	51	26	17	11
18	268	209	135	75	47	23	15	10
19	265	205	131	71	43	21	14	9
20	261	201	127	67	39	19	12	
22	254	193	119	59	32	15	10	
24	246	185	111	51	27	13		
25	242	180	106	47	25	12		
26	238	176	102	43	23			
28	229	167	93	37	20			
30	220	158	83	32	17			
31	216	152	78	30	16			
32	211	148	73	29				
34	201	137	65	25				
36	192	127	58	23				
37	186	120	55	21				
38	181	115	52					
40	171	104	47					
Properties								
Area A (in.^{2})	14.6	11.9	8.40	5.58	4.30	3.17	2.68	2.23
I (in.^{4})	279	161	72.5	28.1	15.2	7.23	4.79	3.02
r (in.)	4.38	3.67	2.94	2.25	1.88	1.51	1.34	1.16
B Bending factor	0.333	0.398	0.500	0.657	0.789	0.987	1.12	1.29
* a	41.7	23.9	10.8	4.21	2.26	1.08	0.717	0.447

* Tabulated values of a must be multiplied by 10^6.
Note: Heavy line indicates Kl/r of 200.

Source: Reprinted from the *Manual of Steel Construction*, 8th ed., with permission of the publishers, the American Institute of Steel Construction.

Referring to Table 6.1, we find that the allowable unit stress F_a for a slenderness ratio of 95 is 13.60 ksi [93.8 MPa]. Thus the allowable axial load is

$$P = A \times F_a = 3.17 \times 13.60 = 43.1 \text{ kips } [192 \text{ kN}]$$

which agrees with the table value of 43 kips.

Problems 6.7.A, B, C, D. Select the minimum nominal-size round pipe column from Table 6.3 for an axial load of 50 kips [222 kN] and the following unbraced heights: (a) 8 ft [2.44 m]; (b) 12 ft [3.66 m]; (c) 18 ft [5.49 m]; (d) 25 ft [7.62 m].

6.8 STRUCTURAL TUBING COLUMNS

Steel columns are fabricated from structural tubing in both square and rectangular shapes. Square tubing is available in sizes of 2 to 16 in. and rectangular sizes range from 3 × 2 to 20 × 12 in. Sections are produced with various wall thicknesses, thus allowing a considerable range of structural capacities. Although round pipe is specified by a nominal outside dimension, tubing is specified by its actual outside dimensions.

The AISC Manual (Ref. 1) contains safe load tables for square and rectangular tubing based on F_y = 46 ksi [317 MPa]. Table 6.4 is a reproduction of one of these tables—for 3 and 4 in. square tubing—and is presented to illustrate the form of the tables.

Example. Using the data given for the example on steel pipe columns in Sec. 6.7 (P = 41 kips and height = 12 ft), select a square structural tubing column from Table 6.4. Compare the two solutions.

Solution. Entering Table 6.4 with an effective length of 12 ft, we find that a load of 43 kips can be supported by a square section 4 × 4 × 3/16. With data from Appendix Table A.1, the properties of the two columns are compared in Table 6.5.

Both pipe and tubing may be available in various steel strengths. We have used the properties in these examples be-

TABLE 6.4 Allowable Column Loads for Steel Structural Tubing[a]

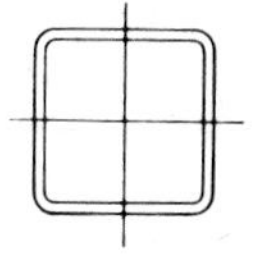

Nominal Size		4 x 4					3 x 3		
Thickness		1/2	3/8	5/16	1/4	3/16	5/16	1/4	3/16
Wt./ft.		21.63	17.27	14.83	12.21	9.42	10.58	8.81	6.87
F_y		46 ksi							
Effective length in feet KL with respect to radius of gyration	0	176	140	120	99	76	86	71	56
	2	168	134	115	95	73	80	67	53
	3	162	130	112	92	71	77	64	50
	4	156	126	108	89	69	73	61	48
	5	150	121	104	86	67	68	57	45
	6	143	115	100	83	64	63	53	42
	7	135	110	95	79	61	57	49	39
	8	126	103	90	75	58	51	44	35
	9	117	97	84	70	55	44	38	31
	10	108	89	78	65	51	37	33	27
	11	98	82	72	60	47	31	27	22
	12	87	74	65	55	43	26	23	19
	13	75	65	58	49	39	22	19	16
	14	65	57	51	43	35	19	17	14
	15	57	49	44	38	30	16	15	12
	16	50	43	39	33	27	14	13	11
	17	44	38	34	29	24	13	11	9
	18	39	34	31	26	21		10	8
	19	35	31	28	24	19			
	20	32	28	25	21	17			
	21	29	25	23	19	16			
	22	26	23	21	18	14			
	23	24	21	19	16	13			
	24		19	17	15	12			
	25				14	11			
Properties									
A (in.^{2})		6.36	5.08	4.36	3.59	2.77	3.11	2.59	2.02
I (in.^{4})		12.3	10.7	9.58	8.22	6.59	3.58	3.16	2.60
r (in.)		1.39	1.45	1.48	1.51	1.54	1.07	1.10	1.13
B Bending factor		1.04	0.949	0.910	0.874	0.840	1.30	1.23	1.17
*a		1.83	1.59	1.43	1.22	0.983	0.533	0.470	0.387

* Tabulated values of a must be multiplied by 10^6.
Note: Heavy line indicates Kl/r of 200.

Source: Reprinted from the *Manual of Steel Construction*, 8th ed. (Ref. 1), with permission of the publishers, the American Institute of Steel Construction.

[a] Loads in kips for axially loaded columns with yield stress of 46 ksi [317 MPa].

TABLE 6.5 Comparison of the Round Pipe and Square Tube

Property	Pipe 4-in. Std	TS 4 × 4 × 0.1875
Area (in.^{2})	3.17	2.77
Weight (lb/ft)	10.79	9.42
Outside dimension (in.)	4.5	4.0
Axial load capacity, 12-ft length (kips)	43	43

cause they appear in the AISC Manual. The choice between round pipe or rectangular tubing for a column is usually made for reasons other than simple structural efficiency. Free-standing columns are often round but when built into wall construction the rectangular shapes are often preferred.

Problem 6.8.A. A structural tubing column TS 4 × 4 × 3/8 of steel with F_y = 46 ksi [317 MPa] is used on an effective length of 12 ft [3.66 m]. Compute the allowable axial load and compare it with the value in Table 6.4.

Problem 6.8.B. Refer to Table 6.4 and select the lightest weight square tubing column with an effective length of 10 ft [3.05 m] that will support an axial load of 64 kips [285 kN].

6.9 DOUBLE-ANGLE STRUTS

Two angle sections separated by the thickness of a connection plate at each end and fastened together at intervals by fillers and welds or bolts are commonly used as compression members in roof trusses (see Fig. 6.3*f*). These members, whether or not in a vertical position, are called struts; their size is determined in accordance with the requirements and formulas for columns in Sec. 6.4. To ensure that the angles act as a unit the intermittent connections are made at intervals such that the slenderness ratio l/r of either angle between fasteners does not exceed the governing slenderness ratio of the built-up member. The least radius of gyration r is used in computing the slenderness ratio of each angle.

The AISC Manual contains safe load tables for struts of two angles with $\frac{3}{8}$-in. separation back-to-back. Three series are given: equal-leg angles, unequal-leg angles with short legs back-to-back, and unequal-leg angles with long legs back-to-back. Table 6.6 has been abstracted from the latter series and lists allowable loads with respect to the X–X and Y–Y axes. The smaller (least) radius of gyration gives the smaller allowable load and, unless the member is braced with respect to the weaker axis, this is the tabular load to be used. The usual practice is to assume K equal to 1.0. The following example shows how the loads in the table are computed.

Example. Two $5 \times 3\frac{1}{2} \times \frac{1}{2}$ angle sections spaced with their long legs $\frac{3}{8}$ in. back-to-back are used as a compression member. If the member is A36 steel and has an effective length of 10 ft, compute the allowable axial load.

Solution. From Appendix Table A.5 we find that the area of the two-angle member is 8.0 in. and that the radii of gyration are $r_x = 1.58$ in. and $r_y = 1.49$ in. Using the smaller r, the slenderness ratio is

$$\frac{Kl}{r} = 1 \times \frac{10 \times 12}{1.49} = 80.5, \text{ say } 81$$

Referring to Table 6.1, we find that $F_a = 15.24$ ksi, making the allowable load

$$P = A \times F_a = 8.0 \times 15.24 = 121.9 \text{ kips}$$

This value is, of course, readily verified by entering Table 6.6 under "*Y–Y Axis*" with an effective length of 10 ft and then proceeding horizontally to the column of loads for the $5 \times 3\frac{1}{2} \times \frac{1}{2}$ angle.

The design of double-angle members for the compression elements in trusses is considered in Chapter 9.

TABLE 6.6 Allowable Axial Compression for Double-Angle Struts[a]

Size (in.)		8 × 6	
Thickness (in.)		3/4	1/2
Weight (lb/ft)		67.6	46.0
Area (in²)		19.9	13.5
r_x (in.)		2.53	2.56
r_y (in.)		2.48	2.44
Effective Length (KL) with Respect to Indicated Axis: *X-X Axis*	0	430	266
	10	370	231
	12	353	222
	14	334	211
	16	315	200
	20	271	175
	24	222	148
	28	168	117
	32	129	90
	36	102	71
Y-Y Axis	0	430	266
	10	368	229
	12	351	219
	14	332	207
	16	311	195
	20	266	169
	24	216	139
	28	162	106
	32	124	81
	36	98	64

Size (in.)		6 × 4		
Thickness (in.)		5/8	1/2	3/8
Weight (lb/ft)		40.0	32.4	24.6
Area (in²)		11.7	9.50	7.22
r_x (in.)		1.90	1.91	1.93
r_y (in.)		1.67	1.64	1.62
Effective Length (KL) with Respect to Indicated Axis: *X-X Axis*	0	253	205	142
	8	214	174	122
	10	200	163	115
	12	185	151	107
	14	168	137	99
	16	150	123	89
	20	110	90	69
	24	76	62	48
	28	56	46	36
Y-Y Axis	0	253	205	142
	6	222	179	125
	8	207	167	117
	10	190	153	108
	12	171	137	97
	14	151	120	86
	16	129	102	74
	20	85	66	49
	24	59	46	34

Size (in.)		5 × 3 1/2	
Thickness (in.)		1/2	3/8
Weight (lb/ft)		27.2	20.8
Area (in²)		8.00	6.09
r_x (in.)		1.58	1.60
r_y (in.)		1.49	1.46
Effective Length (KL) with Respect to Indicated Axis: *X-X Axis*	0	173	129
	4	159	119
	6	150	113
	8	139	105
	10	126	96
	12	113	86
	14	97	75
	16	81	63
	20	52	40
Y-Y Axis	0	173	129
	4	158	118
	6	148	110
	8	136	101
	10	122	91
	12	107	79
	14	90	67
	16	72	53
	20	46	34

Size (in.)		5 × 3		
Thickness (in.)		1/2	3/8	5/16
Weight (lb/ft)		25.6	19.6	16.4
Area (in²)		7.50	5.72	4.80
r_x (in.)		1.59	1.61	1.61
r_y (in.)		1.25	1.23	1.22
Effective Length (KL) with Respect to Indicated Axis: *X-X Axis*	0	162	121	94
	4	149	112	88
	6	141	106	83
	8	130	98	77
	10	119	90	71
	12	106	81	64
	14	92	70	57
	16	76	59	49
	20	49	38	32
Y-Y Axis	0	162	121	94
	4	145	108	85
	6	132	99	78
	8	118	88	69
	10	101	75	60
	12	82	61	49
	14	62	46	38
	16	47	35	29
	20	30	22	19

Source: Abstracted from data in the *Manual of Steel Construction*, 8th ed. (Ref. 1), with permission of the publishers, the American Institute of Steel Construction.

[a] Loads in kips for angles with long legs back to back with $\frac{3}{8}$-in. separation and steel with F_y = 36 ksi [250 MPa].

When designing double angles or structural tees as compression members without the help of safe load tables consideration must be given to the possibility that it may be necessary to reduce the allowable stress when these members have thin parts. This condition is indicated by the presence of a value for Q_s in the tabulated properties in Appendix A. When a value is given for Q_s the safe axial load, as calculated normally, must be multiplied by this value for the true allowable load. Load values given in the safe load tables in the AISC Manual have incorporated this requirement.

TABLE 6.6 *(Continued)*

	4 × 3				3 1/2 × 2 1/2				3 × 2				2 1/2 × 2		
	1/2	3/8	5/16		3/8	5/16	1/4		3/8	5/16	1/4		3/8	5/16	1/4
	22.2	17.0	14.4		14.4	12.2	9.8		11.8	10.0	8.2		10.6	9.0	7.2
	6.50	4.97	4.18		4.22	3.55	2.88		3.47	2.93	2.38		3.09	2.62	2.13
	1.25	1.26	1.27		1.10	1.11	1.12		0.940	0.948	0.957		0.768	0.776	0.784
	1.33	1.31	1.30		1.11	1.10	1.09		0.917	0.903	0.891		0.961	0.948	0.935
0	140	107	90	0	91	77	60	0	75	63	51	0	67	57	46
2	134	103	86	2	86	73	57	2	70	59	48	2	61	52	42
4	126	96	81	4	80	67	53	3	67	57	46	3	58	49	40
6	115	88	74	6	71	60	48	4	63	54	44	4	53	45	37
8	102	78	66	8	61	52	41	6	55	46	38	5	48	41	34
10	88	67	57	10	50	42	34	8	44	38	31	6	42	36	30
12	71	55	47	12	37	31	26	10	32	27	23	8	30	26	21
14	54	42	36	14	27	23	19	12	22	19	16	10	19	16	14
16	41	32	27	16	21	18	15	14	16	14	12	12	13	11	9
18	33	25	22	18	16	14	12								
20	26	20	17												
0	140	107	90	0	91	77	60	0	75	63	51	0	67	57	46
2	135	103	86	2	87	73	57	2	70	59	48	2	63	53	43
4	127	97	81	4	80	67	53	3	67	56	46	3	60	51	41
6	117	89	74	6	72	60	47	4	63	53	43	4	57	48	39
8	105	80	67	8	62	52	41	6	54	45	36	6	49	41	33
10	92	70	58	10	50	42	33	8	43	36	28	8	40	34	27
12	77	58	48	12	37	31	25	10	30	25	20	10	30	24	19
14	61	45	37	14	28	23	18	12	21	17	14	12	21	17	13
16	47	35	29	16	21	17	14	14	15	13	10	14	15	12	10
18	37	27	23	18	17	14	11								
20	30	22	18												

Problem 6.9.A. A double-angle compression member 8 ft [2.44 m] long is composed of two angles 4 × 3 × $\frac{3}{8}$ in. with the long legs $\frac{3}{8}$ in. back-to-back. If the member is fabricated from A36 steel, determine the allowable concentric load.

Problem 6.9.B. Using Table 6.6, select a double-angle compression member that will support an axial load of 50 kips [222 kN] if the effective length is 10 ft [3.05 m].

6.10 COLUMNS WITH BENDING

Many steel columns must sustain bending in addition to the usual axial compression. Figure 6.7 shows three of the most common

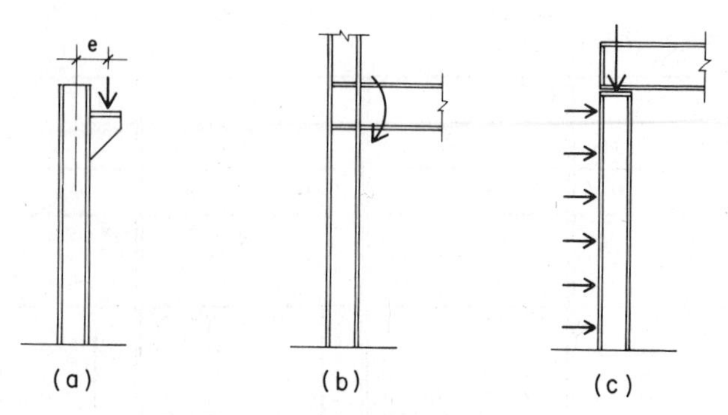

FIGURE 6.7 Columns with bending.

situations that result in this combined effect. When framing members are supported at the column face or on a bracket, the compression load may actually occur with some eccentricity, as shown in Fig. 6.7*a*. When moment-resistive connections are used, and the column becomes a member of a rigid-frame bent, moments will be induced in the ends of the columns, as shown in Fig. 6.7*b*. Columns in exterior walls frequently function as part of the general wall framing; if vertical spanning for wind load is involved, the columns may receive a direct beam loading, as shown in Fig. 6.7*c*.

The fundamental relationship for a column with bending is one of interaction, the simplest form of which is expressed by the straight-line interaction formula:

$$\frac{f_a}{F_a} + \frac{f_b}{F_b} = 1$$

In reality, the problem has the potential for great complexity. The usual problems of dealing with buckling effects of the column must be combined with that of the laterally unsupported beam and some possible synergetic behaviors, such as the *P*-delta effect. While the basic form of the interaction formula is used for investigation, numerous adjustments must often be made for various situations. The AISC Specifications are quite extensive and complex and not very self-explanatory with regard to this problem. For a full treatment of the topic the reader is referred to one of the major textbooks on steel design, such as *Steel Buildings: Analysis and Design,* by S. W. Crawley and R. M. Dillon (Ref. 4).

For use in preliminary design work, or to obtain a first trial section to be used in a more extensive design investigation, a procedure may be used that involves the determination of an equivalent total design load that incorporates the bending effects. This is done by using the bending factors, B_x and B_y, which are listed in Table 6.2. The equivalent design load is determined as

$$P' = P + B_x M_x + B_y M_y$$

where P' = equivalent axial load,
P = actual axial load,
B_x = bending factor for the section's x-axis,
M_x = bending moment about the x-axis,
B_y = bending factor for the section's y-axis,
M_y = bending moment about the y-axis.

The following example illustrates the use of the method.

Example 1. It is desired to use a 10-in. W shape for a column in a situation such as that shown in Fig. 6.8. The axial compression load from above is 120 kips and the beam load is 24 kips, with the beam attached at the column face. The column is 16 ft high and has a K-factor of 1.0. Select a trial section for the column.

Solution. Since bending occurs only about the x-axis, we use only the B_x-factor for this case. Scanning the column of B_x-factors in Table 6.2, we observe that the factor for 10 W sections varies from 0.261 to 0.277. As we have not yet determined the section to be used, it is necessary to make an assumption for the factor and to verify the assumption after the selection is made. Let us assume a B_x of 0.27, with which we find that

$$P' = P + B_x M_x = (120 + 24) + 0.27(24)(5)$$

$$= 144 + 32.4 = 176.4 \text{ kips}$$

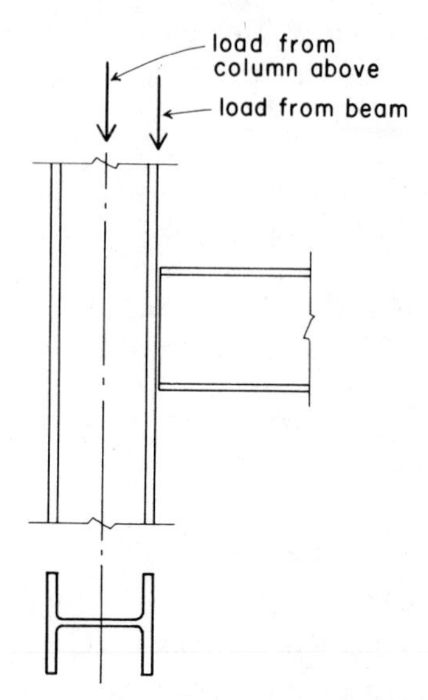

FIGURE 6.8 Load eccentricity with column framing.

From Table 6.2, for a *KL* of 16 ft, we obtain a W 10 × 45. For this shape the B_x-factor is 0.271, which is very close to our assumption.

For most situations use of the bending factors in the manner just demonstrated will result in conservative selections. If the designer intends to use the section thus obtained in a more thorough investigation, it is probably wise to reduce the size slightly before proceeding with the work.

Example 2. It is desired to use a 12-in. W section for a column that sustains an axial compression plus bending about both axes. Select a column for the following data: axial load of 60 kips, bending about the *x*-axis of 40 kip-ft, bending about the *y*-axis of 32 kip-ft, column unbraced height of 12 ft.

Solution. In Table 6.2 we observe that in the midrange of sizes for 12-in. sections approximate values for bending factors are $B_x = 0.215$ and $B_y = 0.63$. Thus

$$P' = P + B_x M_x + B_y M_y$$
$$= 60 + (0.215)(40 \times 12) + (0.63)(32 \times 12)$$
$$= 60 + 103 + 242 = 405 \text{ kips}$$

From Table 6.2, for the 12-ft height, the lightest section is a W 12 × 79, with an allowable load of 431 kips and bending factors of $B_x = 0.217$ and $B_y = 0.648$. As these bending factors slightly exceed those assumed, a new design value for P' should be found to verify the section. Thus

$$P' = 60 + (0.217)(40 \times 12) + (0.648)(32 \times 12) = 413 \text{ kips}$$

and the section is still a valid choice.

A major occurrence of the condition of a column with bending is that of the case of a column in a rigid-frame bent. For steel structures this most often occurs when a steel frame is developed as a moment-resisting space frame in three dimensions with the rigid-frame action being used for lateral bracing. If the frame is made rigid (by using moment-resistive connections between columns and beams), both vertical gravity and horizontal wind or earthquake loads will result in bending and shear in the columns. Multiple-bayed, multistoried frames are highly indeterminate, requiring investigative procedures beyond the scope of the work developed here. Some aspects of frame behavior are discussed in Chapter 7.

Problem 6.8.A. It is desired to use a 12-in. W section for a column to support a beam, as shown in Fig. 6.8. Select a trial size for the column for the following data: column axial load = 200 kips [890 kN], beam reaction = 30 kips [133 kN], unbraced height of column = 14 ft [4.27 m].

Problem 6.10.B. A 14-in. W section is to be used for a column that sustains bending about both axes. Select a trial section for the column for the following data: axial load = 160 kips [712 kN], bending about the x-axis = 65 kip-ft [88 kN-m], bending about the y-axis = 45 kip-ft [61 kN-m], unbraced column height = 16 ft [4.88 m].

6.11 COLUMN FRAMING

Connection details for columns must be developed with considerations of the column shape and size, the shape and orientation of other framing, and the particular structural functions of the joints. Some common connections are shown in Fig. 6.9. When beams sit directly on the top of a column, the usual solution is to fasten a bearing plate on top of the column, with provision for the attachment of the beam to the plate. The plate serves no specific structural purpose in this case, functioning essentially only as an attachment device, assuming that the load transfer is one of simple vertical bearing (see Fig. 6.9*a*).

In many situations beams must frame into the side of a column. If simple transfer of vertical load is all that is required, a common solution is the connection shown in Fig. 6.9*b*, in which a pair of steel angles are used to connect the beam web to the column flange or the column web. This type of connection is discussed in Sec. 10-7. If moment must be transferred between the columns and beams, as in the development of rigid-frame bents, the most common solution involves the use of welding to achieve a direct

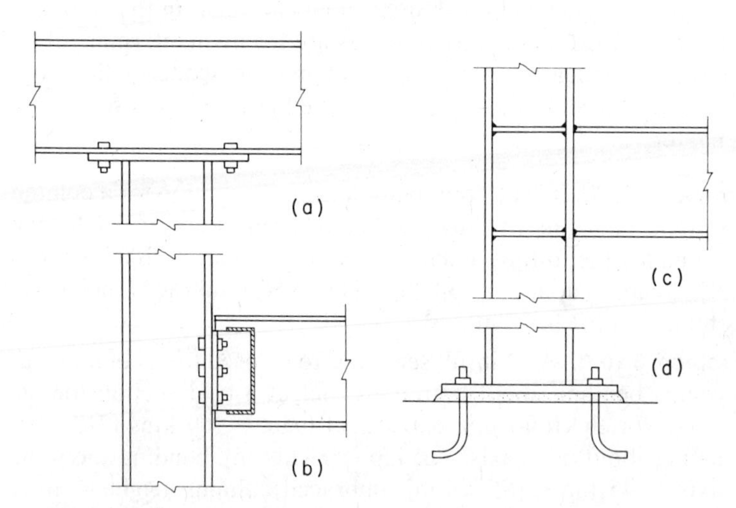

FIGURE 6.9 Typical column connections with lightly loaded frames.

connection between the members, as shown in Fig. 6.9*c*. The details of rigid connections vary considerably, mostly on the basis of the size of the members and the magnitude of forces.

Another common connection is that of the bottom of a column to its supports. Figure 6.9*d* shows a typical solution for transfer of bearing to a concrete footing, consisting of a steel plate welded to the column bottom and bolted to preset anchor bolts. In this case the plate has a definite structural function, serving to transform the highly concentrated, punching effect of the column force into a low-valued bearing stress on the much softer concrete. Thus while the general form of the connection is similar, the role of the plate in Fig. 6.9*a* and *d* are considerably different. In the construction of steel frames of buildings it is common practice to use columns of two-story lengths. This results in a greater cross-sectional area in the upper story than the load requires, but the cost of the excess material is offset by the saving in fabricating costs for the extra splice. When columns of two-story lengths are used the load on the lower story length determines the required cross section of the column.

In order not to conflict with the beam and girder connections, column splices are made 2 ft or more above the floor level. In general, splices are made with plates $\frac{3}{8}$ or $\frac{1}{2}$ thick that are bolted, welded, or riveted to the flanges of the columns, as indicated in Fig. 6.10. The splice plates are not designed to resist compressive

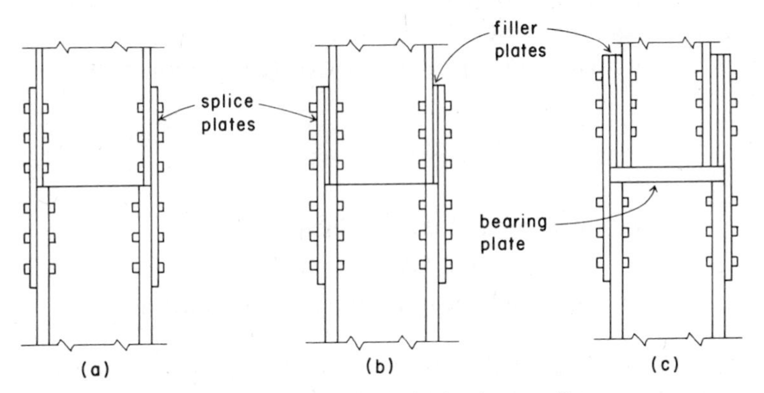

FIGURE 6.10 Typical bolted column splices.

stresses; their function is to hold the column sections in position. Because the upper column transmits its load directly to the column below, the surfaces in contact should be milled to provide full bearing areas. When the upper column has a smaller width than the supporting column filler plates are used (Fig. 6.10*b*). If the difference in width is so great that a full bearing area between the columns is not achieved, a horizontal plate is used as in Fig. 6.10*c*.

Part 4 of the AISC Manual provides extensive data relating to the development of connections for structural steel members. Some of the material is presented in the discussions in this book in Chapter 10. Issues of specific concern are also discussed in some of the design examples in Chapter 12.

6.12 COLUMN BASES

The bottoms of steel columns typically bear on a concrete or masonry support. Force transfer to the support frequently involves only direct compressive bearing, in which case a base plate is typically used to spread the bearing stress out onto the softer support material. Column base plates vary from relatively modest ones for small, lightly loaded columns to huge thick ones for the heavy 14 W shapes in high-strength steel grades. The following procedure is based on the specifications and design illustrations in the AISC Manual (Ref. 1).

The plan area required for the base plate is determined as

$$A_1 = \frac{P}{F_p}$$

where A_1 = plan area of the bearing plate,
P = compression load from the column,
F_p = allowable bearing stress on the concrete.

Allowable bearing is based on the concrete design strength f'_c. It is limited to a value of $0.3f'_c$ when the plate covers the entire area of the support member, which is seldom the case. If the support member has a larger plan area A_2, the allowable bearing stress may be increased by a factor of $\sqrt{A_2/A_1}$, but not greater than two. For modest-sized columns supported on footings, the maximum factor is most likely to be used. For large columns or those supported on pedestals or piers, the adjustment may be less.

For a W-shaped column, the basis for determination of the thickness of the plate due to bending is shown in Fig. 6.11. Once the required value for A_1 is found, the dimensions B and N are established so that the projections m and n are approximately equal. Choice of dimensions must also relate to the locations of anchor bolts and to any details for development of the attachment of the plate to the column. The required plate thickness is determined with the formula

$$t = \sqrt{\frac{3f_p m^2}{F_b}} \quad \text{or} \quad t = \sqrt{\frac{3f_p n^2}{F_b}}$$

where t = thickness of the bearing plate, in inches,
f_p = actual bearing pressure: P/A_1,
F_b = allowable bending stress in the plate: $0.75F_y$.

The following example illustrates the process for a column with a relatively light load.

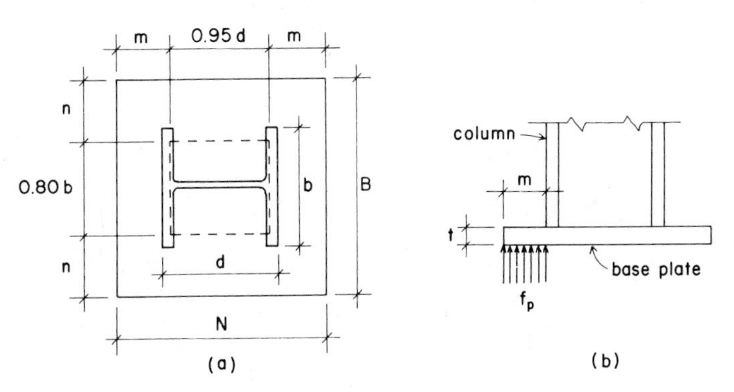

FIGURE 6.11 Reference dimensions for column base plates.

Example 1. Design a base plate of A36 steel for a W 12 × 58 column with a load of 250 kips. The column bears on a concrete footing with $f'_c = 3$ ksi.

Solution. We assume the footing area to be considerably larger than the plate area; thus $F_p = 0.6f'_c = 1.8$ ksi. Then

$$A_1 = \frac{P}{F_p} = \frac{250}{1.8} = 138.9 \text{ in.}^2$$

If the plate is square,

$$B = N = \sqrt{138.9} = 11.8 \text{ in.}$$

Since this is almost the same size as the column, we will assume the plan size layout shown in Fig. 6.12, which allows the welding of the plate to the column and the placing of the anchor bolts. It

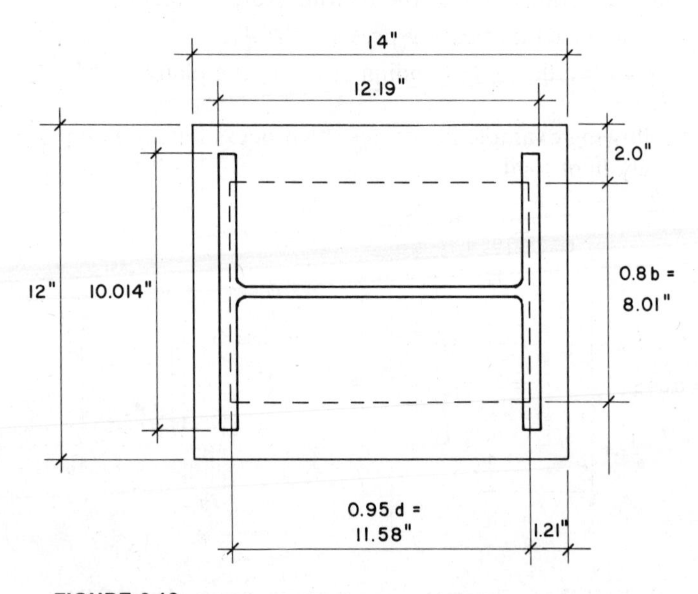

FIGURE 6.12

may be observed that the dimension labeled n in Fig. 6.11 is critical in this case, and the plate thickness is thus found as

$$t = \sqrt{\frac{3f_p n^2}{F_b}}$$

for which

$$f_p = \frac{P}{A_1} = \frac{250}{12(14)} = 1.49 \text{ ksi}$$

and

$$t = \sqrt{\frac{3(1.49)(2)^2}{0.75(36)}} = \sqrt{0.662} = 0.814 \text{ in.}$$

Plates are usually specified in thickness increments of $\frac{1}{8}$ in., so the minimum thickness would be $\frac{7}{8}$ in. (0.875 in.).

When a column transfers only compression force to its support, the stress transfer between the column and the base plate is one of simple bearing. For erection handling, the column and plate usually are attached, but there is no stress computation to be made for this connection. Lightly loaded columns typically have their modest-sized base plates welded to them in the fabricating shop. As the column and base plate get larger for heavily loaded columns, this becomes impractical and the plates are shipped loose and set in place to await the columns; with minor attachment to the column made mostly just to accurately position the column and hold it in place during erection of the building frame.

In some cases, however, the column-to-support force transfer may involve tension (uplift), shear (lateral force), or bending (fixed base–rotationally resistive). These actions all require special detailing, some cases of which are discussed in Chapter 7 and in the design examples in Chapter 12. Some recommended details for simple moment-resistive column bases are given in Part 4 of the AISC Manual.

Problem 6.12.A. Design a column base plate for a W 8 × 31 column that is supported on concrete for which the allowable bearing capacity is 750 psi [5000 kPa]. The load on the column is 178 kips [792 kN].

Problem 6.12.B. Design a column base plate for the W 8 × 31 in Problem 6.12.A if the bearing pressure allowed on the concrete is 1125 psi [7800 kPa].

7

FRAMES AND BENTS

A major use of steel for building structures is in frames consisting of vertical columns and horizontal-spanning members. These systems are often constituted as simple post-and-beam arrangements, with columns functioning as simple, axially loaded compression members and with horizontal members functioning as simple beams. In some cases, however, there may be more complex interactions between the frame members. Such is the case of the rigid frame, in which the members are connected for moment transfer, and the braced frame, in which diagonal members produce truss actions. In this chapter we consider some aspects of behavior and problems of design of rigid and braced frames.

7.1 BENTS

A bent is a planar frame formed to develop resistance to lateral loads, such as those produced by wind or earthquake effects. The simple frame in Fig. 7.1*a* consists of three members connected by pinned joints with pinned joints also at the bottom of the columns. In theory, this frame may be stable under vertical load only, if both the load and the frame are perfectly symmetrical. However, any lateral (in this case, horizontal) load or even a slightly unbalanced vertical load will topple the frame. One means for restoring stability to such a frame is to connect the tops of the columns to the ends of the beam with moment-resistive connections, as shown in Fig. 7.1*b*. If the modification is made, the deformation of the frame under vertical loading will be as shown in Fig. 7.2*a*, with moments developed in the beam and columns as indicated. If the frame is subjected to a lateral load, the frame deformation will be of the form shown in Fig. 7.2*b*.

While the transformation of a frame into a rigid frame bent may be done primarily for the purpose of achieving lateral stability, the form of response to vertical load is also unavoidably altered. Thus in the frame shown in Fig. 7.1*a*, the columns (stable or not) would be subject to only vertical axial compression under vertical loading on the beam, while they are also subject to bending when connected to the beam to produce rigid frame action. Under the combination of vertical and lateral loads, the bent will function as shown in Fig. 7.2*c*.

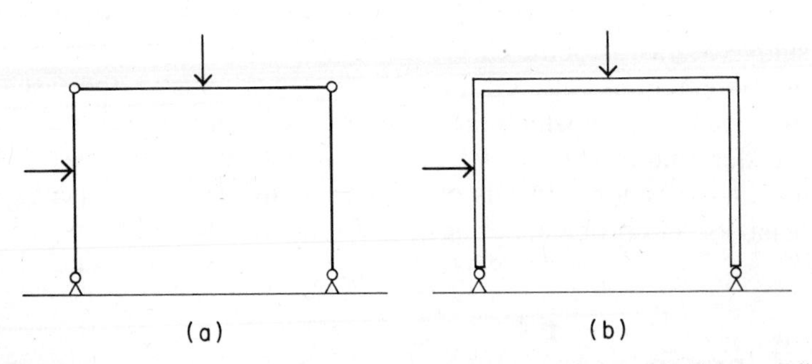

FIGURE 7.1 Single-unit steel frame: (a) with all pinned connections—ordinary post-and-beam construction; (b) rigid frame with moment-resistive beam-to-column connections.

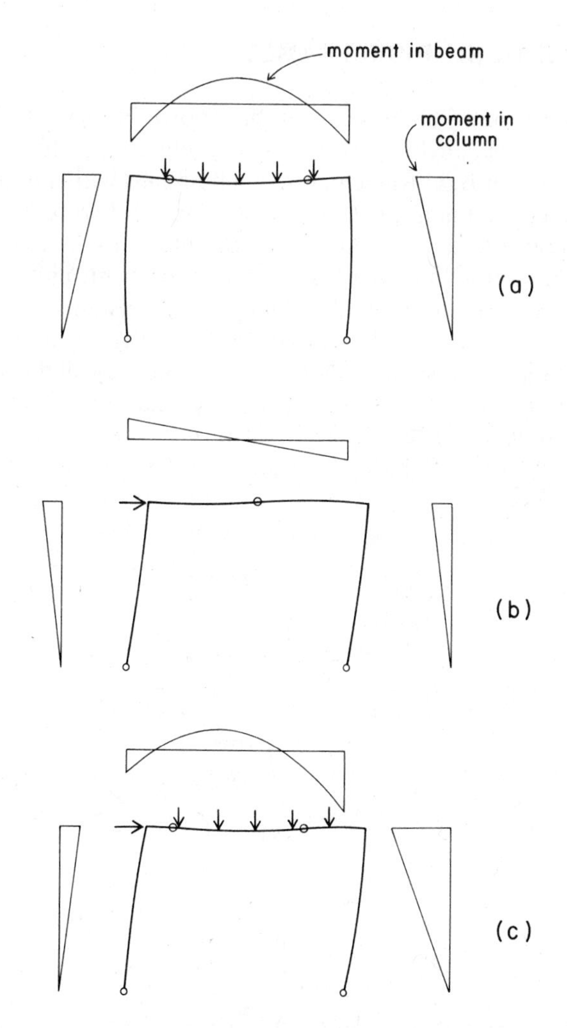

FIGURE 7.2 Actions of the single-unit rigid frame: (a) under gravity load; (b) under lateral load; (c) under combined gravity and lateral loads.

7.2 MULTIUNIT RIGID FRAMES

Single-unit rigid frames, such as that shown in Fig. 7.1, are used frequently for single-space, one-story buildings. However, the greater use for rigid-frame bents is in buildings with multiple horizontal bays and multiple levels (multistoried). Figure 7.3*a* shows the response of a two-story, two-bay rigid frame to lateral loading. Note that all members of the frame are bent, indicating that they are contribute to the development of resistance to the loading. Even when a single member is loaded, such as the single beam in Fig. 7.3*b*, some response is developed by all the members in the frame. This is a major aspect of the nature of such frames.

Multiunit frames are typically also three-dimensional with regard to the total framework. Rigid-frame action may be three-

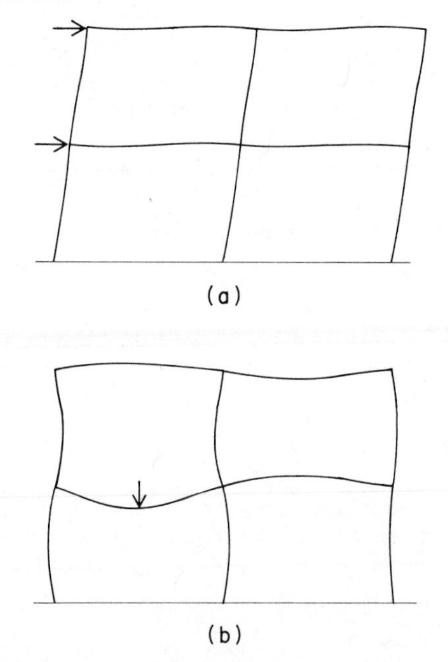

FIGURE 7.3 Actions of a multistory, multiple-span rigid frame: (a) under lateral loads; (b) with gravity load on a single beam.

dimensional or may be limited to the action of selected planar bents. With cast-in-place concrete construction, the three-dimensional rigid-frame action is usually unavoidable. With frames of wood or steel, however, the normal nature of the frame is usually of a simple post-and-beam system, with moment-resistive connections being of a special nature, rather than the ordinary means of connection. Thus it is possible to select which connections to make moment-resistive and which to leave with essentially no moment capacity. In this manner, rigid-frame action may be selectively controlled through the development of individual planar bents within the total frame system. The development of selected bracing bents is illustrated in the design examples in Chapter 12.

In steel frames moment-resistive connections are most often achieved with welding. The term "rigid" as used in referring to a frame actually applies essentially to the connections; implying that the joint resists deformation sufficiently to prevent any significant rotation of one connected member with respect to the other. The term is actually not such a good description of the general nature of the frame with regard to lateral resistance, since the other means for bracing frames against lateral loads (by shear panels or trussing) typically produce more rigid (stiff, deformation-resistive) structures.

Rigid frames are generally statically indeterminate and their investigation and design is beyond the scope of the work in this book. Some of the problems of designing for combined compression and bending are discussed in Chapter 6 and some detailing for moment-resistive connections is discussed in Chapter 10. Approximate design of rigid bents is illustrated in the design examples in Chapter 12.

7.3 BRACED FRAMES

The term "braced frame" is used to describe a frame that is braced by the use of diagonal members. In its simplest form, such a frame is constituted as a vertically cantilevered truss, as shown in Fig. 7.4*a*. With single diagonals, such a trussed frame is capable of being analyzed by statics, as illustrated for the trusses in

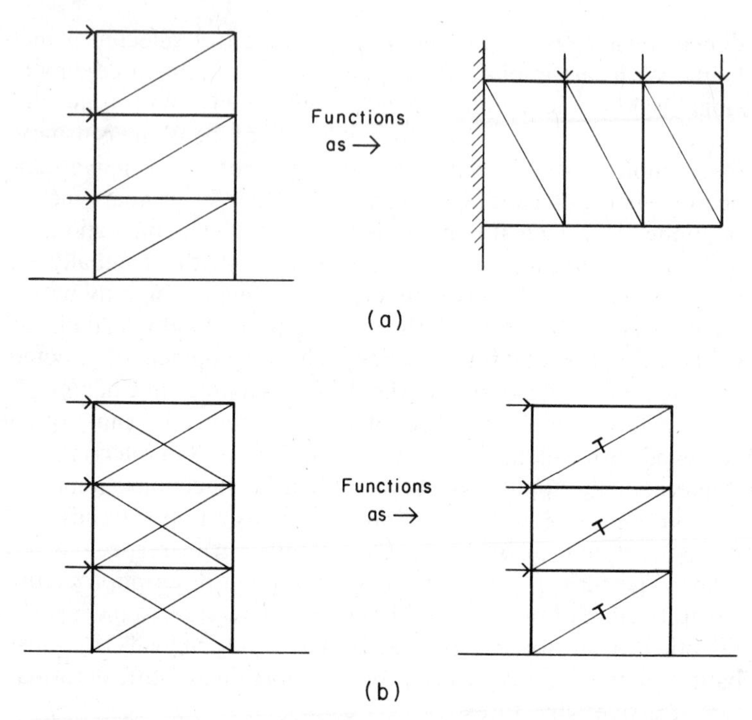

FIGURE 7.4 Frame with concentric bracing: (a) essential cantilever action under lateral loads; (b) assumed behavior with light X-bracing.

Chapter 9. For lateral bracing, however, the single diagonal arrangement is not very common. For light frames in small buildings, a common form used is that of the X-brace, as shown in Fig. 7.4*b*. Although this system is technically indeterminate, it is usually designed by assuming that the diagonals work only in tension. If the diagonals are quite slender (a common case) this may be literally true, as they will buckle under a very small axial compression force, leaving the opposed tension member to sustain the stability of the frame.

A special form of bracing is that described as eccentric bracing. The term derives from the fact that one or both ends of the brace is not connected to a joint of the frame (as in Fig. 7.4). Common forms of eccentric bracing are those shown in Fig. 7.5.

The core-bracing bents of high-rise buildings are mostly of one of these forms. With eccentric bracing, some members of the frame are subject to bending, at the location of the eccentric joints. Thus the action of the eccentrically braced frame has aspects of rigid frame action combined with truss action. When properly designed, this can result in a frame with some of the advantages of both systems; notably the stiffness of the truss and the redundancy and energy capacity of the rigid frame.

7.4 DIAPHRAGM-BRACED FRAMES

The third common way to brace a frame structure is by using rigid, planar elements, called diaphragms. When occurring in a vertical plane, these are called shear walls. When the walls are of sufficient strength, this bracing is usually the most deformation

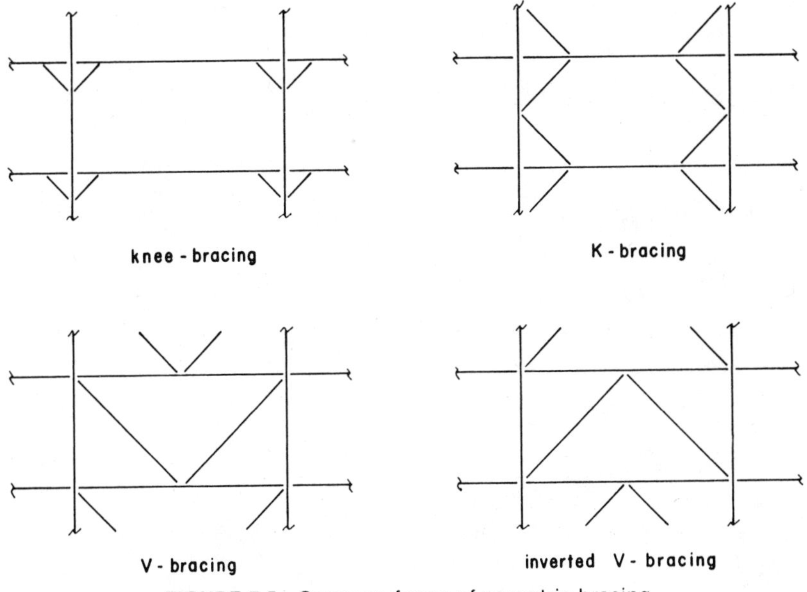

FIGURE 7.5 Common forms of eccentric bracing.

resistive. It is generally the most common bracing system for low-rise construction.

A building must be braced in a three-dimensional manner, and the use of shear walls, rigid frames, or braced frames generally deals only with the bracing of vertical planes. Horizontal planes are most typically braced by forming roof and floor decks as rigid diaphragms, although trussing may also be used horizontally in some cases.

Discussions of the design of shear walls and horizontal diaphragms are given in the examples in Chapter 12.

8

MISCELLANEOUS STEEL COMPONENTS AND SYSTEMS

The work in this book deals principally with structures that use rolled structural products and formed sheet steel units for common structural systems. While the vast majority of steel structures for buildings are formed from this limited inventory of components, there are many other uses of steel for large and small, simple and complex systems. In this chapter we treat briefly some of the other uses of steel for building structures.

8.1 MANUFACTURED SYSTEMS

Various manufacturers produce components that may be used to fashion complete structures, or—in some cases—complete buildings. The "package" building may come totally assembled (as with a mobile home) or in a kit for user assembly. Major use of such construction is made for utilitarian buildings for industrial and agricultural applications. Assurance of cost (not just low price) is a major reason for selection of such construction in many cases.

A certain stigma of architectural tastelessness is often associated with this construction (the army Quonset hut, Butler Brothers corn crib syndrome). Nevertheless, a significant amount of building construction is produced with these systems, including housing, schools, and other not-so-utilitarian uses.

Various factors favor the use of steel for the structures of many of these systems. The ability to achieve light, strong systems with dependable performance is a major asset. The noncombustible nature of the material also permits wider usage.

Many of the components of these systems are priority items, patented and produced by a single manufacturer, making the system a one-of-a-kind, unique product. In a sense, however, most of the elements used for steel structures of any kind are industrial products, produced in some controlled, standardized form. Designing any steel structure is therefore largely a matter of assembling a system from predesigned, prefabricated parts. The form of the end product of design is largely predictable, and not so different in many ways from the "package" building.

8.2 CABLE STRUCTURES

The strongest steels used for structures are generally those used to produce high-strength wire, with ultimate strength up to 300 ksi. At this level of strength, and due to the cold-working effect of the drawing process, the steel has negligible ductility and the wire is quite brittle. Usage is limited to grouping of the thin wires to produce cables. (Note that we use the common name "cable" here, although the products used for structures are often more accurately described as strand or rope.)

Cables represent a high efficiency of mass in the form of strength, but are limited in usage in many ways. First, stress resistance is limited to tension, so development of compression, shear, and bending is not possible. This limits cable use to forms and systems of structure that make major use of tension, and even so, require other elements to develop a full structure (masts, struts, booms, anchors, etc.).

While the high strength of the steel in cables permits development of great force, the modulus of elasticity of the material is not changed. In fact, a cable has a somewhat reduced effective modulus, as the cross section is not solid. Thus a great deal of elongation is required to develop force in the cable, and if loading conditions change significantly and produce change in the force in the cable, major change of length may occur. This sometimes makes it more practical to use lower-strength steel elements for tension members in order to limit deformations of the structure.

Cable structures must generally be quite "honest" in form. That is, the cables must follow the true paths of simple tension force. This requires quite precise determination of the geometry of force resolution for the cable structure.

8.3 COMPOSITE STRUCTURAL ELEMENTS

The term "composite" usually refers to structural elements in which two or more materials with significantly different stress-strain character are combined for mutual resistance to force. The most common example of this is reinforced concrete. Some other uses in building structures are the following:

Composite Steel–Concrete Beams

A common form of construction for floors with steel framing consists of site-cast concrete slabs on top of steel beams (see Fig. 5.13*b*). If the concrete slab is mechanically anchored by some means to the top flanges of the beams, it may be considered to contribute to the development of compression force for a combined resisting moment. Figure 8.1 shows the typical form of anchors (called shear connectors or studs) to achieve the anchorage. The studs are welded to the top of the beam flange and engage the site-cast concrete. While the strength of the steel framing is enhanced, the more significant gain in many cases is in reduced deflections. Part 2 of the AISC Manual contains a section titled "Composite Design for Building Construction" with an example of design of such construction. Tables are also provided in

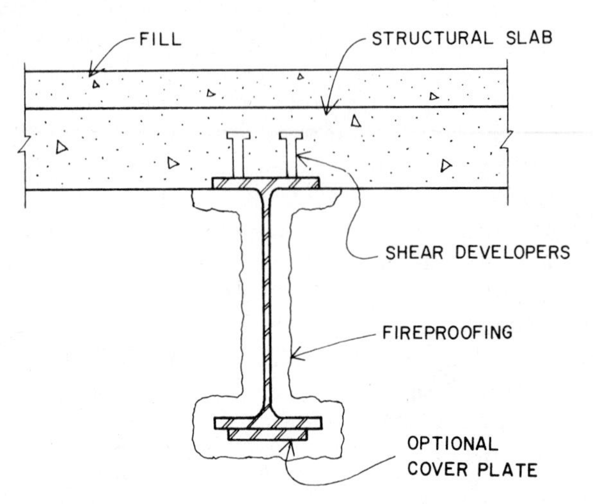

FIGURE 8.1 Composite construction with steel beams and concrete slab.

the manual for values of the "Transformed Section Modulus," which is the index of the beam's increased moment resistance.

Flitched Beams

In construction using solid timber beams, there are some occasions when the deflection of a beam is of critical concern. As heavy timber is generally available only in green wood condition, some long-term sag is inevitable for these beams. One means long used to control this is to combine a timber beam with steel plates in a flitched beam; two forms for which are shown in Fig. 8.2. For large wood sections, the wood member may be sandwiched between two plates (Fig. 8.2*b*), but a more common usage is that shown in Fig. 8.2*a*, with a single plate between two thinner wood members. The discussion that follows presents an example of the design of a flitched beam.

The components of a flitched beam are securely held together with through bolts so that the elements act as a single unit. The computations for determining the strength of such a beam illustrate the phenomenon of two different materials in a beam acting as a unit. The computations are based on the premise that the two materials deform equally. Let

Δ_1 and Δ_2 = deformations per unit length of the outermost fibers of the two materials, respectively,

f_1 and f_2 = unit bending stresses in the outermost fibers of the two materials, respectively,

E_1 and E_2 = moduli of elasticity of the two materials, respectively.

Since by definition the modulus of elasticity of a material is equal to the unit stress divided by the unit deformation, then

$$E_1 = \frac{f_1}{\Delta_1} \quad \text{and} \quad E_2 = \frac{f_2}{\Delta_2}$$

and transposing

$$\Delta_1 = \frac{f_1}{E_1} \quad \text{and} \quad \Delta_2 = \frac{f_2}{E_2}$$

Since the two deformations must be equal,

$$\frac{f_1}{E_1} = \frac{f_2}{E_2} \quad \text{and} \quad f_2 = f_1 \times \frac{E_2}{E_1}$$

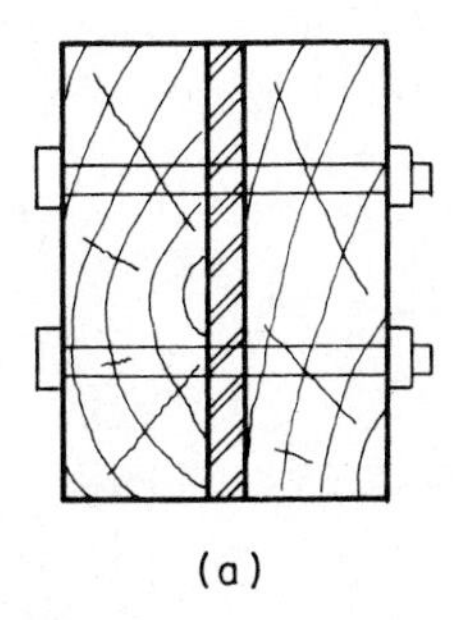

(a)

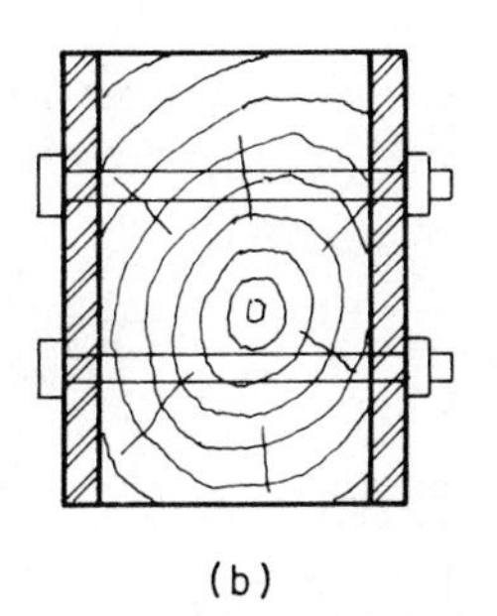

(b)

FIGURE 8.2 Flitched beams; composite steel and wood construction.

This simple equation for the relationship between the stresses in the two materials of a composite beam may be used as the basis for investigation or design of a flitched beam, as is demonstrated in the following example.

Example. A flitched beam is formed as shown in Fig. 8.2*a* consisting of two 2 × 12 planks of Douglas fir, No. 1 grade, and a 0.5 × 11.25-in. [13 × 285 mm] plate of A36 steel. Compute the allowable uniformly distributed load this beam will carry on a simple span of 14 ft [4.2 m]. For the steel: $E = 29{,}000{,}000$ psi [200 GPa], and the maximum allowable bending stress F_b is 22 ksi [150 MPa]. For the wood: $E = 1{,}800{,}000$ psi [12.4 GPA], the maximum allowable bending stress for single-member use is 1500 psi [10.3 MPa], and S for one 2 × 12 = 31.6 in.^{3}

Solution. (1) We first apply the formula just derived to determine which of the two materials limits the beam action. For a trial we assume the stress in the steel plate to be the limiting value and use the formula to find the maximum usable stress in the wood. Thus

$$f_w = f_2 \times \frac{E_w}{E_s} = 22{,}000 \times \frac{1{,}800{,}000}{29{,}000{,}000} = 1366 \text{ psi} \ \ [9.3 \text{ MPa}]$$

As this produces a stress lower than that of the table limit for the wood, our assumption is correct. That is, if we permit a stress higher than 1366 psi in the wood, the steel stress will exceed its limit of 22 ksi.

(2) Using the stress limit just determined for the wood, we now find the capacity of the wood members. Calling the load capacity of the wood W_w, we find that

$$M = \frac{W_w L}{8} = \frac{W_w \times 14 \times 12}{8} = 21\ W_w$$

Then using the S of 31.6 in.^{3} for the 2 × 12, we find

$$M = 21\ W_w = f_w \times S_{2w} = 1366 \times (2 \times 31.6)$$

$$W_w = 4111 \text{ lb} \ \ [18.35 \text{ kN}]$$

(3) For the plate we first must find the section modulus as follows:

$$S_s = \frac{bd^2}{6} = \frac{0.5 \times (11.25)^2}{6} = 10.55 \text{ in.}^3 \ [176 \times 10^3 \text{ mm}^3]$$

Then

$$M = 21\ W_s = f_s \times S_s = 22{,}000 \times 10.55$$
$$W_s = 11{,}052 \text{ lb} \ [50.29 \text{ kN}]$$

and the total capacity of the combined section is

$$W = W_w + W_s = 4111 + 11{,}052 = 15{,}163 \text{ lb} \ [68.64 \text{ kN}]$$

Although the load-carrying capacity of the wood elements is actually reduced in the flitched beam, the resulting total capacity is substantially greater than that of the wood members alone. This significant increase in strength achieved with small increase in size is a principal reason for popularity of the flitched beam. In addition, there is a significant reduction in deflection in most applications, and—most noteworthy—a reduction in sag over time.

Problem 8.3.A. A flitched beam consists of a single 10 × 14 of Douglas fir, Select Structural grade, and two A36 steel plates, each 0.5 × 13.5 in. [13 × 343 mm] (see Fig. 8.2*b*). Compute the magnitude of the concentrated load this flitched beam will support at the center of a 16 ft [4.8 m] simple span. Neglect the weight of the beam. Use a value of 22 ksi for the limiting bending stress in the steel. For the wood, E = 1,600,000 psi, allowable bending stress is 1600 psi, and S = 228.6 in.^{3}.

Composite Decks: Formed Steel plus Concrete

When formed sheet steel decks are used with concrete fill of a structural grade, three possibilities exist for the actions of the steel and concrete elements, as follows:

1. The concrete may be considered to function only as fill material, with the steel deck designed as the structural support.
2. The steel deck units may serve the singular purpose of supporting the wet, freshly poured concrete, with the concrete deck eventually hardening to become an ordinary, reinforced slab.
3. The steel deck may perform the preceding task, but in addition, become interactive with the concrete in composite action, serving the function ordinarily performed by the bottom reinforcing bars in resisting positive bending moments (midspan moments that cause tension in the bottom of the slab).

Steel deck units that are intended for the third purpose just described are typically formed with some lugs or indentations, similar to the raised ridges on ordinary steel reinforcing bars, intended to allow the deck to more fully engage the concrete for the composite action.

Manufacturers of deck units that are intended for use in composite construction usually provide data for the design of structural decks for such action. Safe load tables and other data from these sources may be used for the complete design of the composite decks.

9

STEEL TRUSSES

9.1 GENERAL

Trussing, or triangulated framing, is a means for developing stability with a light frame. It is also a means for producing very light two-dimensional or three-dimensional structural elements for towers, spanning systems, or structures in general. In this chapter we deal with some uses of simple, planar trusses for building structures, with a concentration on roof trusses—an application that generally makes fullest use of the potential lightness and freedom of form of the truss.

A historically common use of the truss is to achieve the simple, double-slope, gabled roof form. This is typically done by use of sloping members and a horizontal bottom member, as shown in Fig. 9.1. Depending on the size of the span, the interior of the simple triangle formed by these three members may be filled by various arrangements of triangulated members. Some of the terminology used for the components of such a truss, as indicated in Fig. 9.1, are as follows:

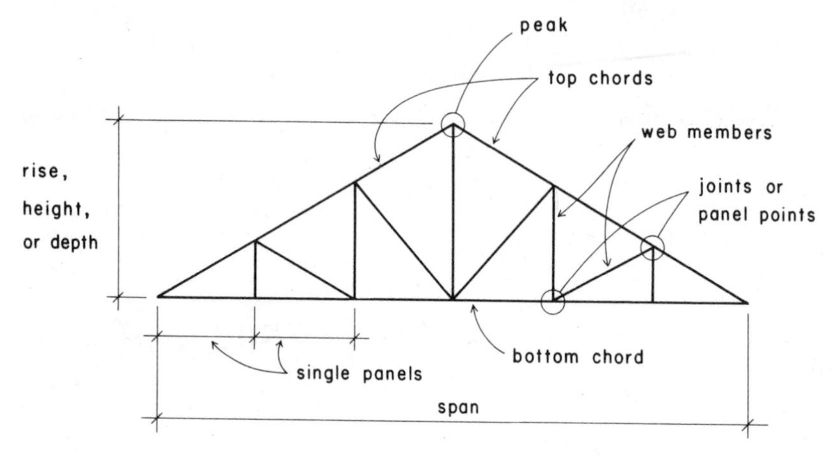

FIGURE 9.1 Elements of a truss.

Chord Members. These are the top and bottom boundary members of the truss, analogous to the top and bottom flanges of a steel beam. For trusses of modest size these members are often made of a single element that is continuous through several joints, with a total length limited only by the maximum ordinarily obtainable for the element selected.

Web Members. The interior members of the truss are called web members. Unless there are interior joints, these members are of a single piece between joints.

Panels. Most trusses have a pattern that consists of some repetitive, modular unit. This unit ordinarily is referred to as the panel of the truss; joints are sometimes referred to as panel points.

A critical dimension of a truss is its overall height, which is sometimes referred to as its rise or its depth. For the truss illustrated, this dimension relates to the establishment of the roof pitch and also determines the length of the web members. A critical concern with regard to the efficiency of the truss as a spanning structure is the ratio of the span of the truss to its height.

Although beams and joists may be functional with span/height ratios as high as 20 to 30, trusses generally require much lower ratios.

Trusses may be used in a number of ways as part of the total structural system for a building. Figure 9.2 shows a series of single-span, planar trusses in the form shown in Fig. 9.1 with the other elements of the building structure that develop the roof system and provide support for the trusses. In this example the trusses are spaced a considerable distance apart. In this situation it is common to use purlins to span between the trusses, supported at the top chord joints of the trusses to avoid bending in the chords. The purlins, in turn, support a series of closely spaced rafters that are parallel to the trusses. The roof deck is then attached to the rafters so that the roof surface actually floats above the level of the top of the trusses.

Figure 9.3 shows a similar structural system of trusses with parallel chords. This system may be used for a floor or a flat roof.

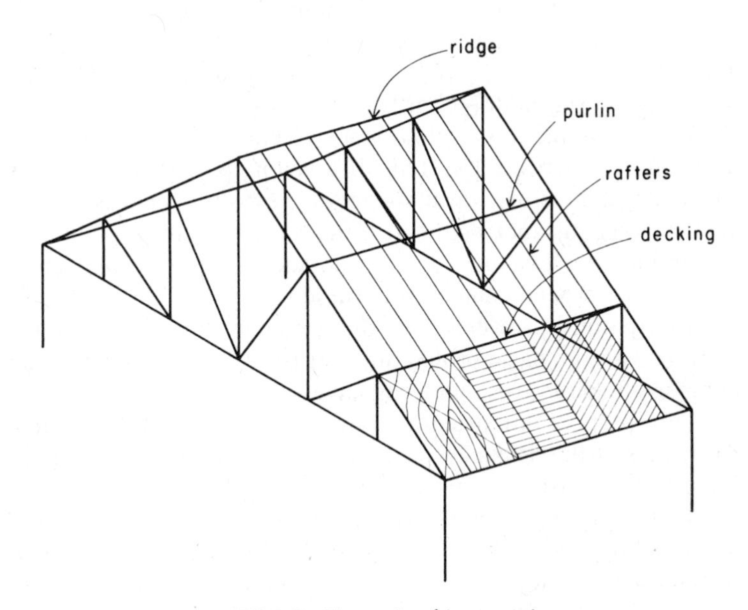

FIGURE 9.2 Elements of truss systems.

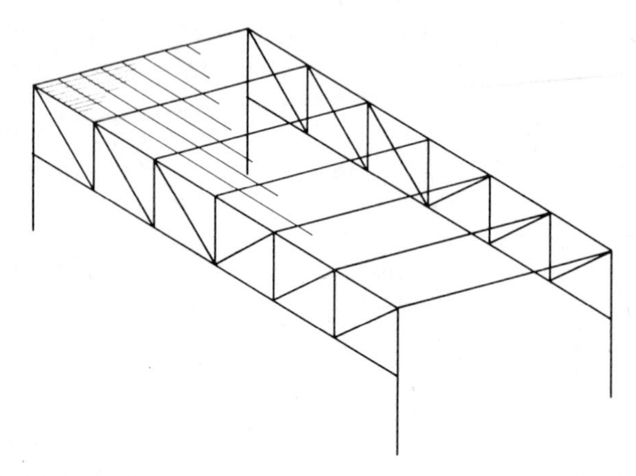

FIGURE 9.3 Structural system with flat, parallel-chorded trusses.

When the trusses are slightly closer together it may be more practical to eliminate the purlins and to increase the size of the top chords to accommodate the additional bending due to the rafters. As an extension of this idea, if the trusses are really close, it may be possible to eliminate the rafters as well and to place the deck directly on the top chords of the trusses.

For various situations additional elements may be required for the complete structural system. If a ceiling is required, another framing system is used at the level of the bottom chords or suspended some distance below it. If the roof and ceiling framing do not provide it adequately, it may be necessary to use some bracing system perpendicular to the trusses.

Truss patterns are derived from a number of considerations, starting with the basic profile of the truss. For various reasons a number of classic truss patterns have evolved and have become standard parts of our structural vocabulary. Some of these carry the names of the designers who first developed them. Several of these common truss forms are shown in Fig. 9.4.

The two most common forms of steel trusses of small to medium size are those shown in Fig. 10.16. In both cases the members may be connected by rivets, bolts, or welds. The most common practice is to use welding for connections that are assembled

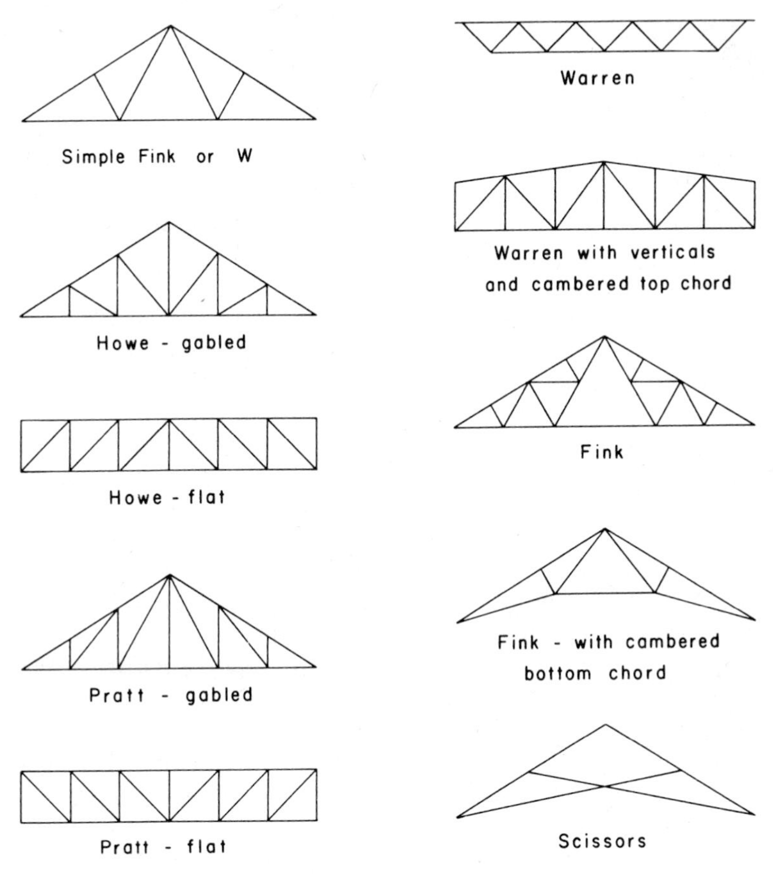

FIGURE 9.4 Truss patterns.

in the fabricating shop and high-strength bolts (torque tensioned) for field connections.

9.2 BRACING FOR TRUSSES

Single planar trusses are very thin structures that require some form of lateral bracing. The compression chord of the truss must be designed for its laterally unbraced length. In the plane of the truss, the chord is braced by other truss members at each joint. However, if there is no lateral bracing, the unbraced length of the chord in a direction perpendicular to the plane of the truss becomes the full length of the truss. Obviously, it is not feasible to design a slender compression member for this unbraced length.

In most buildings other elements of the construction ordinarily provide some or all of the necessary bracing for the trusses. In the structural system shown in Fig. 9.5*a*, the top chord of the truss is braced at each truss joint by the purlins. If the roof deck is a reasonably rigid planar structural element and is adequately attached to the purlins, this constitutes a very adequate bracing of the compression chord–which is the main problem for the truss. However, it is also necessary to brace the truss generally for out-of-plane movement throughout its height. In Fig. 9.5*a* this is done by providing a vertical plane of X-bracing at every other panel point of the truss. The purlin does an additional service by serving as part of this vertical plane of trussed bracing. One panel of this bracing is actually capable of bracing a pair of trusses, so that it would be possible to place it only in alternate bays between the trusses. However, the bracing may be part of the general bracing system for the building, as well as providing for the bracing of the individual trusses. In the latter case, it would probably be continuous.

Light trusses that directly support a deck, as shown in Fig. 9.5*b*, usually are adequately braced by the deck. This constitutes continuous bracing, so that the unbraced length of the chord in this case is actually zero. Additional bracing in this situation often is limited to a series of continuous steel rods or single small angles that are attached to the bottom chords as shown in the illustration.

Another form of bracing that is used is that shown in Fig. 9.5*c*. In this case a horizontal plane of X-bracing is placed between two trusses at the level of the bottom chords. This single braced bay may be used to brace several other bays of trusses by connecting them to the X-braced trusses with horizontal struts. As in the previous example, with vertical planes of bracing, the top chord is braced by the roof construction. It is likely that bracing of this form is also part of the general lateral bracing system for the

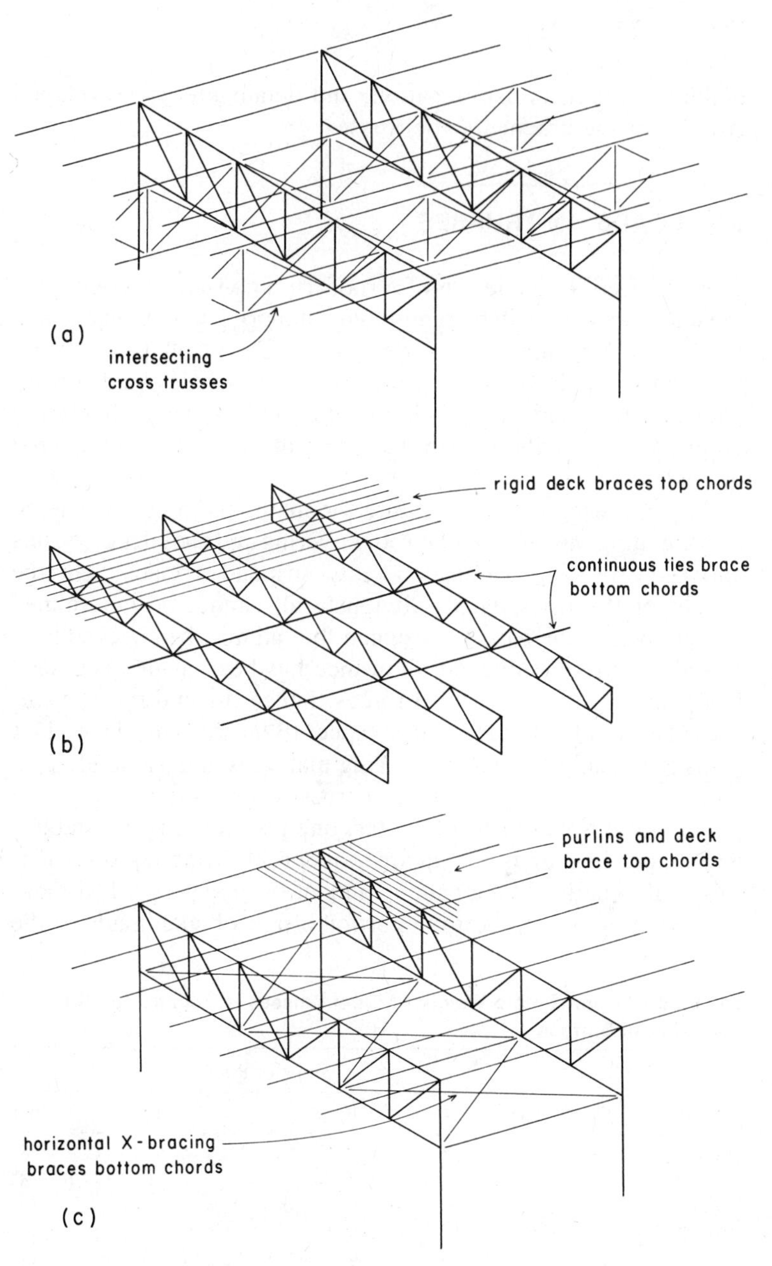

FIGURE 9.5 Forms of lateral bracing for trusses.

building so that its use, location, and details are not developed strictly for the bracing of the trusses.

9.3 LOADS ON TRUSSES

The first step in the design of a roof truss consists of computing the loads the truss will be required to support. These are dead and live loads. The former includes the weight of all construction materials supported by the truss; the latter includes loads resulting from snow and wind, and, on flat roofs, occupancy loads and an allowance for the possible ponding of water due to impaired drainage.

The following items constitute the materials to be considered in computing the dead loads: roof covering and roof deck, purlins and sway bracing, ceiling and any suspended loads, and the weight of the truss itself. Obviously all cannot be determined exactly before the truss is designed, but all may be checked later to see whether a sufficient allowance has been made. The dead loads are downward vertical forces, hence the end reactions of the truss are also vertical with respect to these loads. Table 11.1 gives the weights of certain roofing materials and Table 9.1 provides estimated weights of steel trusses for various spans and pitches. With respect to the latter, one procedure is to establish an estimate in pounds per square foot of roof surface and consider this load as acting at the panel points of the upper chord. A more exact method would be to apportion a part of such loads to the

TABLE 9.1 Approximate Weight of Steel Trusses in Pounds per Square Foot of Roof Surface

	Slope of Roof			
Span (ft)	45°	30°	25°	Flat
Up to 40	5	6	7	8
40–50	6	7	7	8
50–60	7	8	9	10
60–70	7	8	9	10
70–80	8	9	10	11

panel points of the lower chord, but this is customary only in trusses with exceptionally long spans. After the truss has been designed its actual weight may be computed and compared with the estimated weight.

The weight allowance for snow load depends primarily on the geographical location of the structure and the roof slope. Freshly fallen snow may weigh as much as 10 lb per cu ft [0.13 kg/m^3] and accumulations of wet or packed snow may exceed this value. The amount of snow retained on a roof over a given period depends on the type of roofing as well as the slope; for example, snow slides off a metal or slate roof more readily than from a wood shingle surface; also, the amount of insulation in the roof construction will influence the period of retention.

Required design live loads for roofs are specified by local building codes. When snow is a potential problem the load is usually based on anticipated snow accumulation. Otherwise the specified load is intended essentially to provide some capacity for sustaining loads experienced during construction and maintenance of the roof. The basic required load can usually be modified when the roof slope is of some significant angle and on the basis of the total roof surface area supported by the structure. Table 11.2 gives the minimum roof live loads specified by the *Uniform Building Code*, 1988 ed. (Ref. 2), which are based on the situation in which snow load is not the critical concern.

Magnitudes of design wind pressures and various other requirements for wind design are specified by local building codes. The code in force for a specific building location should be used for any design work. For a general explanation of the analysis and design of the effects of wind and earthquake forces on buildings the reader is referred to *Simplified Building Design for Wind and Earthquake Forces* by Ambrose and Vergun (Ref. 6).

9.4 GRAPHICAL ANALYSIS FOR INTERNAL FORCES IN PLANAR TRUSSES

Figure 9.6 shows a single span, planar truss that is subjected to vertical gravity loads. We use this example to illustrate the proce-

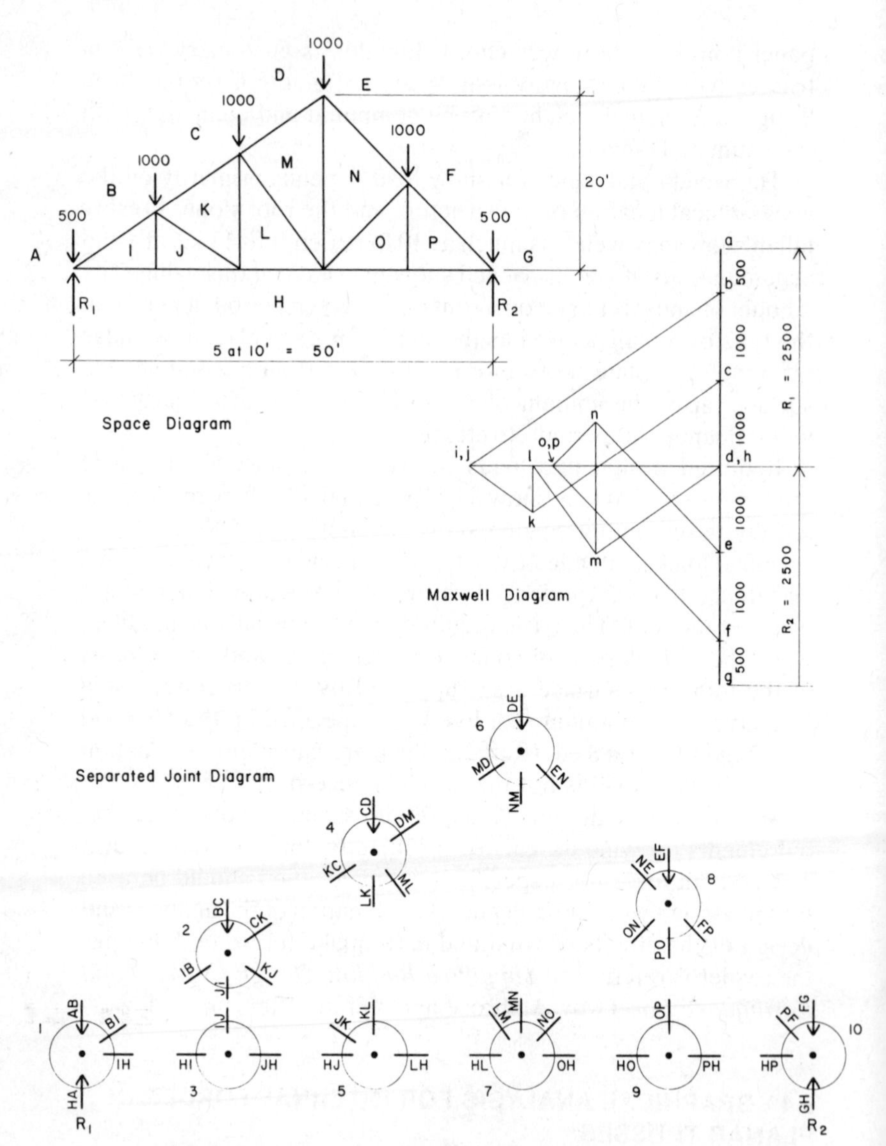

FIGURE 9.6 Examples of graphic diagrams for a planar truss.

dures for determining the internal forces in the truss; that is, the tension and compression forces in the individual members of the truss. The space diagram in the figure shows the truss form, the support conditions, and the loads. The letters on the space diagram identify individual forces at the truss joints. The sequence of placement of the letters is arbitrary, the only necessary consideration being to place a letter in each space between the loads and the individual truss members so that each force at a joint can be identified by a two-letter symbol.

The separated joint diagram in the figure provides a useful means of visualizing the complete force system at each joint as well as the interrelation of the joints through the truss members. The individual forces at each joint are designated by two-letter symbols that are obtained by simple reading around the joint in the space diagram in a clockwise direction. Note that the two-letter symbols are reversed at the opposite ends of each of the truss members. Thus the top chord member at the left end of the truss is designated as *BI* when shown in the joint at the left support (joint 1) and as *IB* when shown in the first interior upper chord joint (joint 2). The purpose of this procedure is demonstrated in the following explanation of the graphical analysis. The third diagram in Figure 9.6 is a composite force polygon for the external and internal forces in the truss. It is called a Maxwell diagram after one of its early users, James Clerk Maxwell, an English engineer. The construction of this diagram constitutes a complete solution for the magnitudes and senses of the internal forces in the truss. The procedure for this construction is as follows.

1. *Construct the Force Polygon for the External Forces.* Before this can be done the values for the reactions must be found. There are graphic techniques for finding the reactions, but it is usually much simpler and faster to find them with an algebraic solution. In this example, although the truss is not symmetrical, the loading is, and it may be observed that each of the reactions is equal to one-half the total load on the truss, or 5000/2 = 2500 lb. Because the external forces in this case are all in a single direction, the

force polygon for the external forces is actually a straight line. Using the two-letter symbols for the forces and starting with letter *A* at the left end, we read the force sequence by moving in a clockwise direction around the outside of the truss. Thus the loads are read as *AB, BC, CD, DE, EF,* and *FG* and the two reactions are read as *GH* and *HA*. By beginning at *A* on the Maxwell diagram the force vector sequence for the external forces is read from *A* to *B*, *B* to *C*, *C* to *D*, and so on, ending back at *A*, which shows that the force polygon closes and the external forces are in the necessary state of static equilibrium. Note that we have pulled the vectors for the reactions off to the side in the diagram to indicate them more clearly. Note also that we have used lowercase letters for the vector ends in the Maxwell diagram, whereas uppercase letters appear on the space diagram. The alphabetic correlation is retained (*A* to *a*) and any possible confusion between the two diagrams is prevented. The letters on the space diagram designate spaces, whereas the letters on the Maxwell diagram designate points of intersection of lines.

2. *Construct the Force Polygons for the Individual Joints.* The graphic procedure for this consists of locating the points on the Maxwell diagram that correspond to the remaining letters, *I* through *P*, on the space diagram. When all the lettered points on the diagram are located, the complete force polygon for each joint may be read on the diagram. To locate these points we use two relationships. The first is that the truss members can resist only those forces that are parallel to the members' positioned directions. Thus we know the directions of all the internal forces. The second relationship is a simple one from plane geometry: A point may be located as the intersection of two lines. Consider the forces at joint 1, as shown in the separated joint diagram in Figure 9.6. Note that there are four forces and that two of them are known (the load and the reaction) and two are unknown (the internal forces in the truss members). The force polygon for this joint, as shown on the Maxwell diagram, is read as *ABIHA*. *AB* represents the load, *BI*, the

force in the upper chord member, *IH*, the force in the lower chord member, and *HA*, the reaction. Thus the location of point *I* on the Maxwell diagram is determined by noting that *I* must be in a horizontal direction from *H* (corresponding to the horizontal position of the lower chord) and in a direction from *B* parallel to the position of the upper chord.

The remaining points on the Maxwell diagram are found by the same process, using two known points on the diagram to project lines of known direction whose intersection will determine the location of another point. Once all the points are located the diagram is complete and can be used to find the magnitude and sense of each internal force. The process for construction of the Maxwell diagram typically consists of moving from joint to joint along the truss. Once one of the letters for an internal space is determined on the Maxwell diagram it may be used as a known point for finding the letter for an adjacent space on the space diagram. The only limitation of the process is that it is not possible to find more than one unknown point on the Maxwell diagram for any single joint. Consider joint 7 on the separated joint diagram in Figure 9.6. If we attempt to solve this joint first, knowing only the locations of letters *A* through *H* on the Maxwell diagram, we must locate four unknown points: *L*, *M*, *N*, and *O*. This is three more unknowns than we can determine in a single step, and we must first solve for three of the unknowns by using other joints.

Solving for a single unknown point on the Maxwell diagram corresponds to finding two unknown forces at a joint because each letter on the space diagram is used twice in the force identifications for the internal forces. Thus for joint 1 in the previous example the letter *I* is part of the identity for forces *BI* and *IH*, as shown on the separated joint diagram. The graphic determination of single points on the Maxwell diagram is therefore analogous to finding two unknown quantities in an algebraic solution. As discussed previously, two unknowns are the maximum that can be solved for in the equilibrium of a coplanar, concurrent force system, which is the condition of the individual joints in the truss.

When the Maxwell diagram is completed the internal forces can be read from the diagram as follows:

1. The magnitude is determined by measuring the length of the line in the diagram with the scale that was used to plot the vectors for the external forces.

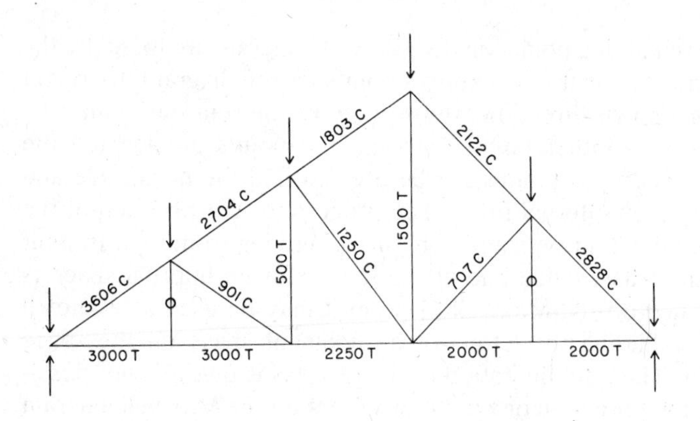

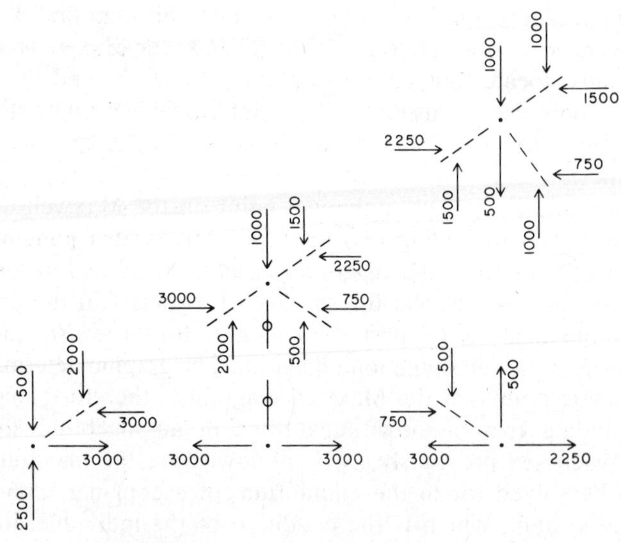

FIGURE 9.7 Internal forces for the truss.

2. The sense of individual forces is determined by reading the forces in clockwise sequence around a single joint in the space diagram and tracing the same letter sequences on the Maxwell diagram.

The degree of accuracy attainable from a graphical analysis depends on the size of the construction and the accuracy of the drafting. The results of the analysis of the truss shown in Fig. 9.6 are displayed on the truss form in Fig. 9.7; *C* indicates compression and *T* indicates tension. Zero force members are indicated by placing a zero directly on the member. The values shown were actually determined from an algebraic analysis, since four place accuracy is not attainable from a graphical construction.

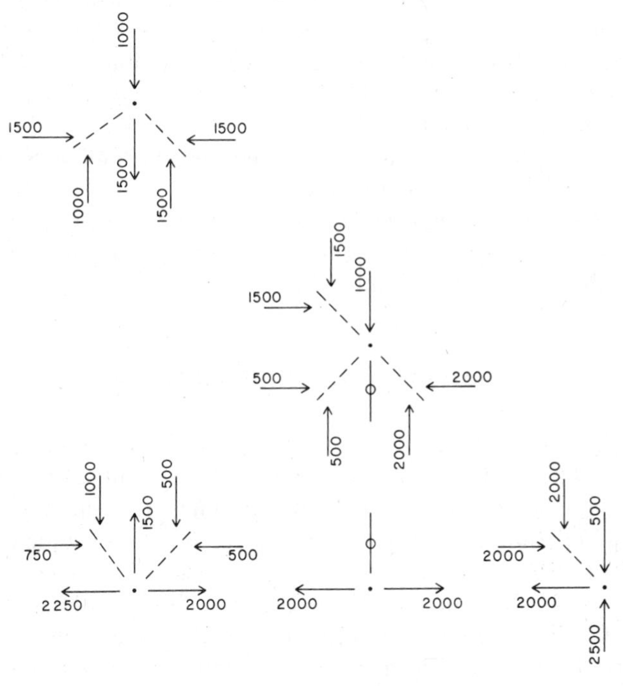

FIGURE 9.7 *(Continued)*

9.5 DESIGN FORCES FOR TRUSS MEMBERS

The primary concern in analysis of trusses is the determination of the critical forces for which each member of the truss must be designed. The first step in this process is the decision about which combinations of loading must be considered. In some cases the potential combinations may be quite numerous. When both wind and seismic actions are potentially critical and more than one type of live loading occurs (e.g., roof loads plus hanging loads) the theoretically possible combinations of loadings can be overwhelming. However, designers are usually able to exercise judgment in reducing the sensible combinations to a reasonable number; for example, it is statistically improbable that a violent windstorm will occur simultaneously with a major earthquake shock.

Once the required design loading conditions are established the usual procedure is to perform separate analyses for each of the loadings. The values obtained can then be combined at will for each member to ascertain the particular combination that establishes the critical result for the member. This means that in some cases certain members will be designed for one combination and others, for different combinations.

In most cases design codes permit an increase in allowable stress for design of members when the critical loading includes forces due to wind or seismic loads.

9.6 COMBINED STRESS IN TRUSS MEMBERS

When analyzing trusses the usual procedure is to assume that the loads will be applied to the truss joints. This results in the members themselves being loaded only through the joins and thus having only direct tension or compression forces. In some cases, however, truss members may be directly loaded; for example, when the top chord of a truss supports a roof deck without benefit of joists. Thus the chord member is directly loaded with a linear uniform load and functions as a beam between its end joints.

The usual procedure in these situations is to accumulate the loads at the truss joints and analyze the truss as a whole for the typical joint loading arrangement. The truss members that sustain the direct loading are then designed for the combined effects of the axial force caused by the truss action and the bending caused by the direct loading.

A typical situation for a roof truss is one in which the actual loading consists of the roof load distributed continuously along the top chords and a ceiling loading distributed continuously along the bottom chords. The top chords are thus designed for a combination of axial compression and bending and the bottom chords for a combination of axial tension plus bending. This will of course result in somewhat larger members being required for both chords and any estimate of the truss weight should account for this anticipated additional requirement.

9.7 INTERNAL FORCES FOUND BY COEFFICIENTS

Figure 9-8 shows a number of simple trusses of both parallel-chorded and gable form. Table 9.2 lists coefficients that may be used to find the values for the internal forces in these trusses. For the gable-form trusses coefficients are given for three different slopes of the top chord: 4 in 12, 6 in 12, and 8 in 12. For the parallel-chorded trusses coefficients are given for two different ratios of the truss depth to the truss panel length: 1 to 1 and 3 to 4. Loading results from vertical gravity loads and is assumed to be applied symmetrically to the truss; internal panel point loads are equal to W.

The table values are based on a value of $W = 1.0$. To use the tables it is necessary only to find the true value of the panel point loading and multiply it by the table coefficient to find the force in a truss member. Note that because of the symmetry of the trusses and the loads, the internal forces in the members are the same on each half of the truss; therefore we have given the coefficients for only the left half of each truss.

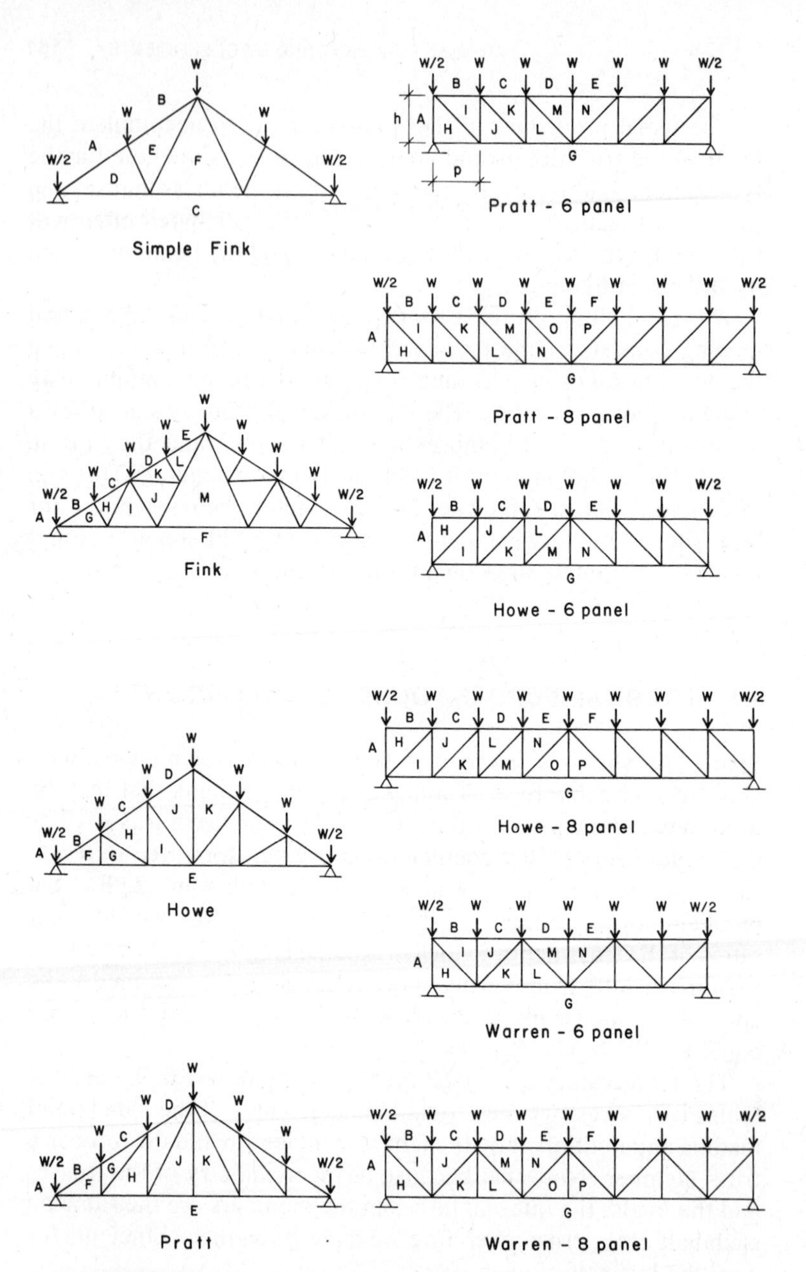

FIGURE 9.8 Simple trusses of parallel-chorded and gable form.

TABLE 9.2 Coefficients for Internal Forces in Simple Trusses

Force in members = (table coefficient) X (panel load, W)

T indicates tension, C indicates compression

Gable Form Trusses

Truss Member	Type of Force	Roof Slope 4/12	Roof Slope 6/12	Roof Slope 8/12
Truss 1 - Simple Fink				
AD	C	4.74	3.35	2.70
BE	C	3.95	2.80	2.26
DC	T	4.50	3.00	2.25
FC	T	3.00	2.00	1.50
DE	C	1.06	0.90	0.84
EF	T	1.06	0.90	0.84
Truss 2 - Fink				
BG	C	11.08	7.83	6.31
CH	C	10.76	7.38	5.76
DK	C	10.44	6.93	5.20
EL	C	10.12	6.48	4.65
FG	T	10.50	7.00	5.25
FI	T	9.00	6.00	4.50
FM	T	6.00	4.00	3.00
GH	C	0.95	0.89	0.83
HI	T	1.50	1.00	0.75
IJ	C	1.90	1.79	1.66
JK	T	1.50	1.00	0.75
KL	C	0.95	0.89	0.83
JM	T	3.00	2.00	1.50
LM	T	4.50	3.00	2.25
Truss 3 - Howe				
BF	C	7.90	5.59	4.51
CH	C	6.32	4.50	3.61
DJ	C	4.75	3.35	2.70
EF	T	7.50	5.00	3.75
EI	T	6.00	4.00	3.00
GH	C	1.58	1.12	0.90
HI	T	0.50	0.50	0.50
IJ	C	1.81	1.41	1.25
JK	T	2.00	2.00	2.00
Truss 4 - Pratt				
BF	C	7.90	5.59	4.51
CG	C	7.90	5.59	4.51
DI	C	6.32	4.50	3.61
EF	T	7.50	5.00	3.75
EH	T	6.00	4.00	3.00
EJ	T	4.50	3.00	2.25
FG	C	1.00	1.00	1.00
GH	T	1.81	1.41	1.25
HI	C	1.50	1.50	1.50
IJ	T	2.12	1.80	1.68

Flat - Chorded Trusses

Truss Member	Type of Force	6 Panel Truss $\frac{h}{p} = 1$	6 Panel Truss $\frac{h}{p} = \frac{3}{4}$	8 Panel Truss $\frac{h}{p} = 1$	8 Panel Truss $\frac{h}{p} = \frac{3}{4}$
Truss 5 - Pratt					
BI	C	2.50	3.33	3.50	4.67
CK	C	4.00	5.33	6.00	8.00
DM	C	4.50	6.00	7.50	10.00
EO	C	–	–	8.00	10.67
GH	O	0	0	0	0
GJ	T	2.50	3.33	3.50	4.67
GL	T	4.00	5.33	6.00	8.00
GN	T	–	–	7.50	10.00
AH	C	3.00	3.00	4.00	4.00
IJ	C	2.50	2.50	3.50	3.50
KL	C	1.50	1.50	2.50	2.50
MN	C	1.00	1.00	1.50	1.50
OP	C	–	–	1.00	1.00
HI	T	3.53	4.17	4.95	5.83
JK	T	2.12	2.50	3.54	4.17
LM	T	0.71	0.83	2.12	2.50
NO	T	–	–	0.71	0.83
Truss 6 - Howe					
BH	O	0	0	0	0
CJ	C	2.50	3.33	3.50	4.67
DL	C	4.00	5.33	6.00	8.00
EN	C	–	–	7.50	10.00
GI	T	2.50	3.33	3.50	4.67
GK	T	4.00	5.33	6.00	8.00
GM	T	4.50	6.00	7.50	10.00
GO	T	–	–	8.00	10.67
AH	C	0.50	0.50	0.50	0.50
IJ	T	1.50	1.50	2.50	2.50
KL	T	0.50	0.50	1.50	1.50
MN	T	0	0	0.50	0.50
OP	O	–	–	0	0
HI	C	3.53	4.17	4.95	5.83
JK	C	2.12	2.50	3.54	4.17
LM	C	0.71	0.83	2.12	2.50
NO	C	–	–	0.71	0.83
Truss 7 - Warren					
BI	C	2.50	3.33	3.50	4.67
DM	C	4.50	6.00	7.50	10.00
GH	O	0	0	0	0
GK	T	4.00	5.33	6.00	8.00
GO	T	–	–	8.00	10.67
AH	C	3.00	3.00	4.00	4.00
IJ	C	1.00	1.00	1.00	1.00
KL	O	0	0	0	0
MN	C	1.00	1.00	1.00	1.00
OP	O	–	–	0	0
HI	T	3.53	4.17	4.95	5.83
JK	C	2.12	2.50	3.54	4.17
LM	T	0.71	0.83	2.12	2.50
NO	C	–	–	0.71	0.83

9.8 DESIGN OF A STEEL ROOF TRUSS

The following example is used to illustrate several of the issues and the general process in the design of short-to-medium-span roof trusses. The form of the truss is shown in Fig. 9.9. The principal loading consists of the concentrated panel point loads

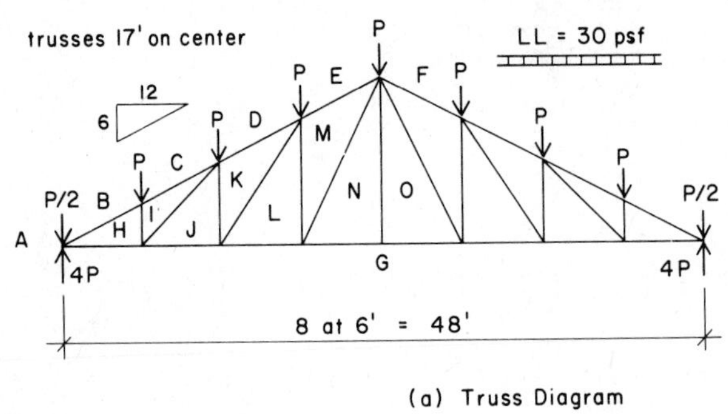

(a) Truss Diagram

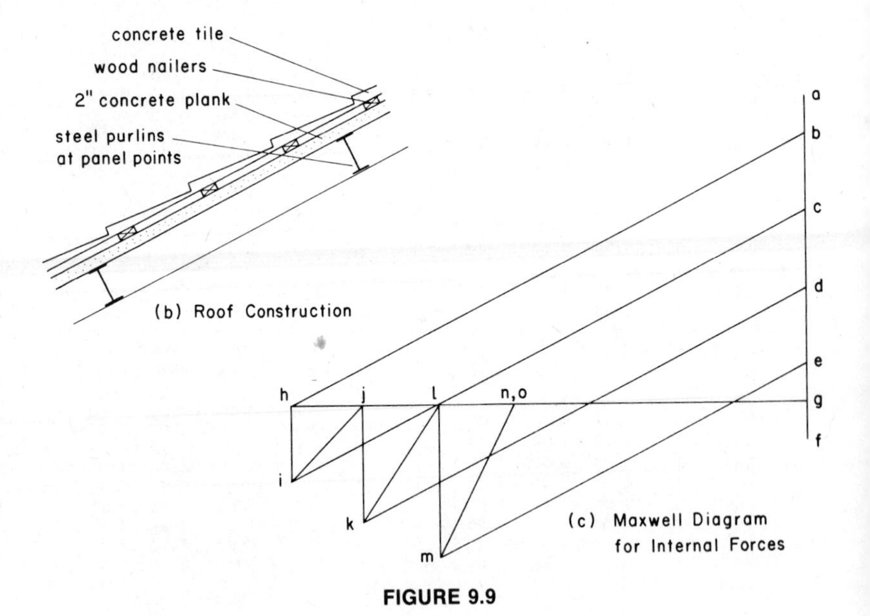

(b) Roof Construction

(c) Maxwell Diagram for Internal Forces

FIGURE 9.9

on the top, delivered by steel purlins that span between the trusses. The purlins support a roof of cement-bonded wood fiber deck units, wood nailing strips, and cement tile roofing. The dead weight of this roof construction is estimated for design:

Concrete tile	10 psf	0.48 kN/m^2
Wood nailing strips	1	0.05
Deck units	8	0.38
Purlins (estimate)	3	0.14
Total dead load	22 psf	1.05 kN/m^2

9.9 DESIGN OF THE PURLIN

Selection of the shape to be used for the purlin depends on a number of considerations, including the need for attachment of the deck to the purlins and the purlins to the trusses. Several possibilities for the choice of a purlin are shown in Fig. 9.10. If a W shape is placed directly on the top chord of the truss as shown in Fig. 9.10*a*, a problem that must be considered is the bending of the purlin on its weak axis because of its tilted position. Two modifications that may be used to eliminate this problem are shown at *b* and *c* in Fig. 9.10. Sag rods and channels were common in earlier times but other options are now considered more favorable.

A consideration that must be made is whether the deck provides lateral bracing for the purlins. The type of deck described for this construction probably does not offer good bracing and the best choice for the purlin may be the tube shown in Fig. 9.10*d*.

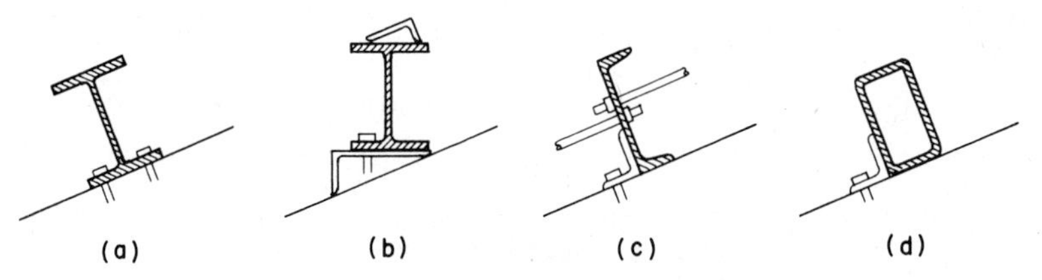

FIGURE 9.10 Alternatives for the steel purlins.

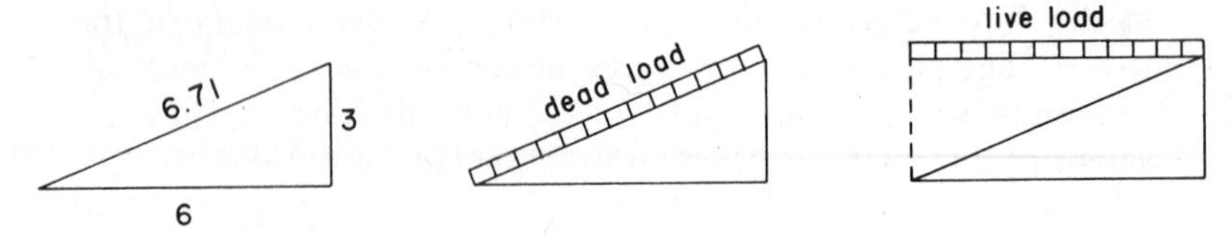

FIGURE 9.11 Roof gravity load distribution on the truss.

Without further belaboring the decision we choose the tube and proceed with its design for the biaxial bending.

As shown in Fig. 9.11, the dead weight of the roof is distributed over the actual roof surface. The load carried by a single interior purlin is thus

$$w_d = 22 \times 6.71 \times 17 = 2510 \text{ lb } [11.16 \text{ kN}]$$

The live load is distributed on the horizontal plane and the purlin load is

$$w_L = 30 \times 6 \times 17 = 3060 \text{ lb } [13.61 \text{ kN}]$$

The total load is 2510 + 3060 = 5570 lb [24.77 kN].

If the member is placed in a tilted position, as shown at *a*, *c*, and *d* in Fig. 9.10, this vertical load must be divided into two components that relate to the major and minor axes of the member. The bending moments developed by these two load components are determined and the combined effect is investigated by use of the combined action formula

$$\frac{f_{bx}}{F_{bx}} + \frac{f_{by}}{F_{by}} \geq 1.0$$

where f_{bx} = actual bending stress about the *x*-axis,
F_{bx} = allowable bending stress for the major axis of the shape,
f_{by} = actual bending stress about the *y*-axis,
F_{by} = allowable bending stress for the minor axis of the shape.

For W and S shapes a higher allowable stress is permitted for the minor axis; $F_{by} = 0.75F_y$. For the tube, however, the allowable stresses are the same.

For a first trial we consider the use of a rectangular steel tube, TS 6 × 4 × 0.3125, for which Table A.7 gives $S_x = 8.72$ in.[3] and $S_y = 6.92$ in.[3] Using the components of the load as shown in Fig. 9.12, we find

$$M_x = \frac{WL}{8} = \frac{4982 \times 17}{8} = 10{,}587 \text{ lb-ft } [14.35 \text{ kN-m}]$$

$$M_y = \frac{WL}{8} = \frac{2491 \times 17}{8} = 5293 \text{ lb-ft } [7.17 \text{ kN-m}]$$

The corresponding maximum bending stresses are

$$f_{bx} = \frac{M_x}{S_x} = \frac{10{,}587 \times 12}{8.72} = 14{,}572 \text{ psi } [100.4 \text{ MPa}]$$

$$f_{by} = \frac{M_y}{S_y} = \frac{5293 \times 12}{6.92} = 9179 \text{ psi } [63.3 \text{ MPa}]$$

Using the formula for the combined action analysis, we obtain

$$\frac{f_{bx}}{F_{bx}} + \frac{f_{by}}{F_{by}} = \frac{14{,}572}{24{,}000} + \frac{9179}{24{,}000} = 0.607 + 0.382 = 0.989$$

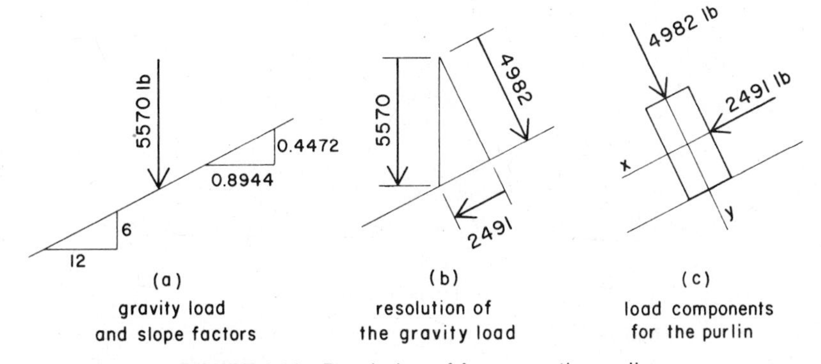

FIGURE 9.12 Resolution of forces on the purlin.

Because this is less than 1.0, the section is adequate.

We note from Table A.7 that this tube weighs 19.08 lb/ft, which means that the average weight of the purlins is 19.08/6 = 3.18 psf. This is close enough to our estimate of 3 psf to warrant no revision in the design loading.

9.10 DETERMINATION OF THE TRUSS REACTIONS AND INTERNAL FORCES

The loading on the truss is assumed to be that shown in Fig. 9.9. A concentrated load of P occurs at each of the interior panel points of the top chord and a load of one-half P occurs at the truss ends. The total load is thus $8P$ and the reactions for the symmetrical truss are each $8P/2 = 4P$. The load P consists of the total load

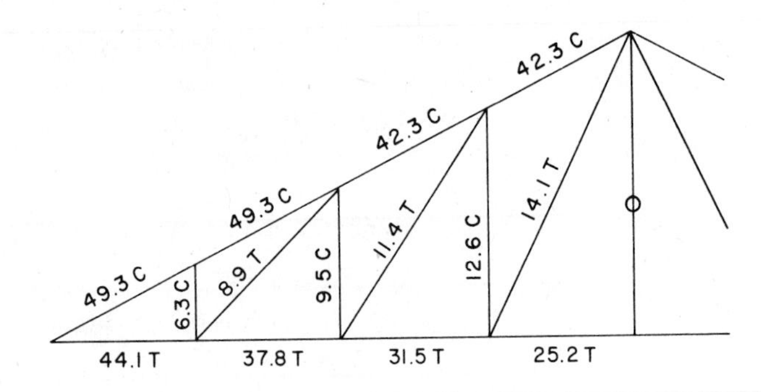

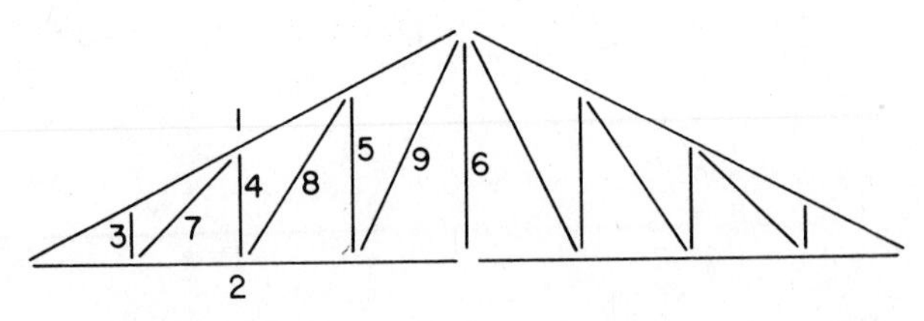

FIGURE 9.13 Internal forces and truss member layout.

on one purlin plus a portion of the weight of the truss. From Table 9.1 we determine that the estimated weight should be 7 psf for the supported area. Thus the panel load increment becomes $7 \times 6 \times 17 = 714$ lb. By adding this to the total purlin load found in Sec. 9.9 we obtain the value of P as $714 + 5570 = 6284$ lb.

For the truss analysis this value of P is rounded off to 6.3 kips. The Maxwell diagram for one-half the truss is shown in Fig. 9.9. The results of the analysis of internal forces are recorded on the truss diagram in Fig. 9.13*a*. For a truss of this size the top chords on each side of the peak would probably be made of a single piece. The bottom chord is a bit long for a single piece and would likely be made with at least one splice. A scheme for the truss member configuration on the basis of these assumptions is shown in Fig. 9.13*b*.

Because of the relatively low slope of the roof and the magnitude of the dead load, it is not likely that this truss will be critically affected by the wind load unless an unusually high wind pressure must be sustained. We proceed with the assumption that design for gravity load is adequate.

9.11 SELECTION OF THE TRUSS MEMBERS

For a truss of this size the construction is most likely to be one of the following:

Top and bottom chords of structural tees with webs of double angles welded directly to the stems of the tees.

All double angles with joints using steel gusset plates; the angles are connected to the plates by welds or high strength bolts.

Another possibility is a truss of welded steel pipe or tube members, which is usually quite expensive and therefore not so favorable unless a neat appearance is considered important for trusses that are exposed to view.

For purpose of illustration we choose the members on the basis of an all-welded construction with double angles and gusset plates. Before proceeding to a consideration of the design of individual members some general limitations and requirements should be established:

1. *Minimum Thickness of Angle Leg.* This is based on the minimum weld that is used. For this size truss it is advisable to select a minimum size of $\frac{1}{4}$ in. [6.35 mm] for the fillet welds. For this weld the minimum dimension recommended for the angle leg is $\frac{5}{16}$ in. [7.94 mm].
2. *Minimum Radius of Gyration.* Based on the recommendations of Sec. 1.8.4 of the AISC Specification, the limits for slenderness ratios are $L/r = 200$ for compression and $L/240$ for tension members.
3. *Unbraced Length of Members.* For web members this is their actual length in either direction. For the chords the unbraced length in the plane of the truss is the panel length. However, the unbraced length perpendicular to the plane of the truss depends on the lateral bracing provided for the truss.

Considering the last point raised, we take the lateral unbraced length to be the panel length for the top chord because the purlins occur at the panel points. For the bottom chord we anticipate that the trusses will be braced at alternate panel points by some form of horizontal bracing. The critical unsupported length of the bottom chord, in a direction perpendicular to the plane of the truss, is therefore 12 ft.

The critical design factors for the truss members are summarized in Table 9.3. The members are referred to by the numbers shown on the proposed truss layout in Fig. 9.13. Members may be designed directly from the table data, subject to the limitations listed. For tension members the critical concern is the stress on the effective area discussed in Sec. 10.3. For compression members the slenderness ratios are usually quite high, requiring a design with the reduced value of compression stress as discussed in Sec. 6.5. The design of compression members is usually made easier with the use of column load tables, such as those in Chapter 6.

TABLE 9.3 Data for the Truss Members

Truss Member (see Fig. 9.13)	Design Force (kips)	Slenderness Considerations		Selection and Properties				
		Critical Length and Axis (ft)	Minimum r and Axis[a] (in.)	Angles Long Legs Back-to-Back	r_x (in.)	r_y (in.)	Weight (lb/ft)	Capacity[b] (kips)
1	49.3 C	6.71	0.40	$3\frac{1}{2} \times 2\frac{1}{2} \times \frac{5}{16}$	1.11	1.10	12.2	56
2	44.1 T	6 − x	0.30 − x	$3\frac{1}{2} \times 2\frac{1}{2} \times \frac{5}{16}$	1.11	1.10	12.2	48
		12 − y	0.60 − y					
3	6.3 C	3	0.18	$2 \times 2 \times \frac{5}{16}$	0.601	1.0	7.84	40
4	9.5 C	6	0.36	$2 \times 2 \times \frac{5}{16}$	0.601	1.0	7.84	24
5	12.6 C	9	0.54	$2 \times 2 \times \frac{5}{16}$	0.601	1.0	7.84	11
6	0	12	0.60	$2 \times 2 \times \frac{5}{16}$	0.601	1.0	7.84	27 T
7	8.9 T	8.48	0.42	$2 \times 2 \times \frac{5}{16}$	0.601	1.0	7.84	27
8	11.4 T	10.8	0.52	$2 \times 2 \times \frac{5}{16}$	0.601	1.0	7.84	27
9	14.1 T	13.4	0.65	$2\frac{1}{2} \times 2 \times \frac{5}{16}$	0.776	0.948	9.0	34

[a] Required for both axes if separate values are not given.
[b] Column load for compression members. Area times 22 ksi for tension members, using the area of the connected legs only.

There are three possibilities for combinations of double angles: equal leg angles, unequal leg angles with long legs back-to-back, and unequal leg angles with short legs back-to-back. The unequal leg angles with long legs back-to-back are often used because they tend to produce a combination that has its stiffness and radius of gyration on both axes closer in value. They also tend to be stronger in bending on the x-axis, offering more resistance to sag when occurring in other than a vertical position. For the bottom chord in this example problem, however, it may be more desirable to use unequal leg angles with short legs back-to-back because they will have greater stiffness in the direction perpendicular to the plane of the truss.

Although other choices are possible, the angle combinations shown in Table 9.3 satisfy the various data and special requirements discussed in this example. The total weight of the truss members, determined by multiplying the member weights by the member lengths, is 2162 lb, to which must be added the weight of gusset plates, lacing, support connections, purlin seat connections, and cross bracing. The total weight is thus likely to be as much as 50% more than that of the members alone. If this is true, the weight per sq ft of supported area is

$$w = \frac{1.5 \times 2162}{48 \times 17} = 3.97 \text{ psf}$$

which is somewhat less than the value of 7 psf obtained for estimation of the truss weight from Table 9.1. This is due in part to the fact that the truss is welded; bolted connections are generally somewhat heavier.

9.12 DESIGN OF THE TRUSS JOINTS

We now consider the design of the welded truss joints. Although there are 10 separate joints, we illustrate the process by considering only two: the left support and the bottom chord joint at 12 ft from the left support.

Figure 9.14 shows a possible layout for the bottom chord joint. Four truss members intersect at this joint: GJ, JK, KL, and LG. However, because the chord is made continuous through the joint, only three sets of angles are connected to the gusset plate.

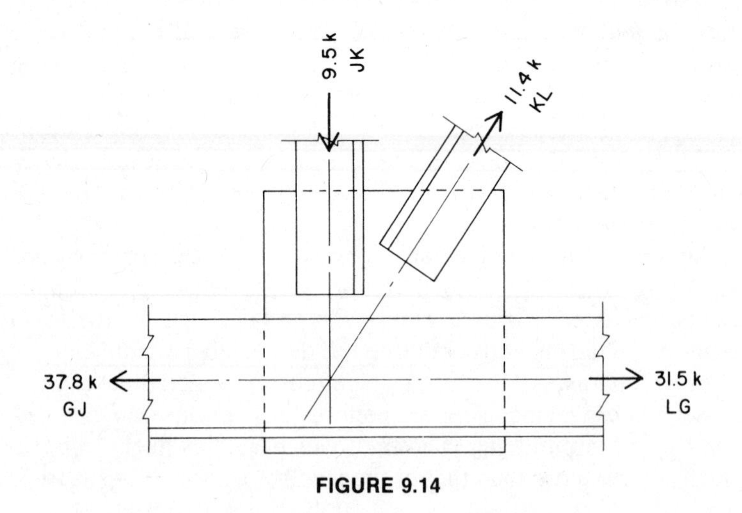

FIGURE 9.14

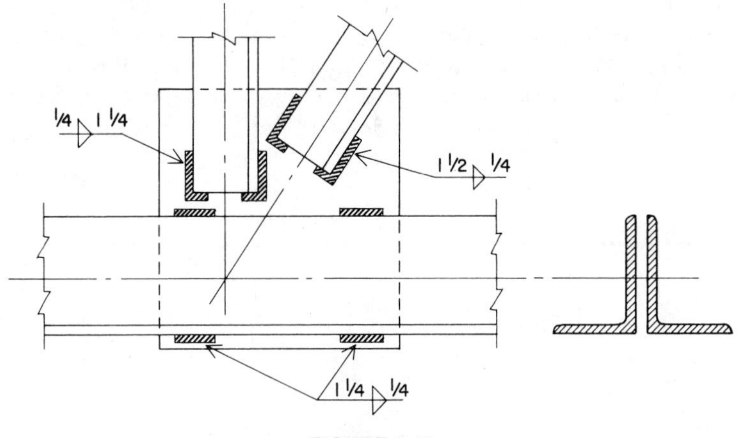

FIGURE 9.15

A basic consideration in the layout of this joint is the need to avoid twisting, which is generally accomplished by having the action lines of the axial forces in the members intersect at a common point: the panel point of the truss. If the stresses in the members are uniformly distributed on the full cross sections of the members, the action lines will coincide with the centroids of the double-angle combinations. The connections to the gusset plates, however, do not fully develop the stresses in the outstanding legs of the angles. Therefore, some designers prefer to consider that the forces at the joint occur symmetrically with respect to the connected legs, as indicated in the layout in Fig. 9.14.

If the forces are axial to the full cross sections of the members, the joint design will develop as illustrated in Fig. 10.22 and explained in Sec. 10.13. If the forces at the joint are symmetrical with respect to the connected leg, the welds will be placed symmetrically, as shown in Fig. 9.15. The design of the former is already explained in Sec. 10.13; therefore, we describe the latter approach in this example.

Using the previously selected size of $\frac{1}{4}$ in. for the fillet welds and assuming welds made with E 60 XX electrodes, we find in Table 10.7 that the weld has a capacity of 3.2 k/in. [0.56 kN/mm]. The total length of weld required for the attachment of an individual truss member is determined by dividing the internal force in the member by the unit capacity of the weld. Because the bottom chord is continuous at the joint under consideration, the attachment of these angles need develop only the difference in force between the chords on each side of the joint. The weld lengths required for this joint are

$$\text{Bottom chord:} \quad L = \frac{37.8 - 31.5}{3.2} = 1.97 \text{ in.}$$

$$\text{Vertical web:} \quad L = \frac{9.5}{3.2} = 3.17 \text{ in.}$$

$$\text{Diagonal web:} \quad L = \frac{11.4}{3.2} = 3.56 \text{ in.}$$

Before selecting the size and layout of individual welds, we must consider a number of other factors;

1. *Minimum Weld Length.* As discussed in Sec. 10.12, the minimum length is four times the weld size, plus an extra $\frac{1}{4}$ in. for starting and stopping. We consider the use of a minimum length of 1.25 in. for an individual weld.
2. Section 1.15 of the AISC Specification requires that connections be designed for a minimum of 6 kips or 50% of the capacity of the connected member, whichever is greater.
3. Connections should be arranged on the gusset plates to maximize the general stability of the joint.

Considering point 2 and using the actual member capacities given in Table 9.3, we find the following alternative lengths for the web members:

$$\text{Vertical web:} \quad L = \frac{0.50 \times 24}{3.2} = 3.75 \text{ in.}$$

$$\text{Diagonal web:} \quad L = \frac{0.50 \times 27}{3.2} = 4.22 \text{ in.}$$

Because both are slightly larger than the lengths determined for the actual internal forces, they become the critical requirements for the joint.

Consideration of point 3 reveals that it is desirable to place the welds for the bottom chords at the edges of the gusset plate (Fig. 9.15). It should be observed that this arrangement produces greater twisting resistance for the gusset plate, both in and out of the plane of the truss.

Note that the welds shown in Fig. 9.15 occur on both sides of the gusset plate and that there is actually twice as much total weld for each truss member.

9.13 TRUSS JOINT WITH TEE CHORD

Figure 9.16 shows a construction that is often used with light steel trusses. In this case the double-angle chord is replaced by a single structural tee and the web members are welded directly to the stem of the tee, thus eliminating the need for the gusset plate. A critical concern in the selection of the tee is the need for sufficient height of the stem to accommodate the welds for the web members. For this type of joint it may be desirable to cut the ends of diagonal members at an angle or to arrange the welds in a different manner to make the joint layout more compact. In Figure 9.16

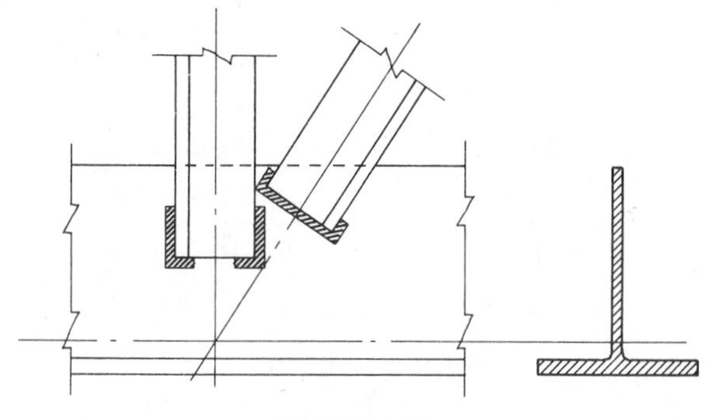

FIGURE 9.16

the weld is placed completely across the end of the diagonal member, which permits the use of a short return weld on the two sides.

When structural tees are used as truss members it is necessary to ensure that the width/thickness ratio of the flange and the depth/thickness ratio of the web comply with the minimum requirements of the AISC Specification. This does not apply to members that sustain only tension force but it does when compression or bending must be resisted.

9.14 TRUSS SUPPORT JOINTS

Figure 9.17 is a detail for the truss joint at the left support. In this layout the joint has been developed to facilitate support in the form of direct vertical bearing on top of a wall, on the top flange of a girder, or on the cap plate of a steel column. The joint has also been developed to allow for the extension of the top chord to form a roof overhang beyond the support. The internal forces are

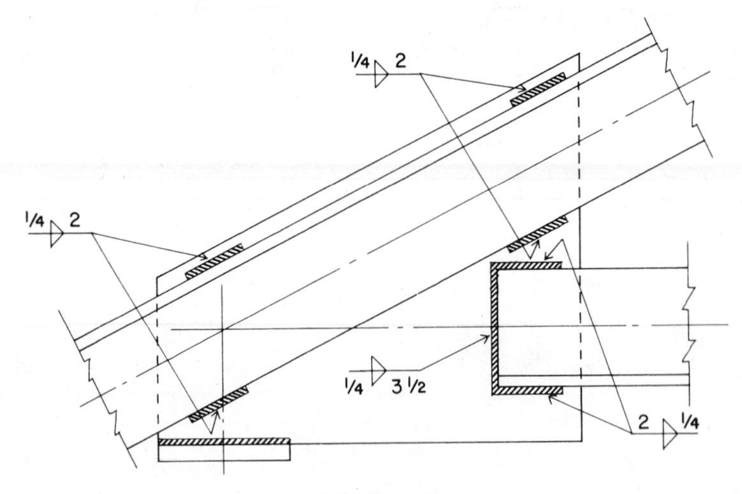

FIGURE 9.17

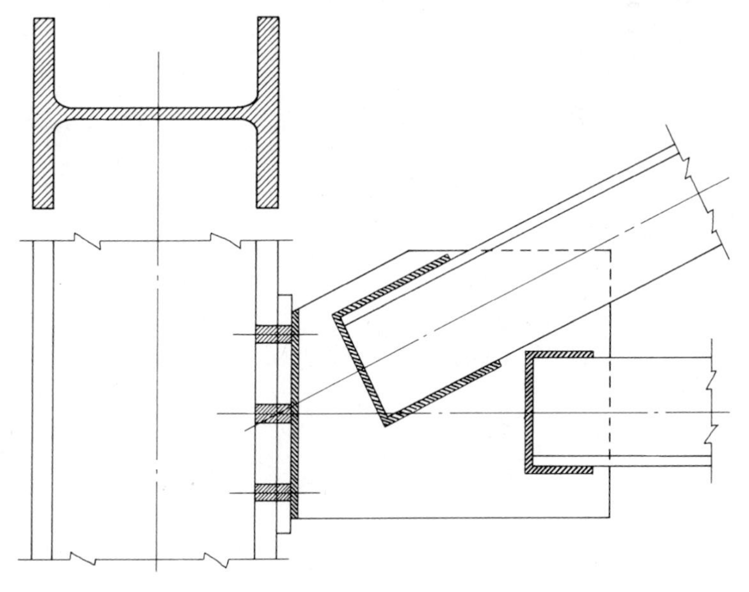

FIGURE 9.18

much greater in this joint than in the joint in Sec. 9.13; thus the amount of weld required is considerably larger.

An alternative detail for the truss support is shown in Fig. 9.18. In this case the joint is developed to facilitate attachment to the face of a steel column with a bolted shear connection.

10

STEEL FASTENINGS

The making of a steel structure typically involves the connecting of many parts. The form and size of the parts, the functional needs of the connections, and the technology available for achieving the connections is all subject to considerable variety. In this chapter we deal with the connection of structural elements, with an emphasis on ordinary means and ordinary structural requirements.

10.1 STRUCTURAL BOLTS

Elements of structural steel are often connected by mating flat parts with common holes and inserting a pin-type device to hold them together. In times past the pin device was a rivet; today it is usually a bolt. A great number of types and sizes of bolt are available, as are many connections in which they are used. The material in this chapter deals with a few of the common bolting methods used in building structures. The diagrams in Fig. 10.1

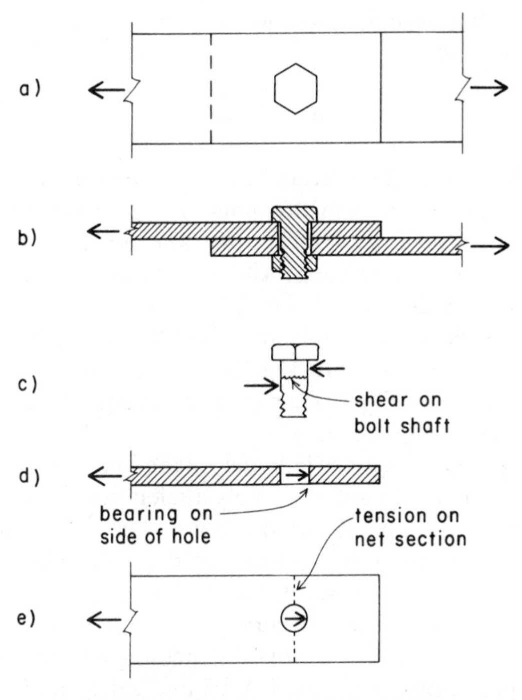

FIGURE 10.1 Actions of bolted joints.

show a simple connection between two steel bars that functions to transfer a tension force from one bar to another. Although this is a tension-transfer connection, it is also referred to as a shear connection because of the manner in which the connecting device (the bolt) works in the connection (see Fig. 10.1*b*). If the bolt tension (due to tightening of the nut) is relatively low, the bolt serves primarily as a pin in the matched holes, bearing against the sides of the holes, as shown in Fig. 10.1*d*. In addition to these functions, the bars develop tension stress that will be a maximum at the section through the bolt holes.

In the connection shown in Fig. 10.1, the failure of the bolt involves a slicing (shear) failure that is developed as a shear stress on the bolt cross section. The resistance of the bolt can be expressed as an allowable shear stress F_v times the area of the bolt cross section, or

$$R = F_v \times A$$

With the size of the bolt and the grade of steel known, it is a simple matter to establish this limit. In some types of connections, it may be necessary to slice the same bolt more than once to separate the connected parts. This is the case in the connection shown in Fig. 10.2, in which it may be observed that the bolt must be sliced twice to make the joint fail. When the bolt develops shear on only one section (Fig. 10.1), it is said to be in *single shear*; when it develops shear on two sections (Fig. 10.2), it is said to be in *double shear*.

When the bolt diameter is larger or the bolt is made of strong steel, the connected parts must be sufficiently thick if they are to develop the full capacity of the bolts. The maximum bearing stress permitted for this situation by the AISC Specification is $F_p = 1.5F_u$, where F_u is the ultimate tensile strength of the steel in the part in which the hole occurs.

Bolts used for the connection of structural steel members come in two types. Bolts designated A307 and called *unfinished* have the lowest load capacity of the structural bolts. The nuts for these bolts are tightened just enough to secure a snug fit of the attached parts; because of this, plus the oversizing of the holes, there is some movement in the development of full resistance. These bolts are generally not used for major connections, especially

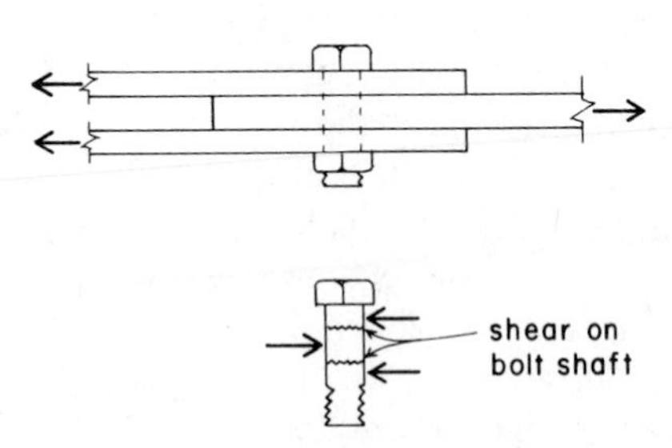

FIGURE 10.2 Bolted joint with double shear.

when joint movement or loosening under vibration or repeated loading may be a problem.

Bolts designated A325 or A490 are called *high-strength bolts*. The nuts of these bolts are tightened to produce a considerable tension force, which results in a high degree of friction resistance between the attached parts. High-strength bolts are further designated as F, N, or X. The F designation denotes bolts for which the limiting resistance is that of friction. The N designation denotes bolts that function ultimately in bearing and shear but for which the threads are not excluded from the bolt shear planes. The X designation denotes bolts that function like the N bolts but for which the threads are excluded from the shear planes.

When bolts are loaded in tension, their capacities are based on the development of the ultimate resistance in tension stress at the reduced section through the threads. When loaded in shear, bolt capacities are based on the development of shear stress in the bolt shaft. The shear capacity of a single bolt is further designated as *S* for single shear (Fig. 10.1) or *D* for double shear (Fig. 10.2). The capacities of structural bolts in both tension and shear are given in Table 10.1. The size range given in the table—$\frac{5}{8}$–$1\frac{1}{2}$ in.—is that listed in the AISC Manual. However, the most commonly used sizes for structural steel framing are $\frac{3}{4}$ and $\frac{7}{8}$ in.

Bolts are ordinarily installed with a washer under both head and nut. Some manufactured high-strength bolts have specially formed heads or nuts that in effect have self-forming washers, eliminating the need for a separate, loose washer. When a washer is used, it is sometimes the limiting dimensional factor in detailing for bolt placement in tight locations, such as close to the fillet (inside radius) of angles or other rolled shapes.

For a given diameter of bolt, there is a minimum thickness required for the bolted parts in order to develop the full shear capacity of the bolt. This thickness is based on the bearing stress between the bolt and the side of the hole, which is limited to a maximum of $F_p = 1.5F_u$. The stress limit may be established by either the bolt steel or the steel of the bolted parts.

Steel rods are sometimes threaded for use as anchor bolts or tie rods. When they are loaded in tension, their capacities are usually limited by the stress on the reduced section at the threads. Tie

TABLE 10.1 Capacity of Structural Bolts (kips)

ASTM Designation	Connection Type[a]	Loading Condition[b]	Nominal Diameter (in.)							
			$\frac{5}{8}$	$\frac{3}{4}$	$\frac{7}{8}$	1	$1\frac{1}{8}$	$1\frac{1}{4}$	$1\frac{3}{8}$	$1\frac{1}{2}$
			Area, Based on Nominal Diameter (in.^{2})							
			0.3068	0.4418	0.6013	0.7854	0.9940	1.227	1.485	1.767
A307		S	3.1	4.4	6.0	7.9	9.9	12.3	14.8	17.7
		D	6.1	8.8	12.0	15.7	19.9	24.5	29.7	35.3
		T	6.1	8.8	12.0	15.7	19.9	24.5	29.7	35.3
A325	F	S	5.4	7.7	10.5	13.7	17.4	21.5	26.0	30.9
		D	10.7	15.5	21.0	27.5	34.8	42.9	52.0	61.8
	N	S	6.4	9.3	12.6	16.5	20.9	25.8	31.2	37.1
		D	12.9	18.6	25.3	33.0	41.7	51.5	62.4	74.2
	X	S	9.2	13.3	18.0	23.6	29.8	36.8	44.5	53.0
		D	18.4	26.5	31.1	47.1	59.6	73.6	89.1	106.0
	All	T	13.5	19.4	26.5	34.6	43.7	54.0	65.3	77.7
A490	F	S	6.7	9.7	13.2	17.3	21.9	27.0	32.7	38.9
		D	13.5	19.4	26.5	34.6	43.7	54.0	65.3	77.7
	N	S	8.6	12.4	16.8	22.0	27.8	34.4	41.6	49.5
		D	17.2	24.7	33.7	44.0	55.7	68.7	83.2	99.0
	X	S	12.3	17.7	24.1	31.4	39.8	49.1	59.4	70.7
		D	24.5	35.3	48.1	62.8	79.5	98.2	119.0	141.0
	All	T	16.6	23.9	32.5	42.4	53.7	66.3	80.2	95.4

Source: Reproduced from data in the *Manual of Steel Construction*, 8th ed. (Ref. 1), with permission of the publishers, the American Institute of Steel Construction.

[a] F = friction; N = bearing, threads not excluded; X = bearing, threads excluded.

[b] S = single shear; D = double shear; T = tension.

rods are sometimes made with *upset ends*, which consist of larger diameter portions at the ends. When these enlarged ends are threaded, the net section at the thread is the same as the gross section in the remainder of the rods; the result is no loss of capacity for the rod.

10.2 LAYOUT OF BOLTED CONNECTIONS

Design of bolted connections generally involves a number of considerations in the dimensional layout of the bolt-hole patterns for the attached structural members. Although we cannot develop all the points necessary for the production of structural steel construction and fabrication details, the material in this section presents basic factors that often must be included in the structural calculations.

Figure 10.3 shows the layout of a bolt pattern with bolts placed in two parallel rows. Two basic dimensions for this layout are limited by the size (nominal diameter) of the bolt. The first is the center-to-center spacing of the bolts, usually called the *pitch*. The AISC Specification limits this dimension to an absolute minimum of $2\frac{2}{3}$ times the bolt diameter. The preferred minimum, however, which is used in this book, is 3 times the diameter.

The second critical layout dimension is the *edge distance*, which is the distance from the center line of the bolt to the nearest

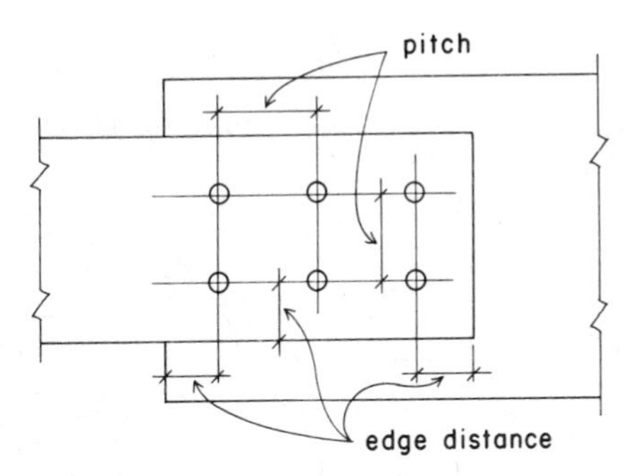

FIGURE 10.3 Pitch and edge distances for bolts.

TABLE 10.2 Pitch and Edge Distances for Bolts

Rivet or Bolt Diameter, d (in.)	Minimum Edge Distance for Punched, Reamed, or Drilled Holes (in.)		Minimum Recommended Pitch, Center to Center (in.)	
	At Sheared Edges	At Rolled Edges of Plates, Shapes, or Bars, or Gas-Cut Edges[a]	$2\frac{2}{3}d$	$3d$
$\frac{5}{8}$	1.125	0.875	1.67	1.875
$\frac{3}{4}$	1.25	1	2	2.25
$\frac{7}{8}$	1.5[b]	1.125	2.33	2.625
1	1.75[b]	1.25	2.67	3

Source: Reproduced from data in the *Manual of Steel Construction*, 8th ed. (Ref. 1), with permission of the publishers, the American Institute of Steel Construction.

[a] May be reduced $\frac{1}{8}$ in. when the hole is at a point where stress does not exceed 25% of the maximum allowed in the connected element.

[b] May be $1\frac{1}{4}$ in. at the ends of beam connection angles.

edge. There is also a specified limit for this as a function of bolt size. This dimension may also be limited by edge tearing, which is discussed in Sec. 10.4.

Table 10.2 gives the recommended limits for pitch and edge distance for the bolt sizes used in ordinary steel construction.

In some cases bolts are staggered in parallel rows (Fig. 10.4). In this case the diagonal distance, labeled m in the illustration, must also be considered. For staggered bolts the spacing in the direction of the rows is usually referred to as the pitch; the spac-

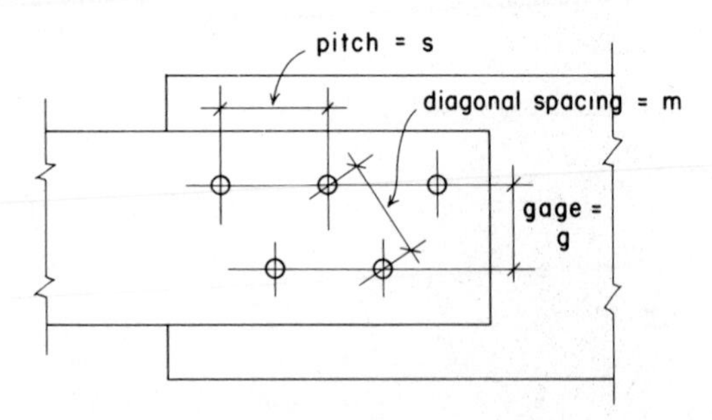

FIGURE 10.4 Standard reference dimensions for layout of bolted joints.

TABLE 10.3 Minimum Pitch to Maintain Three Diameters Center to Center of Holes

Diameter of Bolt	m	Distance, g (in.)								
		1	$1\frac{1}{4}$	$1\frac{1}{2}$	$1\frac{3}{4}$	2	$2\frac{1}{4}$	$2\frac{1}{2}$	$2\frac{3}{4}$	3
$\frac{5}{8}$	$1\frac{7}{8}$	$1\frac{5}{8}$	$1\frac{3}{8}$	$1\frac{1}{8}$	$\frac{5}{8}$	0				
$\frac{3}{4}$	$2\frac{1}{4}$	2	$1\frac{7}{8}$	$1\frac{5}{8}$	$1\frac{3}{8}$	1	0			
$\frac{7}{8}$	$2\frac{5}{8}$	$2\frac{1}{2}$	$2\frac{3}{8}$	$2\frac{1}{8}$	2	$1\frac{3}{4}$	$1\frac{3}{8}$	$\frac{3}{4}$	0	
1	3	$2\frac{7}{8}$	$2\frac{3}{4}$	$2\frac{5}{8}$	$2\frac{1}{2}$	$2\frac{1}{4}$	2	$1\frac{3}{8}$	$1\frac{1}{8}$	0

Source: Reproduced from data in the *Manual of Steel Construction*, 8th ed. (Ref. 1), with permission of the publishers, the American Institute of Steel Construction (see Fig. 10.4).

ing of the rows is called the gage. The reason for staggering the bolts is that sometimes the rows must be spaced closer (gage spacing) than the minimum spacing required for the bolts selected. Table 10.3 gives the pitch required for a given gage spacing to keep the diagonal spacing m within the recommended diameter limit.

Location of bolt lines is often related to the size and type of structural members being attached. This is especially true of bolts placed in the legs of angles or in the flanges of W, M, S, C, and structural tee shapes. Figure 10.5 shows the placement of bolts in the legs of angles. When a single row is placed in a leg, its recommended location is at the distance labeled g from the back of the angle. When two rows are used, the first row is placed at the distance g_1, and the second row is spaced a distance g_2 from the first. Table 10.4 gives the recommended values for these distances.

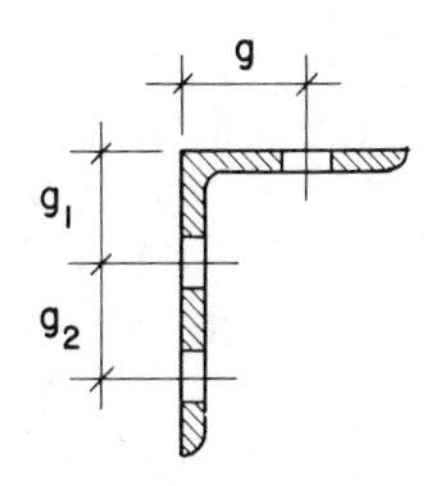

FIGURE 10.5 Gage dimensions for steel angles.

TABLE 10.4 Usual Gage Dimensions for Angles (in.)

Gage Dimension	Width of Angle Leg								
	8	7	6	5	4	$3\frac{1}{2}$	3	$2\frac{1}{2}$	2
g	$4\frac{1}{2}$	4	$3\frac{1}{2}$	3	$2\frac{1}{2}$	2	$1\frac{3}{4}$	$1\frac{3}{8}$	$1\frac{1}{8}$
g_1	3	$2\frac{1}{2}$	$2\frac{1}{4}$	2					
g_2	3	3	$2\frac{1}{2}$	$1\frac{3}{4}$					

Source: Reproduced from data in the *Manual of Steel Construction*, 8th ed. (Ref. 1), with permission of the publishers, the American Institute of Steel Construction.

When placed at the recommended locations in rolled shapes, bolts will end up a certain distance from the edge of the part. Based on the recommended edge distance for rolled edges given in Table 10.2, it is thus possible to determine the maximum size of bolt that can be accommodated. For angles, the maximum fastener may be limited by the edge distance, especially when two rows are used: however, other factors may in some cases be more critical. The distance from the center of the bolts to the inside fillet of the angle may limit the use of a large washer where one is required. Another consideration may be the stress on the net section of the angle, especially if the member load is taken entirely by the attached leg.

10.3 TENSION CONNECTIONS

When tension members have reduced cross sections, two stress investigations must be considered. This is the case for members with holes for bolts or for bolts or rods with cut threads. For the member with a hole (Fig. 10.1*d*), the allowable tension stress at the reduced cross section through the hole is $0.50F_u$, where F_u is the ultimate tensile strength of the steel. The total resistance at this reduced section (also called the net section) must be compared with the resistance at other, unreduced sections at which the allowable stress is $0.60F_y$.

For threaded steel rods the maximum allowable tension stress at the threads is $0.33F_u$. For steel bolts the allowable stress is specified as a value based on the type of bolt. The load capacity of various types and sizes of bolt is given in Table 10.1.

When tension elements consist of W, M, S, and tee shapes, the tension connection is usually not made in a manner that results in the attachment of all the parts of the section (e.g., both flanges plus the web for a W). In such cases the AISC Specification requires the determination of a reduced effective net area A_e that consists of

$$A_e = C_t A_n$$

where A_n = actual net area of the member,
C_t = reduction coefficient.

Unless a larger coefficient can be justified by tests, the following values are specified:

1. For W, M, or S shapes with flange widths not less than two-thirds the depth and structural tees cut from such shapes, when the connection is to the flanges and has at least three fasteners per line in the direction of stress, $C_t = 0.90$.
2. For W, M, or S shapes not meeting the above conditions and for tees cut from such shapes, provided the connection has not fewer than three fasteners per line in the direction of stress, $C_t = 0.85$.
3. For all members with connections that have only two fasteners per line in the direction of stress, $C_t = 0.75$.

Angles used as tension members are often connected by only one leg. In a conservative design, the effective net area is only that of the connected leg, less the reduction caused by bolt holes. Rivet and bolt holes are punched larger in diameter than the nominal diameter of the fastener. The punching damages a small amount of the steel around the perimeter of the hole; consequently the diameter of the hole to be deducted in determining the net section is $\frac{1}{8}$ in. greater than the nominal diameter of the fastener.

When only one hole is involved, as in Fig. 10.1, or in a similar connection with a single row of fasteners along the line of stress, the net area of the cross section of one of the plates is found by multiplying the plate thickness by its net width (width of member minus diameter of hole).

When holes are staggered in two rows along the line of stress (Fig. 10.6), the net section is determined somewhat differently. The AISC Specification reads:

> In the case of a chain of holes extending across a part in any diagonal or zigzag line, the net width of the part shall be obtained by deducting from the gross width the sum of the diameters of all the holes in the chain and adding, for each gage space in the chain, the quantity $s^2/4g$, where
>
> s = longitudinal spacing (pitch) in inches or any two successive holes
>
> and
>
> g = transverse spacing (gage) in inches for the same two holes
>
> The critical net section of the part is obtained from that chain which gives the least net width.

The AISC Specification also provides that in no case shall the net section through a hole be considered as more than 85% of the corresponding gross section.

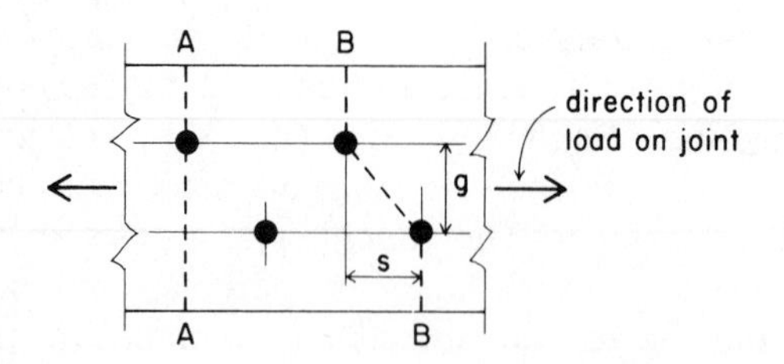

FIGURE 10.6 Determination of net cross-sectional area.

10.4 TEARING IN BOLTED CONNECTIONS

One possible form of failure in a bolted connection is that of tearing out the edge of one of the attached members. The diagrams in Fig. 10.7 show this potentiality in a connection between two plates. The failure in this case involves a combination of shear and tension to produce the torn-out form shown. The total tearing force is computed as the sum required to cause both forms of failure. The allowable stress on the net tension area is specified as $0.50F_u$, where F_u is the maximum tensile strength of the steel. The allowable stress on the shear areas is specified as $0.30F_u$. With the edge distance, hole spacing, and diameter of the holes known, the net widths for tension and shear are determined and multiplied by the thickness of the part in which the tearing occurs. These areas are then multiplied by the appropriate stresses to find the total tearing force that can be resisted. If this force is greater than the connection design load, the tearing problem is not critical.

Another case of potential tearing is shown in Fig. 10.8. This is the common situation for the end framing of a beam in which support is provided by another beam, whose top is aligned with that of the supported beam. The end portion of the top flange of the supported beam must be cut back to allow the beam web to extend to the side of the supporting beam. With the use of a bolted connection, the tearing condition shown is developed.

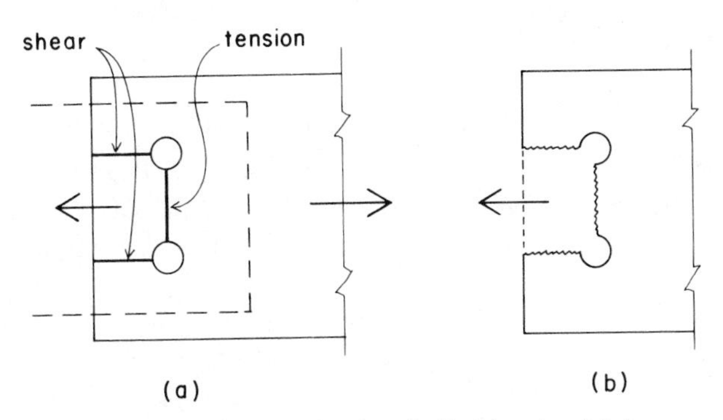

FIGURE 10.7 Tearing in a bolted tension joint.

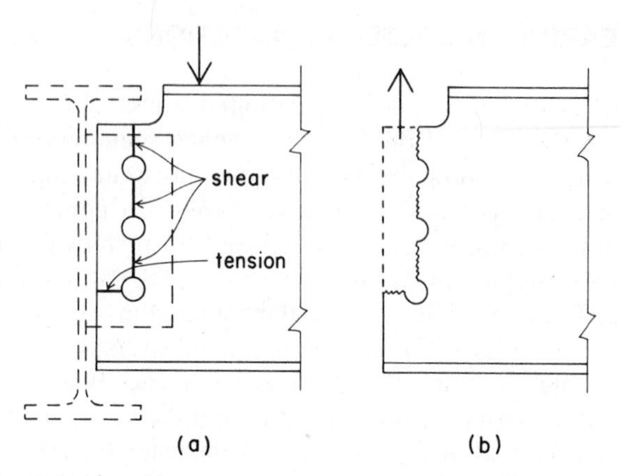

FIGURE 10.8 Tearing in a bolted beam connection.

10.5 DESIGN OF A BOLTED TENSION CONNECTION

The issues raised in several of the preceding sections are illustrated in the following design example. Before proceeding with the problem data, we should consider some of the general requirements for this joint.

If friction-type bolts are used, the surfaces of the connected parts must be cleaned and made reasonably true. If high-strength bolts are used, the determination to exclude threads from the shear failure planes must be established.

The AISC Specification has a number of general requirements for connections:

1. Need for a minimum of two bolts per connection.
2. Need for a minimum connection capacity of 6 kips.
3. Need for the connection to develop at least 50% of the full potential capacity of the member (for trusses only).

Although a part of the design problem may be the selection of the type of fastener or the required strength of steel for the attached parts, we provide this as given data in the example problem.

Example. The connection shown in Fig. 10.9 consists of a pair of narrow plates that transfer a load of 100 kips [445 kN] in tension to a single 10-in. [254-mm] wide plate. The plates are A36 steel with $F_u = 58$ ksi [400 MPa] and are attached with $\frac{3}{4}$-in. A325F bolts placed in two rows. Determine the number of bolts required, the width and thickness of the narrow plates, the thickness of the wide plate, and the layout of the bolts.

Solution. From Table 10.1 we find the double-shear (D) capacity for one bolt is 15.5 kips [69 kN]. The required number of bolts is thus

$$n = \frac{\text{connection load}}{\text{bolt capacity}} = \frac{100}{15.5} = 6.45$$

and the minimum number for a symmetrical connection is eight.

With eight bolts used, the load on one bolt is

$$P = \frac{100}{8} = 12.5 \text{ kips} \ \ [55.6 \text{ kN}]$$

According to Table 10.2, the $\frac{3}{4}$-in. bolts require a minimum edge distance of 1.25 in. (at a sheared edge) and a recommended

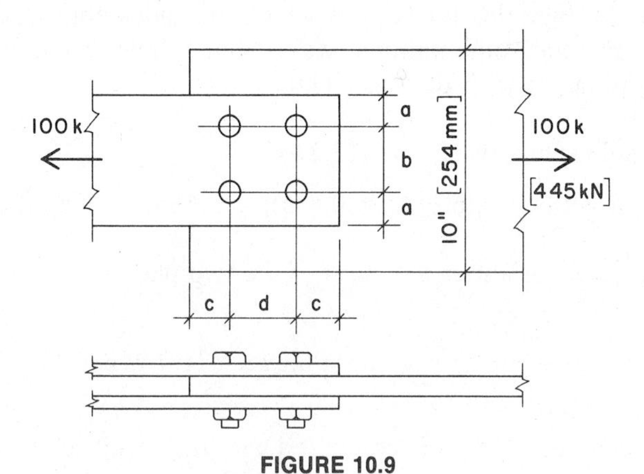

FIGURE 10.9

pitch of 2.25 in. The minimum width for the narrow plates is therefore (see Fig. 10.9)

$$w = b + 2(a) = 2.25 + 2(1.25) = 4.75 \text{ in. } [121 \text{ mm}]$$

With no other constraining conditions given, we arbitrarily select a width of 6 in. [152.4 mm] for the narrow plates. Checking first for the requirement of a maximum tension stress of $0.60F_y$ on the gross area, we find

$$F_t = 0.60F_y = 0.60(36) = 21.6 \text{ ksi } [149 \text{ MPa}]$$

(*Note:* AISC permits rounding off to 22 ksi.)

$$A_{req} = \frac{100}{22} = 4.55 \text{ in.}^2 \ [2928 \text{ mm}^2]$$

and the required thickness with the width selected is

$$t = \frac{4.55}{2(6)} = 0.38 \text{ in. } [9.6 \text{ mm}]$$

We therefore select a minimum thickness of $\frac{7}{16}$ in. (0.4375 in.) [11 mm]. The next step is to check the stress condition on the net section through the holes, for which the allowable stress is $0.50F_u$. For the computations, we assume a hole diameter $\frac{1}{8}$ in. [3.18 mm] larger than the bolt. Thus

$$\text{hole size} = 0.875 \text{ in. } [22.23 \text{ mm}]$$

$$\text{net width} = 2\{6 - (2 \times 0.875)\} = 8.5 \text{ in. } [216 \text{ mm}]$$

and the stress on the net section of the two plates is

$$f_t = \frac{100}{0.4375 \times 8.5} = 26.89 \text{ ksi } [187 \text{ MPa}]$$

This computed stress is compared with the specified allowable stress of

$$F_t = 0.50F_u = 0.50 \times 58 = 29 \text{ ksi} \ [200 \text{ MPa}]$$

Bearing stress is computed by dividing the load on a single bolt by the product of the bolt diameter and the plate thickness. Thus

$$f_p = \frac{12.5}{2 \times 0.75 \times 0.4375} = 19.05 \text{ ksi} \ [146 \text{ MPa}]$$

This is compared with the allowable stress of

$$F_p = 1.5F_u = 1.5 \times 58 = 87 \text{ ksi} \ [600 \text{ MPa}]$$

For the middle plate the procedure is essentially the same except that, in this case, the plate width is given. As before, on the basis of stress on the unreduced section, we determine that the total area required is 4.55 in.^{2} [2928 mm^2]. Thus the thickness required is

$$t = \frac{4.55}{10} = 0.455 \text{ in.} \ [11.6 \text{ mm}]$$

We therefore select a minimum thickness of $\frac{1}{2}$ in. (0.50 in.) [13 mm]. We then proceed as before to check the stress on the net width. The net width through the two holes is

$$w = 10 - (2 \times 0.875) = 8.25 \text{ in.} \ [210 \text{ mm}]$$

and the tension stress on this net cross section is

$$f_t = \frac{100}{8.25 \times 0.5} = 24.24 \text{ ksi} \ [177 \text{ MPa}]$$

which is less than the allowable stress of 29 ksi [200 MPa] determined previously.

The computed bearing stress on the wide plate is

$$f_p = \frac{12.5}{0.75 \times 0.50} = 33.3 \text{ ksi} \ [243 \text{ MPa}]$$

which is considerably less than the allowable determined before, $F_p = 87$ ksi [600 MPa].

In addition to the layout restrictions given in Sec. 10.2, the AISC Specification requires that the minimum spacing in the direction of the load be

$$\frac{2P}{F_u t} + \frac{D}{2} \qquad (D \text{ is bolt diameter})$$

and that the minimum edge distance in the direction of the load be

$$\frac{2P}{F_u t} \qquad (\text{dimension } c \text{ in Fig. 10.9})$$

where P = force transmitted by one fastener to the critical connected part,

F_u = specified minimum (ultimate) tensile strength of the connected part,

t = thickness of the critical connected part.

For our case

$$\frac{2P}{F_u t} = \frac{2 \times 12.5}{58 \times 0.5} = 0.862 \text{ in.}$$

which is considerably less than the specified edge distance listed in Table 10.2 for a $\frac{3}{4}$-in. bolt at a sheared edge: 1.25 in.

For the spacing

$$\frac{2P}{F_u t} + \frac{D}{2} = 0.862 + 0.375 = 1.237 \text{ in.}$$

which is also not critical.

A final problem that must be considered is the potential of tearing out the two bolts at the ends of the plates. Because the combined thickness of the two outer plates is greater than that of the middle plate, the critical case in this connection is that of the

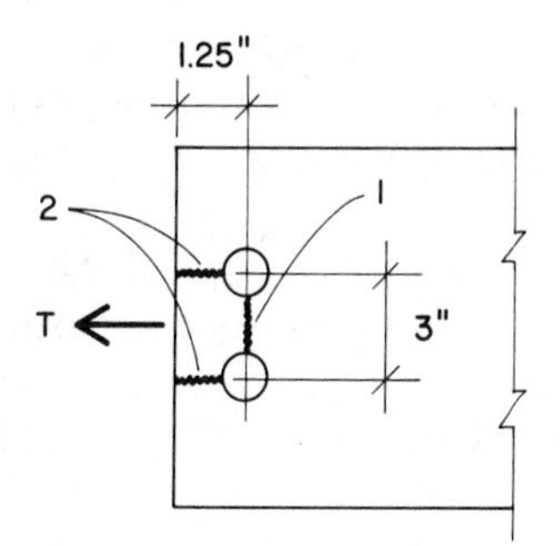

FIGURE 10.10

middle plate. Figure 10.10 shows the condition for the tearing, which involves tension on the section labeled "1" and shear on the two sections labeled "2."

For the tension section

$$w_{(net)} = 3 - 0.875 = 2.125 \text{ in. } [54.0 \text{ mm}]$$

$$F_t = 0.50F_u = 29 \text{ ksi } [200 \text{ MPa}]$$

For the shear sections

$$w_{(net)} = 2\left(1.25 - \frac{0.875}{2}\right) = 1.625 \text{ in. } [41.3 \text{ mm}]$$

$$F_v = 0.30F_u = 17.4 \text{ ksi } [120 \text{ MPa}]$$

The total resistance to tearing is

$$T = (2.125 \times 0.5 \times 29) + (1.625 \times 0.5 \times 17.4)$$
$$= 44.95 \text{ kips } [205 \text{ kN}]$$

Because this is greater than the combined load of 25 kips [111.2 kN] on the two bolts, the problem is not critical.

Connections that transfer compression between the joined parts are essentially the same with regard to the bolt stresses and bearing on the parts. Stress on the net section is less likely to be critical because the compression members will usually be designed for column action, with a considerably reduced value for the allowable compression stress.

Problem 10.5.A. A bolted connection of the general form shown in Fig. 10.9 is to be used to transmit a tension force of 200 kips [890 kN] by using $\frac{7}{8}$-in. A490N bolts and plates of A36 steel. The outer plates are to be 8 in. [200 mm] wide, and the center plate is to be 12 in. [300 mm] wide. Find the required thicknesses of the plates and the number of bolts needed if the bolts are placed in two rows. Sketch the bolt layout with the necessary dimensions.

Problem 10.5.B. Design a connection for the data in Problem 10.5.A except that the bolts are 1-in. A325N, the outside plates are 9 in. wide, and the bolts are placed in three rows.

10.6 BOLTED FRAMING CONNECTIONS

The joining of structural steel members in a structural system generates a wide variety of situations, depending on the form of the connected parts, the type of connecting device used, and the nature and magnitude of the forces that must be transferred between the members. Figure 10.11 shows a number of common connections that are used to join steel columns and beams consisting of rolled shapes.

In the joint shown in Fig. 10.11*a*, a steel beam is connected to a supporting column by the simple means of resting it on top of a steel plate that is welded to the top of the column. The bolts in this case carry no computed loads if the force transfer is limited to that of the vertical end reaction of the beam. The only computed stress condition that is likely to be of concern in this situation is that of crippling the beam web (Sec. 5.10). This is a situation in which the use of unfinished bolts is indicated.

The remaining details in Fig. 10.11 illustrate situations in which the beam end reactions are transferred to the supports by attachment to the beam web. This is, in general, an appropriate form of force transfer because the vertical shear at the end of the

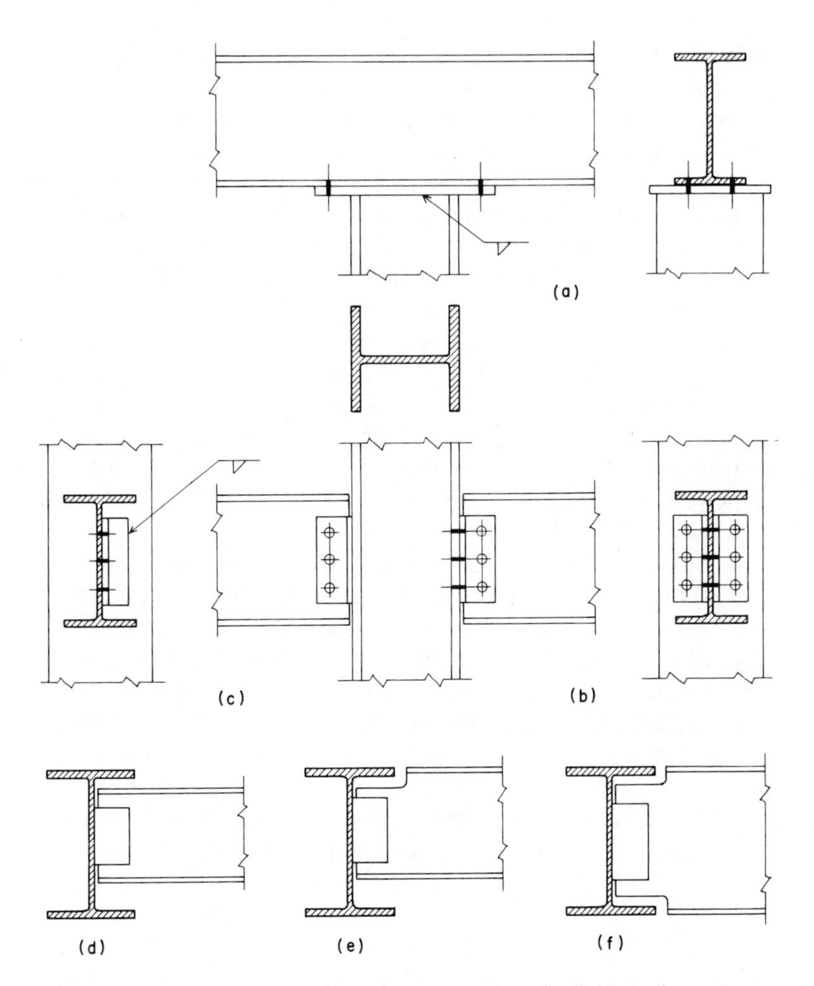

FIGURE 10.11 Typical bolted framing connections for light steel structures with rolled shapes.

beam is resisted primarily by the beam web. The most common form of connection is that which uses a pair of angles (Fig. 10.11*b*). The two most frequent examples of this type of connection are the joining of a steel beam to the side of a column (Fig. 10.11*b*) or to the side of another beam (Fig. 10.11*d*). A beam may also be joined to the web of a W-shaped column in this manner if the column depth provides enough space for the angles.

An alternative to this type of connection is shown in Fig. 10.11*c*, where a single plate is welded to the side of a column, and the beam web is bolted to one side of the plate. This is generally acceptable only when the magnitude of the load on the beam is low because the one-sided connection experiences some torsion.

When the two intersecting beams must have their tops at the same level, the supported beam must have its top flange cut back, as shown at Fig. 10.11*e*. This is to be avoided, if possible, because it represents an additional cost in the fabrication and also reduces the shear capacity of the beam. Even worse is the situation in which the two beams have the same depth and which requires cutting both flanges of the supported beam (see Fig. 10.11*f*). When these conditions produce critical shear in the beam web, it will be necessary to reinforce the beam end.

Alignment of the tops of beams is usually done to simplify the installation of decks on top of the framing. When steel deck is used it may be possible to adopt some form of the detail shown in Fig. 10.12, which permits the beam tops to be offset by the depth of the deck ribs. Unless the flange of the supporting beam is quite thick, it will probably provide sufficient space to permit the connection shown, which does not require cutting the flange of the supported beam.

Figure 10.13 shows additional framing details that may be used in special situations. The technique described in Fig. 10.13*a* is sometimes used when the supported beam is shallow. The verti-

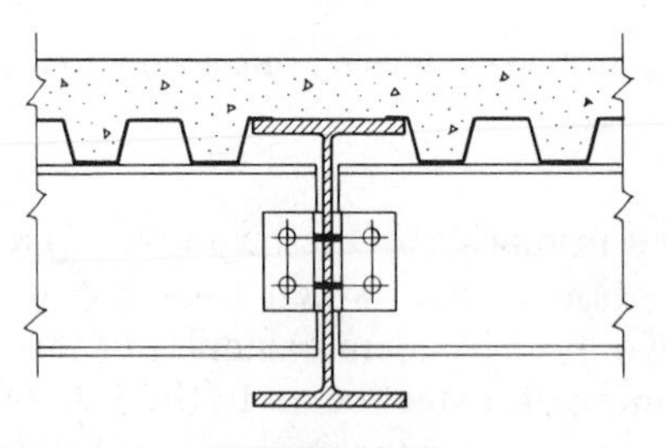

FIGURE 10.12

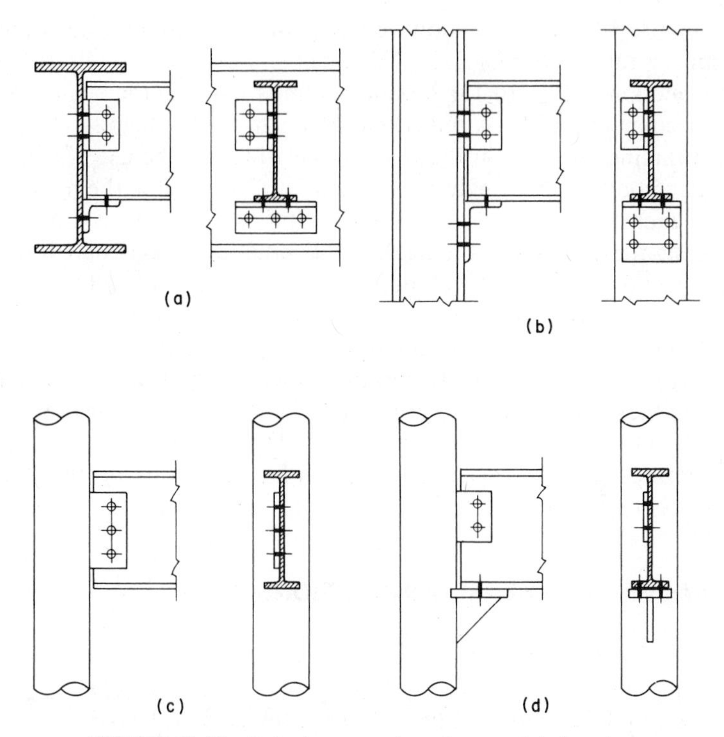

FIGURE 10.13 Bolted connections for special situations.

cal load in this case is transferred through the seat angle, which may be bolted or welded to the carrying beam. The connection to the web of the supported beam merely provides additional resistance to roll-over, or torsional rotation, on the part of the beam. Another reason for favoring this detail is the possibility that the seat angle may be welded in the shop and the web connection made with small unfinished bolts in the field, which greatly simplifies the field work.

Figure 10.13*b* shows the use of a similar connection for joining a beam and column. For heavy beam loads the seat angle may be braced with a stiffening plate. Another variation of this detail involves the use of two plates rather than the angle which may be used if more than four bolts are required for attachment to the column.

Figure 10.13*c* and *d* shows connections commonly used when pipe or tube columns carry the beams. Because the one-sided connection in Fig. 10.13*c* produces some torsion in the beam, the seat connection is favored when the beam load is high.

Framing connections quite commonly involve the use of welding and bolting in a single connection, as illustrated in the figures. In general, welding is favored for fabrication in the shop and bolting for erection in the field. If this practice is recognized, the connections must be developed with a view to the overall fabrication and erection process and some decision made regarding what is to be done where. With the best of designs, however, the contractor who is awarded the job may have some of his own ideas about these procedures and may suggest alterations in the details.

Development of connection details is particularly critical for structures in which a great number of connections occur. The truss is one such structure.

10.7 FRAMED BEAM CONNECTIONS

The connection shown in Fig. 10.11*b* is the type used most frequently in the development of structures that consist of I-shaped beams and H-shaped columns. This device is referred to as a *framed beam connection,* for which there are several design considerations:

1. *Type of Fastening.* This may be accomplished with rivets or with any of the several types of structural bolt. The angles may also be welded in place; the most common practice being to weld the angles to the beam web in the fabricating shop and to bolt them to the supports in the field.
2. *Number of Fasteners.* This refers to the number of bolts used on the beam web; there are twice this number in the outstanding legs of the angles. The capacities are matched, however, because the web bolts are in double shear, the others in single shear.
3. *Size of the Angles.* This depends on the size of the fasteners, the magnitude of the loads, and the size of the support,

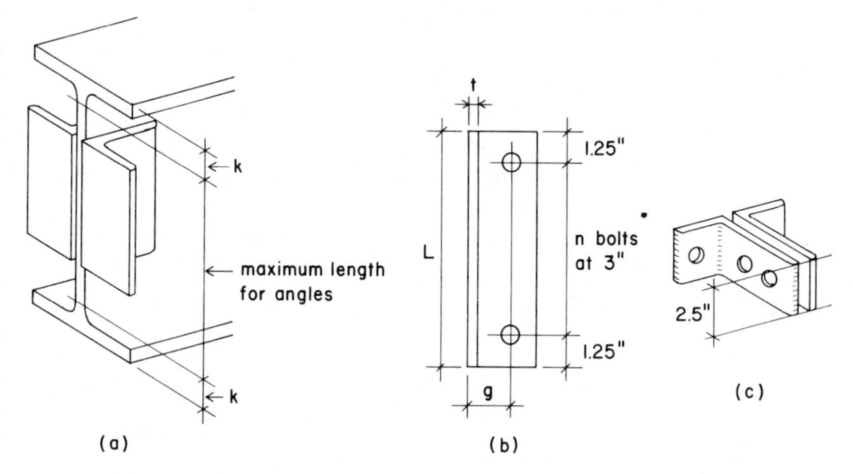

FIGURE 10.14 Framed beam connections for rolled shapes, using steel connection angles.

if it is a column with a particular limiting dimension. Two sizes used frequently are 4 × 3 in. and 5 × $3\frac{1}{2}$ in. Thickness of the angle legs is usually based on the size and type of the fastener.

4. *Length of the Angles.* This is primarily a function of the size of the fasteners. As shown in Fig. 10.14, typical dimensions are an end distance of 1.25 in. and a pitch of 3 in. In special situations, however, smaller dimensions may be used with bolts of 1 in. or smaller diameter.

The AISC Manual (Ref. 1) provides considerable information to assist in the design of this type of connection in both the bolted and welded versions. A sample for bolted connections that use A325 F bolts and angles of A36 steel is given in Table 10.5. The angle lengths in the table are based on the standard dimensions, as shown in Fig. 10.14. For a given beam shape the maximum size of connection (designated by the number of bolts) is limited by the dimension of the flat portion of the beam web. By referring to Appendix Table A.1 we can determine this dimension for any beam designation.

TABLE 10.5 Framed Beam Connections with A325F Bolts and A36 Angles

No. of Bolts, n^a	Angle Length, L^a (in.)	Total Shear Capacity of Bolts (kips)			Use with the Following Rolled Shapes
		Bolt Diameter, d (in.)			
		$\frac{3}{4}$	$\frac{7}{8}$	1	
		Usual Angle Thickness, t (in.)			
		$\frac{1}{4}$	$\frac{5}{16}$	$\frac{1}{2}$	
10	$29\frac{1}{2}$	155	210	275	W 36
9	$26\frac{1}{2}$	139	189	247	W 36, 33
8	$23\frac{1}{2}$	124	168	220	W 35, 33, 30
7	$20\frac{1}{2}$	108	147	192	W 36, 33, 30, 27, 24, S 24
6	$17\frac{1}{2}$	92.8	126	165	W 36, 33, 30, 27, 24, 21, S 24
5	$14\frac{1}{2}$	77.3	105	137	W 30, 27, 24, 21, 18, S 24, 20, 18, C 18
4	$11\frac{1}{2}$	61.9	84.2	110	W 24, 21, 18, 16, S 24, 20, 18, 15, C 18, 15
3	$8\frac{1}{2}$	46.4	61.9[b]	82.5	W 18, 16, 14, 12, 10, S 18, 15, 12, 10, C 18, 15, 12, 10
2	$5\frac{1}{2}$	30.9	39.4[b]	55.0	W 12, 10, 8, S 12, 10, 8, C 12, 10, 9, 8
1	$2\frac{1}{2}$	15.4	21.0	27.5	W 6, 5, M 6, 5, C 7, 6, 5

Source: Adapted from data in the *Manual of Steel Construction*, 8th ed. (Ref. 1), with permission of the publishers, the American Institute of Steel Construction.

[a] See Fig. 10.14.

[b] Limited by shear on the angles.

Although there is no specified limit for the minimum size of a framed connection to be used with a beam, the general rule is to choose one with an angle length of at least one-half the beam depth. This is intended in the most part to ensure some rotational stability for the beam end.

The one-bolt connection with an angle length of only 2.5 in. (Fig. 10.14*c*) is the shortest. This special connection has double-gage spacing of bolts in the beam web to ensure its stability.

The following example illustrates the general design procedure for a framed beam connection. In practice, this process can be shortened because experience permits judgments that will eventually make some of the steps unnecessary. Other design aids in joints must be relatively easy to produce and economical, especially if there are many trusses of a single type in the building structural system. Considerations involved in the design of connections for the joints include the truss configuration, member shapes and sizes, and the fastening method—usually welding or high-strength bolts.

In most cases the preferred method of fastening for connections made in the fabricating shop is welding. In most cases trusses will be shop-fabricated in the largest units possible, which means the whole truss for modest span trusses or the maximum-sized unit that can be transported for large trusses. Bolting is mostly used for connections made at the building site. For the small truss, the only bolting is usually done for the connections to supports and to supported elements or bracing. For the large truss, bolting may also be done at splice points between shop-fabricated units. All of this is subject to many considerations relating to the nature of the rest of the building structure, the particular location of the site, and the practices of local fabricators and erectors.

Two common forms for light steel trusses are shown in Fig. 10.16. In Fig. 10.16*a* the truss members consist of pairs of angles and the joints are achieved by using steel gusset plates to which the members are attached. For top and bottom chords the angles are often made continuous through the joint, reducing the number of connectors required and the number of separate cut pieces of

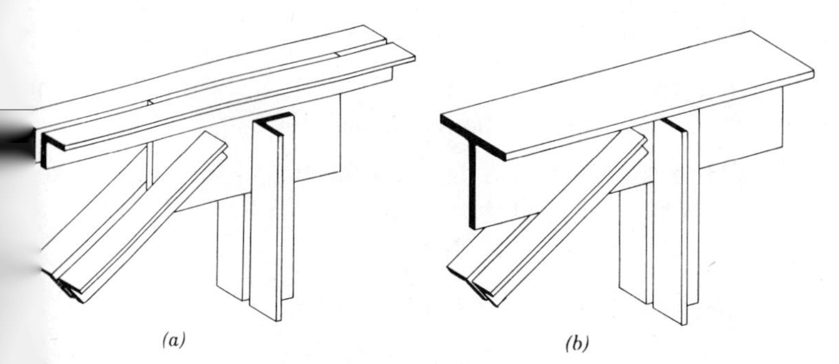

FIGURE 10.16 Typical framing details for light steel trusses.

the AISC Manual (Ref. 2) will shorten the work for some computations.

Example. A beam consists of a W 27 × 84 of A36 steel with F_u of 58 ksi [400 MPa] that is needed to develop an end reaction of 80 kips [356 kN]. Design a standard framed beam connection with A325F bolts and angles of A36 steel.

Solution. A scan of Table 10.5 reveals that the range of possible connections for a W 27 is $n = 5$ to $n = 7$. For the required load possible choices are

$$n = 6,\ \tfrac{3}{4}\text{-in. bolts, angle } t = \tfrac{1}{4}\text{ in., load} = 92.8\text{ kips [413 kN]}$$

$$n = 5,\ \tfrac{7}{8}\text{-in. bolts, angle } t = \tfrac{5}{16}\text{ in., load} = 105\text{ kips [467 kN]}$$

Bolt size is ordinarily established for a series of framing rather than for each element. Having no other criterion, we make an arbitrary choice of the connection with $\frac{7}{8}$-in. bolts.

The bolt capacity in double shear is the primary consideration in the development of data in Table 10.5. We must make a separate investigation of the bearing on the beam web because it is not incorporated in the table data. It is actually seldom a problem except in heavily loaded beams, but the following procedure should be used:

From Appendix Table A.1 the thickness of the beam web is 0.460 in. [11.7 mm]. The total bearing capacity of the five bolts is

$$V = n \times (\text{bolt diameter}) \times (\text{web } t) \times 1.5F_u$$

$$= 5 \times 0.875 \times 0.460 \times 87 = 175.1\text{ kips [780 kN]}$$

which is considerably in excess of the required load of 80 kips [356 kN].

Another concern in the typical situation is that for the shear stress through the net section of the web, reduced by the chain of bolt holes. If the connection is made as shown in Fig. 10.11*b* or *d*, this section is determined as the full web width (beam depth), less the sum of the hole diameters, times the web thickness, and the allowable stress is specified as $0.40F_y$. From Appendix Table A.1 for the W 27, $d = 26.92$ in. [684 mm]. The net shear width through the bolt holes is thus

$$w = 26.92 - (5 \times 0.9375) = 22.23 \text{ in. } [565 \text{ mm}]$$

and the computed stress due to the load is

$$f_v = \frac{80}{22.23 \times 0.460} = 7.82 \text{ ksi } [54 \text{ MPa}]$$

which is less than the allowable of $0.40 \times 36 = 14.4$ ksi [100 MPa].

If the top flange of the beam is cut back to form the type of connection shown in Fig. 10.11*e*, a critical condition that must be investigated is that of tearing out the end portion of the beam web, as discussed in Sec. 10.4. This is also called *block shear*, which refers to the form of the failed portion (Fig. 10.15). If the angles are placed with the edge distances shown in Fig. 10.15, this failure block will have the dimensions of 14 × 2.25 in. [356 × 57 mm]. The tearing force V is resisted by a combination of tension stress and shearing stress. The allowable stresses for this situation are $0.30F_u$ for shear and $0.50\,F_u$ for tension. Next find the net widths of the sections, multiply them by the web thickness to

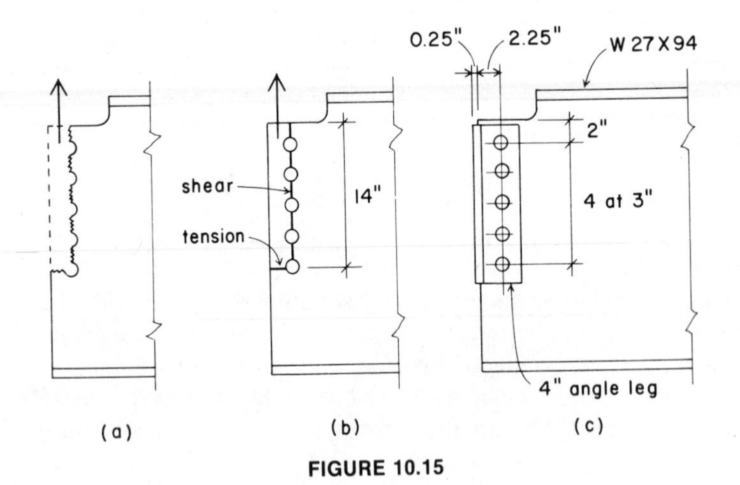

FIGURE 10.15

obtain the areas, and multiply by the allowable stresses to ob the total resisting forces.

For the tension resistance

$$w = 2.25 - \frac{0.9375}{2} = 1.78 \text{ in. } [45.1 \text{ mm}]$$

For the shear resistance

$$w = 14 - (4\tfrac{1}{2} \times 0.9375) = 9.78 \text{ in. } [249 \text{ mm}]$$

For the total resisting force

$$V = (\text{tension } w \times t_w) \times (0.50F_u) + (\text{shear } w \times t_w) \times ($$

$$= (1.78 \times 0.46 \times 29) + (9.78 \times 0.46 \times 17.4)$$

$$= 23.7 + 78.3 = 102 \text{ kips}$$

Because this potential total resistance exceeds the loac the tearing is not critical.

Problem 10.7.A. A W 30 × 108 of A36 steel with F_u = MPa] is required to develop an end reaction of 120 kip Determine the possible choices for a framed beam with A325F bolts and A36 angles. Investigate the res the smallest bolts to bearing and shear if the beam fla back, and the resistance to tearing if the connection i shown in Fig. 10.15.

Problem 10.7.B. Proceed as in Problem 10.7.A ex beam is a W 24 × 69 and the reaction is 100 kips [

Problem 10.7.C. Proceed as in Problem 10.7.A e beam is a W 16 × 45 and the reaction is 50 kips [

10.8 BOLTED TRUSS CONNECTIONS

A major factor in the design of trusses is the de truss joints. Since a single truss typically has s

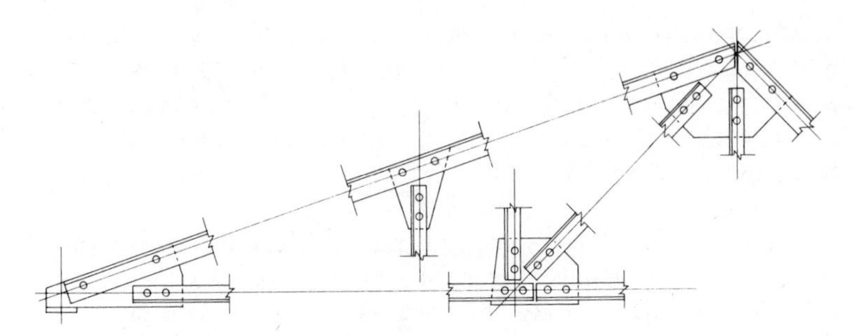

FIGURE 10.17 Typical light steel truss with bolted joints.

the angles. For flat-profiled, parallel-chorded trusses of modest size, the chords are sometimes made from tees, with interior members fastened to the tee web (Fig. 10.16*b*).

Figure 10.17 shows a layout for several joints of a light roof truss, employing the system shown in Fig. 10.16*a*. This is a form commonly used in the past for roofs with high slopes, with many short-span trusses fabricated in a single piece in the stop, usually with riveted joints. Trusses of this form are now mostly welded or use high-strength bolts as shown in Fig. 10.17.

Development of the joint designs for the truss shown in Fig. 10.17 would involve many considerations, including:

1. *Member Size and Load Magnitude.* This determines primarily the size and type of connector (bolt) required, based on individual connector capacity.
2. *Angle Leg Size.* This relates to the maximum diameter of bolt that can be used, based on minimum edge distances.
3. *Thickness and Profile Size of Gusset Plates.* The preference is to have the lightest weight added to the structure (primarily for the cost per pound of the steel), which is achieved by reducing the plates to a minimum thickness and general minimum size.
4. *Layout of Members at Joints.* The general attempt is to have the action lines of the forces (vested in the rows of bolts) all meet at a single point, thus avoiding twisting in the joint.

Many of the points mentioned are determined by data. Minimum edge distances for bolts (Table 10.2) can be matched to usual gage dimensions for angles (Table 10.4). Forces in members can be related to bolt capacities in Table 10.1, the general intent being to keep the number of bolts to a minimum in order to make the required size of the gusset plate smaller.

Other issues involve some judgment or skill in the manipulation of the joint details. For really tight or complex joints, it is often necessary to study the form of the joint with carefully drawn large-scale layouts. Actual dimensions and form of the member ends and the gusset plates may be derived from these drawings.

The truss shown in Fig. 10.17 has some features that are quite common for small trusses. All member ends are connected by only two bolts, the minimum required by the specifications. This simply indicates that the minimum-sized bolt chosen has sufficient capacity to develop the forces in all members with only two bolts. At the top chord joint between the support and the peak, the top chord member is shown as being continuous (uncut) at the joint. This is quite common where the lengths of members available are greater than the joint-to-joint distances in the truss, a cost savings in member fabrication as well as connection.

If there are only one or a few of the trusses as shown in Fig. 10.17 to be used in a building, the fabrication may indeed be as shown in the illustration. However, if there are many such trusses, or the truss is actually a manufactured, standardized product, it is much more likely to be fabricated with joints as described previously, employing welding for shop work and bolting only for field connections.

10.9 WELDED CONNECTIONS

Welding is in some instances an alternative means of making connections in a structural joint, the other principal option being structural bolts. A common situation is that of a connecting device (bearing plate, framing angles, etc.) that is welded to one member in the shop and fastened by bolting to a connecting member in the field. However, there are also many instances of joints that are fully welded, whether done in the shop or at the site of the building construction. For some situations the use of welding may be the only reasonable means of making an attachment for a joint. As in many other situations, the design of welded joints requires considerable awareness of the problems encountered by the welder and the fabricator of the welded parts. In the following sections we present some of the problems and potential uses of welding in building structures.

One advantage of welding is that it offers the possibility for direct connection of members, often eliminating the need for intermediate devices, such as gusset plates or framing angles. Another advantage is the lack of need for holes (required for bolts), which permits development of the capacity of the unreduced cross section of tension members. Welding also offers the possibility of developing exceptionally rigid joints, an advantage in moment-resistive connections or generally nondeforming connections.

10.10 ELECTRIC ARC WELDING

Although there are many welding processes, electric arc welding is the one generally used in steel building construction. In this type of welding, an electric arc is formed between an electrode and the two pieces of metal that are to be joined. The intense heat melts a small portion of the members to be joined, as well as the end of the electrode or metallic wire. The term *penetration* is used to indicate the depth from the original surface of the base metal to the point at which fusion ceases. The globules of melted metal from the electrode flow into the molten seat and, when cool, are united with the members that are to be welded together. *Partial penetration* is the failure of the weld metal and base metal to fuse at the root of a weld. It may result from a number of items, and such incomplete fusion produces welds that are inferior to those of full penetration (called *complete penetration* welds).

10.11 FORMS OF WELDED JOINTS

When two members are to be joined, the end may or may not be grooved in preparation for welding. In general, there are three classifications of joints: *butt joints, tee joints,* and *lap joints*. The selection of the type of weld to use depends on the magnitude of the load requirement, the manner in which it is applied, and the cost of preparation and welding. Several joints are shown in Fig. 10.18. The type of joint and preparation permit a number of variations. In addition, welding may be done from one or both sides. The scope of this book prevents a detailed discussion of the many joints and their uses and limitations.

A weld commonly used for structural steel in building construction is the *fillet* weld. It is approximately triangular in cross section and is formed between the two intersecting surfaces of the joined members (see Fig. 10.19*a* and *b*). The *size* of a fillet weld is the leg length of the largest inscribed isosceles right triangle, *AB* or *BC* (see Fig. 10.19*a*). The *root* of the weld is the point at the

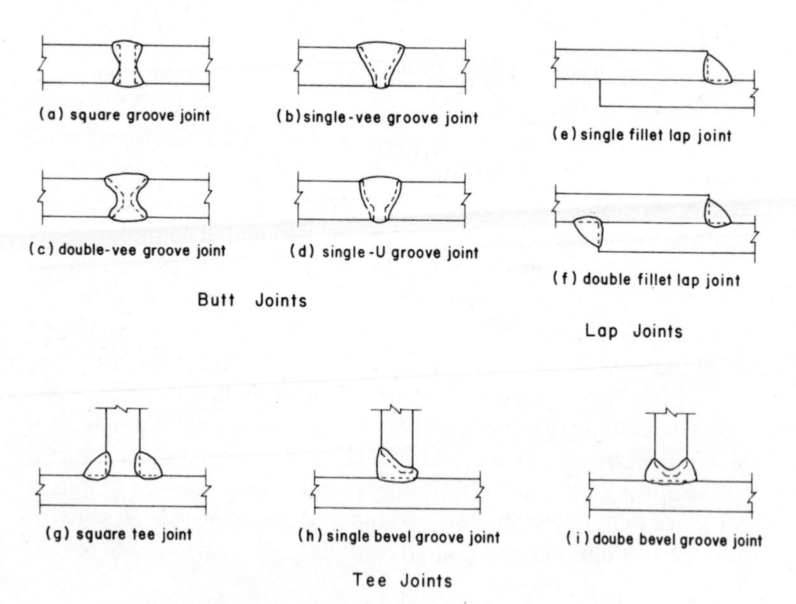

FIGURE 10.18 Typical forms for welded joints.

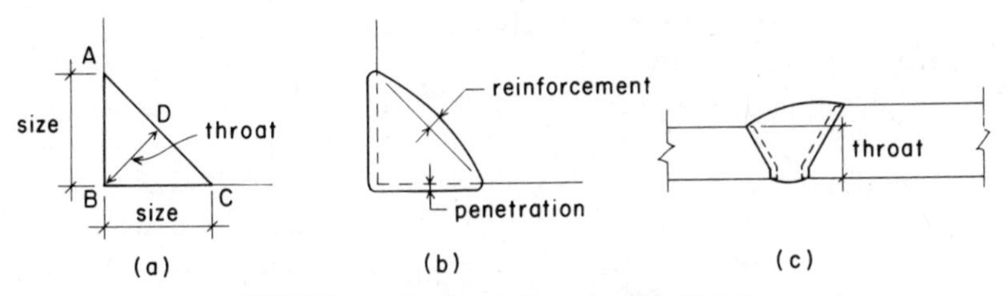

FIGURE 10.19 Properties of welded joints.

bottom of the weld, point B in Fig. 10.19*a*. The *throat* of a fillet weld is the distance from the root to the hypotenuse of the largest isosceles right triangle that can be inscribed within the weld cross section, distance BC in Fig. 10.19*a*. The exposed surface of a weld is not the plane surface indicated in Fig. 10.19*a* but is usually somewhat convex, as shown in Fig. 10.19*b*. Therefore the actual throat may be greater than that shown in Fig. 10.19*a*. This additional material is called *reinforcement*. It is not included in determining the strength of a weld.

A single-vee groove weld between two members of unequal thickness is shown in Fig. 10.19*c*. The *size* of a butt weld is the thickness of the thinner part joined, with no allowance made for the weld reinforcement.

10.12 STRESSES IN FILLET WELDS

If the dimension (size) of AB in Fig. 10.20*a* is one unit in length, $(AD)^2 + (BD)^2 = 1^2$. Because AD and BD are equal, $2(BD)^2 = 1^2$,

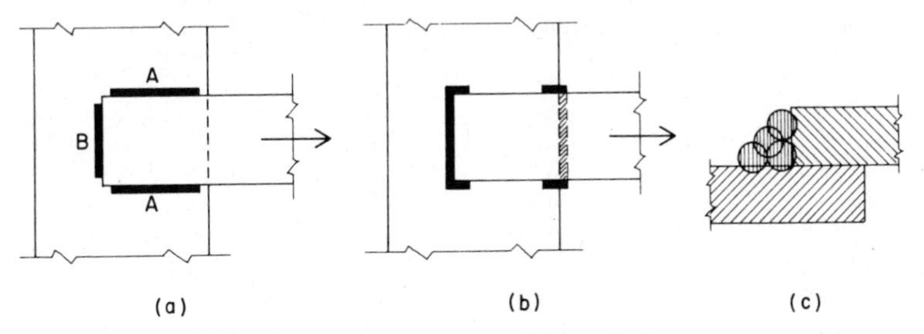

FIGURE 10.20 Welding of lapped plates.

and $BD = \sqrt{0.5}$, or 0.707. Therefore the throat of a fillet weld is equal to the *size* of the weld multiplied by 0.707. As an example, consider a $\frac{1}{2}$-in. fillet weld. This would be a weld with dimensions AB or BC equal to $\frac{1}{2}$ in. In accordance with the above, the throat would be 0.5×0.707, or 0.3535 in. Then, if the allowable unit shearing stress on the throat is 21 ksi, the allowable working strength of a $\frac{1}{2}$-in. fillet weld is $0.3535 \times 21 = 7.42$ kips *per lin in. of weld*. If the allowable unit stress is 18 ksi, the allowable working strength is $0.3535 \times 18 = 6.36$ kips *per lin in. of weld*.

The permissible unit stresses used in the preceding paragraph are for welds made with E 70 XX- and E 60 XX-type electrodes on A36 steel. Particular attention is called to the fact that *the stress in a fillet weld is considered as shear on the throat, regardless of the direction of the applied load*. Neither plug nor slot welds shall be assigned any values in resistances other than shear. The allowable working strengths of fillet welds of various sizes are given in Table 10.6 with values rounded to $\frac{1}{10}$ kip.

The stresses allowed for the metal of the connected parts (known as the *base metal*) apply to complete penetration groove welds that are stressed in tension or compression parallel to the axis of the weld or are stressed in tension perpendicular to the effective throat. They apply also to complete or partial penetra-

TABLE 10.6 Allowable Working Strength of Fillet Welds

	Allowable Load (kips/in.)		Allowable Load (kN/mm)		
Size of Weld (in.)	E 60 XX Electrodes $F_{vw} = 18$ (ksi)	E 70 XX Electrodes $F_{vw} = 21$ (ksi)	E 60 XX Electrodes $F_{vw} = 124$ (MPa)	E 70 XX Electrodes $F_{vw} = 145$ (MPa)	Size of Weld (mm)
$\frac{3}{16}$	2.4	2.8	0.42	0.49	4.76
$\frac{1}{4}$	3.2	3.7	0.56	0.65	6.35
$\frac{5}{16}$	4.0	4.6	0.70	0.81	7.94
$\frac{3}{8}$	4.8	5.6	0.84	0.98	9.52
$\frac{1}{2}$	6.4	7.4	1.12	1.30	12.7
$\frac{5}{8}$	8.0	9.3	1.40	1.63	15.9
$\frac{3}{4}$	9.5	11.1	1.66	1.94	19.1

TABLE 10.7 Relation Between Material Thickness and Minimum Size of Fillet Welds

Material Thickness of the Thicker Part Joined		Minimum Size of Fillet Weld	
in.	mm	in.	mm
To $\frac{1}{4}$ inclusive	To 6.35 inclusive	$\frac{1}{8}$	3.18
Over $\frac{1}{4}$ to $\frac{1}{2}$	Over 6.35 to 12.7	$\frac{3}{16}$	4.76
Over $\frac{1}{2}$ to $\frac{3}{4}$	Over 12.7 to 19.1	$\frac{1}{4}$	6.35
Over $\frac{3}{4}$	Over 19.1	$\frac{5}{16}$	7.94

tion groove welds stressed in compression normal to the effective throat and in shear on the effective throat. Consequently, allowable stresses for butt welds are the same as for the base metal.

The relation between the weld size and the maximum thickness of material in joints connected only by fillet welds is shown in Table 10.7. The maximum size of a fillet weld applied to the square edge of a plate or section that is $\frac{1}{4}$ in. or more in thickness should be $\frac{1}{16}$ in. less than the nominal thickness of the edge. Along edges of material less than $\frac{1}{4}$ in. thick, the maximum size may be equal to the thickness of the material.

The effective area of butt and fillet welds is considered to be the effective length of the weld multiplied by the effective throat thickness. The minimum effective length of a fillet weld should not be less than four times the weld size. For starting and stopping the arc, approximately $\frac{1}{4}$ in. should be added to the design length of fillet welds.

Figure 10.20*a* represents two plates connected by fillet welds. The welds marked *A* are longitudinal; *B* indicates a transverse weld. If a load is applied in the direction shown by the arrow, the stress distribution in the longitudinal weld is not uniform, and the stress in the transverse weld is approximately 30% higher per unit of length.

Added strength is given to a transverse fillet weld that terminates at the end of a member, as shown in Fig. 10.20*b*, if the weld is returned around the corner for a distance not less than twice the weld size. These end returns, sometimes called *boxing*, afford considerable resistance to the tendency of tearing action on the weld.

The $\frac{1}{4}$-in. fillet weld is considered to be the minimum practical size, and a $\frac{5}{16}$-in. weld is probably the most economical size that can be obtained by one pass of the electrode. A small continuous weld is generally more economical than a larger discontinuous weld if both are made in one pass. Some specifications limit the single-pass fillet weld to $\frac{5}{16}$ in. Large fillet welds require two or more passes (multipass welds) of the electrode, as shown in Fig. 10.20*c*.

10.13 DESIGN OF WELDED JOINTS

The most economical weld to use for a given condition depends on several factors. It should be borne in mind that members to be connected by welding must be firmly clamped or held rigidly in position during the welding process. When welding a beam to a column, you must provide a support to keep the beam in position for the welding work. The seat angle is not considered as adding strength to the connection. The designer must have in mind the actual conditions during erection and must provide for economy and ease in working the welds. Seat angles or similar members used to facilitate erection are *shop-welded* before the material is sent to the site. The welding done during erection is called *field welding*. In preparing welding details, the designer indicates shop or field welds on the drawings. Conventional welding symbols are used to identify the type, size, and position of the various welds. Only engineers or architects experienced in the design of welded connections should design or supervise welded construction. It is apparent that a wide variety of connections is possible; experience is the best aid in determining the most economical and practical connection.

The following examples illustrate the basic principles on which welded connections are designed.

Example 1. A bar of A36 steel, $3 \times \frac{7}{16}$ in. [76.2 × 11 mm] in cross section, is to be welded with E 70 XX electrodes to the back of a channel so that the full tensile strength of the bar may be developed. What is the size of the weld? (See Fig. 10.21.)

Solution. The area of the bar is 3 × 0.4375 = 1.313 in.^{2} [76.2 × 11 = 838.2 mm^2]. Because the allowable unit tensile stress of the steel is 22 ksi (Table 2.1), the tensile strength of the bar is $F_t \times A$ = 22 × 1.313 = 28.9 kips [152 × 838.2/10^3 = 127 kN]. The weld must be of ample dimensions to resist a force of this magnitude.

A $\frac{3}{8}$-in. [9.52-mm] fillet weld will be used. Table 10.7 gives the allowable working strength as 5.6 kips/in. [0.98 kN/mm]. Hence the required length of weld to develop the strength of the bar is 28.9 ÷ 5.6 = 5.16 in. [127 ÷ 0.98 = 130 mm]. The position of the weld with respect to the bar has several options, three of which are shown in Fig. 10.21*a*, *c*, and *d*.

Example 2. A $3\frac{1}{2} \times 3\frac{1}{2} \times \frac{5}{16}$-in. [89 × 89 × 7.94-mm] angle of A36 steel subjected to a tensile load is to be connected to a plate by fillet welds, using E 70 XX electrodes. What should the dimensions of the welds be to develop the full tensile strength of the angle?

Solution. We shall use a $\frac{1}{4}$-in. fillet weld which has an allowable working strength of 3.7 kips/in. [0.65 kN/mm] (Table 10.7). From Appendix Table A.3 the cross-sectional area of the angle is 2.09 in.^{2} [1348 mm^2]. By using the allowable tension stress of 22 ksi [152 MPa] for A36 steel (Table 2.1), the tensile strength of the angle is 22 × 2.09 = 46 kips [152 × 1348/10^3 = 205 kN]. Therefore the required total length of weld to develop the full strength of the angle is 46 ÷ 3.7 = 12.4 in. [205 ÷ 0.65 = 315 mm].

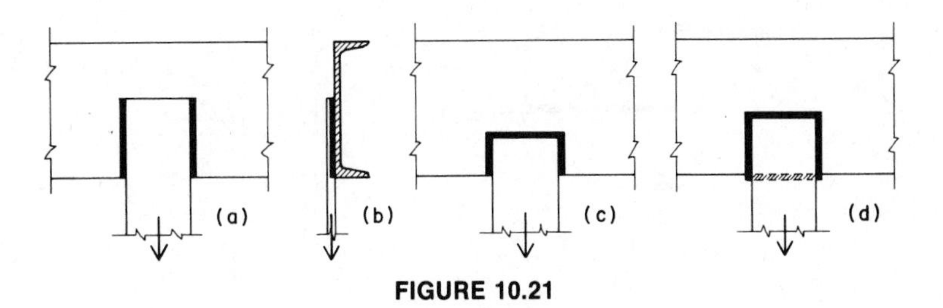

FIGURE 10.21

An angle is an unsymmetrical cross section, and the welds marked L_1 and L_2 in Fig. 10.22 are made unequal in length so that their individual resistance will be proportioned in accordance to the distributed area of the angle. From Appendix Table A.3 we find that the centroid of the angle section is 0.99 in. [25 mm] from the back of the angle; hence the two welds are 0.99 in. [25 mm] and 2.51 in. [64 mm] from the centroidal axis, as shown in Fig. 10.22. The lengths of welds L_1 and L_2 are made inversely proportional to their distances from the axis, but the sum of their lengths is 12.4 in. [315 mm]. Therefore,

$$L_1 = \frac{2.51}{3.5} \times 12.4 = 8.9 \text{ in. [227 mm]}$$

and

$$L_2 = \frac{0.99}{3.5} \times 12.4 = 3.5 \text{ in. [88 mm]}$$

These are the design lengths required, and as noted earlier, each weld would actually be made $\frac{1}{4}$ in. [6.4 mm] longer than its computed length.

When angle shapes are used as tension members and connected by fastening only one leg, it is questionable to assume a stress distribution of equal magnitude on the entire cross section. Some designers therefore prefer to ignore the stress in the unconnected leg and to limit the capacity of the member in tension to the force obtained by multiplying the allowable stress by the area

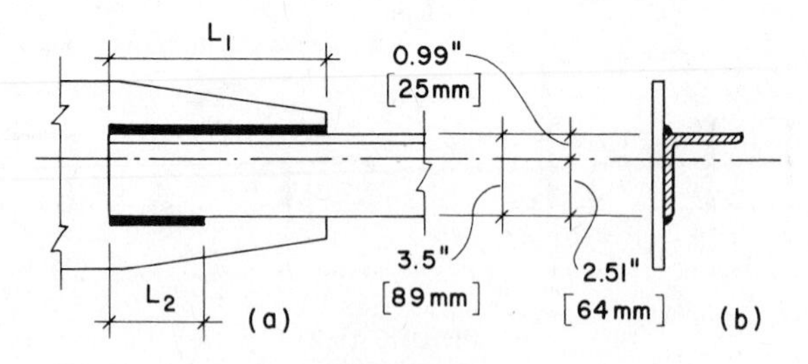

FIGURE 10.22

of the connected leg only. If this is done, it is logical to use welds of equal length on each side of the leg, as in Example 1.

Problem 10.13.A. A 4 × 4 × $\frac{1}{2}$-in. angle of A36 steel is to be welded to a plate with E 70 XX electrodes to develop the full tensile strength of the angle. Using $\frac{3}{8}$-in. fillet welds, compute the design lengths L_1 and L_2, as shown in Fig. 10.22, assuming the development of tension on the entire cross section of the angle.

Problem 10.13.B. Redesign the welded connection in Problem 10.13.A assuming that the tension force is developed only by the connected leg of the angle.

10.14 BEAMS WITH CONTINUOUS ACTION

As noted earlier, one of the advantages of welding is that beams having continuous action at the supports (type 1 construction) are readily provided for. The usual bolted or riveted connections of type 2 construction are assumed to offer no rigidity at the supports and the bending moment throughout the length of the beam is positive. By the use of welding, however, a beam may be connected at its supports in such a manner that the beam is *fixed* or *restrained* and a negative bending moment results. For the same span and loading the maximum bending moment for a continuous beam is smaller than for a simple beam and a lighter beam section is required.

When beams are rigidly connected by means of moment-resisting connections the fibers in the upper flange *at the supports* are in tension and the lower flange is in compression. This is shown diagrammatically in Fig. 10.23*a*. Therefore, in designing the welds for beams that have continuous action, we must provide for tension and compression at the supports in the upper and lower flanges, respectively. A wide variety of welds is possible. The following example illustrates the principles by which they are designed:

Example. A W 12 × 40 framing into column flanges at its ends is to be connected by welding to provide continuous action. For erection of the beam it is necessary that its length be slightly

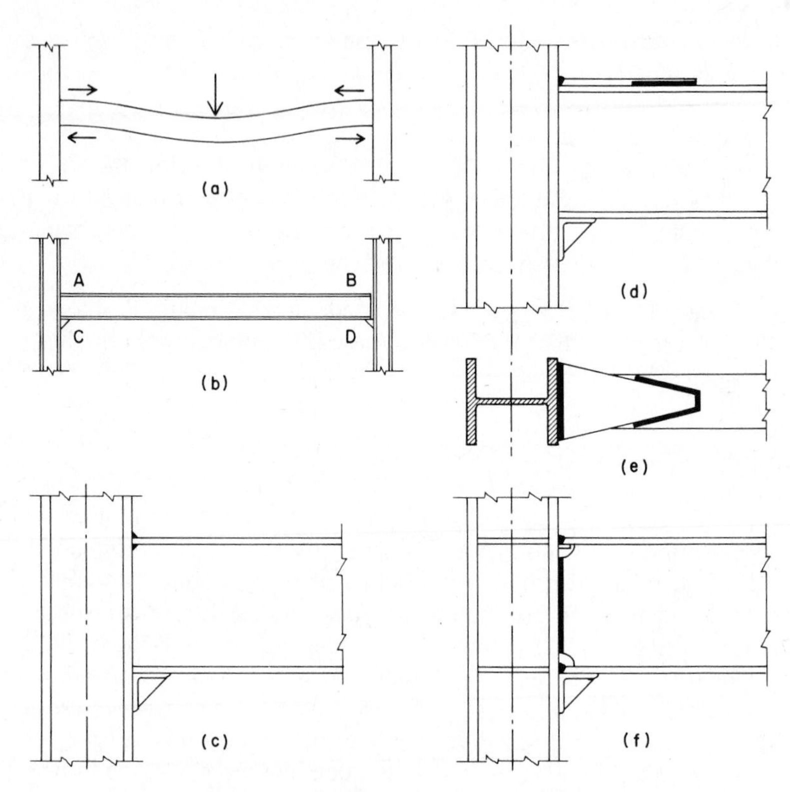

FIGURE 10.23 Welded joints for restrained beams.

shorter than the distance between the flanges of the columns to which it will be welded. Seat angles are shop-welded to the columns, and the beam is supported on them during field welding of the connection. For this example consider that the left end of the beam is tightly held against the column flange; this leaves a short space between the right end of the beam and the column on the right, as shown in Fig. 10.23*b*. Because of this difference in end conditions, each weld must receive individual consideration. As a means of identification, the different welds are referred to as *A*, *B*, *C*, and *D*, as indicated in Fig. 10.23*b*.

The negative bending moment in the beam at the supports is 1150 kip-in. [130 kN-m] and E 70 XX electrodes are used. Design the welds.

Solution. The tensile and compressive forces in the beam flanges constitute a *mechanical couple*, which consists of two equal parallel forces, opposite in direction but not having the same line of action. The moment of a couple is the magnitude of one of the forces multiplied by the perpendicular distance between their lines of action. Therefore, if the negative bending moment is 1150 kip-in. [130 kN-m] and the distance between the two flange forces is approximately 12 in. [305 mm], the forces are each 1150 ÷ 12 = 96 kips [130 ÷ 0.305 = 426 kN].

From Appendix Table A.1 we find that the beam has a flange width of 8.005 in. [203 mm] and a flange thickness of 0.515 in. [13.08 mm]. Suppose we run a $\frac{3}{8}$-in. [9.52-mm] fillet weld across the upper flange for weld A. This weld has an allowable working load of 5.6 kips per in. [0.98 kN per mm] (Table 10.6), and, because the flange is 8 in. [203 mm] wide, its resistance is 8 = 5.6 = 44.8 kips [0.98 × 203 = 199 kN]. If $\frac{3}{8}$-in. fillet welds are to be used at this joint, the total length of the weld must be 96 ÷ 5.6 = 17.2 in. [426 ÷ 0.98 = 435 mm]. However, the flange is only 8 in. wide, and welds on the upper and under surfaces of the flange, as shown in Fig. 10.23*c*, do not provide sufficient length. A solution for this condition would be to investigate larger welds, but the weld at the underside of the flange requires overhead welding which should be avoided whenever possible.

An alternate detail for the joint at *A* is shown in Fig. 10.23*d* and *e*. This joint has a tapered plate which is welded to the column face with a large butt weld and to the beam flange with fillet welds on the edges of the plate. This joint could also be used at *B* because it does not require the end of the beam to be placed against the column face.

At joints *C* and *D* the force to be transferred from the beam flange to the column face is one of compression. If the beam end is placed flush against the column face, this transfer may be achieved by direct bearing, with no welding required. However, fillet welds probably should be used on the upper surface of the lower beam flange at the column face or at the edges of the beam flange on top of the seat angle to ensure that the beam will not slip with respect to the column. This connection cannot be used at *D*, however, because the beam is not in contact with the column.

The best connection for use in developing a moment close to the limit for the beam is probably that shown in Fig. 10.23*f*. In this joint the beam flanges are cut to permit the placing of a full penetration groove weld, which will develop the capacity of the flange fully in both tension and compression. In the illustration the beam web is shown welded to the column face to develop the vertical shear force, or end reaction, of the beam. The seat angle in this case is used strictly as a temporary erection device to hold the beam in place while the welds are being made. If desired, it could be removed after the welds are completed.

The forces from the beam flanges in connections like those shown in Fig. 10.23 exert considerable bending on the column flanges, making reinforcing plates (Fig. 10.23*f*) necessary. Although generally desirable, this reinforcement may not be required when the beam flanges are considerably narrower than the column or the column flanges are quite thick.

10.15 MISCELLANEOUS WELDED CONNECTIONS

Plug and Slot Welds

One method of connecting two overlapping plates uses a weld in a hole made in one of the two plates (see Fig. 10.24). Plug and slot welds are those in which the entire area of the hole or slot receives weld metal. The maximum and minimum diameters of plug and slot welds and the maximum length of slot welds are shown in Fig. 10.24. If the plate containing the hole is not more than $\frac{5}{8}$ in. thick, the hole should be filled with weld metal. If the plate is more than $\frac{5}{8}$ in. thick, the weld metal should be at least one-half the thickness of the material but not less than $\frac{5}{8}$ in.

The stress in a plug or slot weld is considered to be shear on the area of the weld at the plane of contact of the two plates being connected. The allowable unit shearing stress, when E 70 XX electrodes are used, is 21 ksi [145 MPa].

A somewhat similar weld consists of a continuous fillet weld at the circumference of a hole, as shown in Fig. 10.24*c*. This is not a plug or slot weld and is subject to the usual requirements for fillet welds.

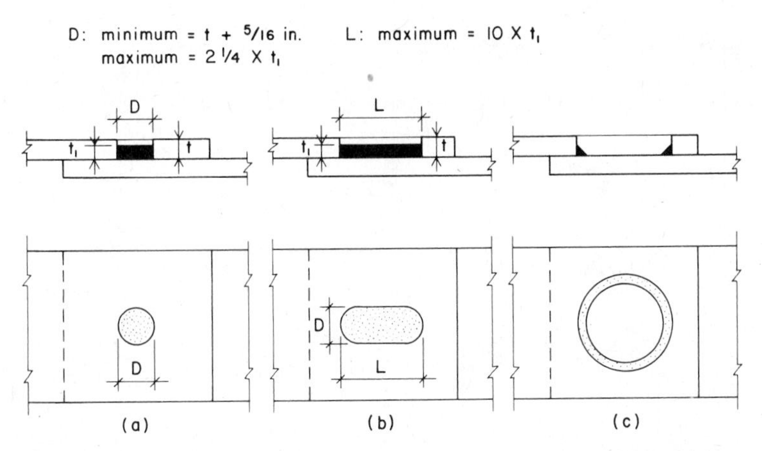

FIGURE 10.24 Welds in holes: (a) plug weld; (b) slot weld; (c) fillet weld in a large hole.

Framing Connections

Part 4 of the AISC Manual contains a series of tables that pertain to the design of welded connections. The tables cover free-end as well as moment-resisting connections. In addition, suggested framing details are shown for various situations.

A few common connections are shown in Fig. 10.25. As an aid to erection, certain parts are welded together in the shop before being sent to the site. Connection angles may be shop-welded to beams and the angles field-welded or field-bolted to girders or columns. The beam connection in Fig. 10.25*a* shows a beam supported on a seat that has been shop-welded to the column. A small connection plate is shop-welded to the lower flange of the beam, and the plate is bolted to the beam seat. After the beams have been erected and the frame plumbed, the beams are field-welded to the seat angles. This type of connection provides no degree of continuity in the form of moment transfer between the beam and column.

The connections shown in Fig. 10.25*b* and *c* are designed to develop some moment transfer between the beam and its supporting column. Auxiliary plates are used to make the connection at the upper flanges.

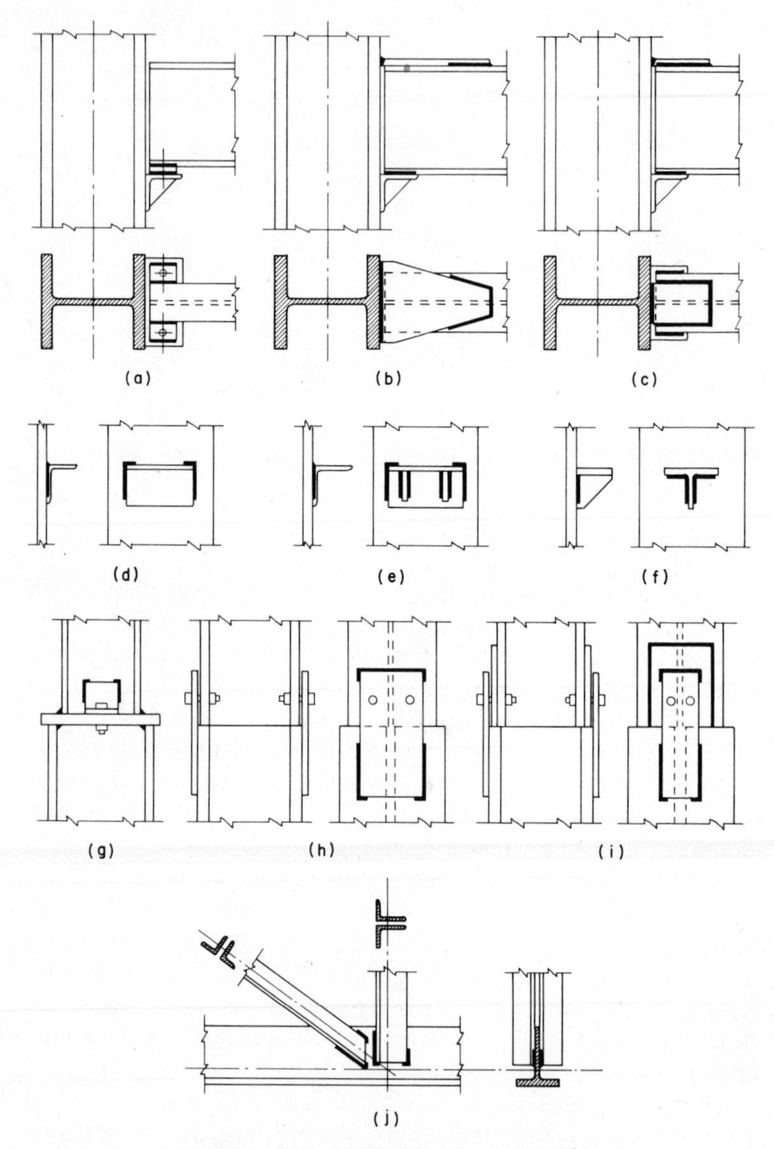

FIGURE 10.25 Welded framing connections.

Beam seats shop-welded to columns are shown in Fig. 10.25*d–f*. A short length of angle welded to the column with no stiffeners is shown in Fig. 10.25*d*. Stiffeners consisting of triangular plates are welded to the legs of the angles shown in Fig. 10.25*e* and add materially to the strength of the seat. Another method of forming a seat, using a short piece of structural tee, is shown in Fig. 10.25*f*.

Various types of column splice are shown in Fig. 10.25*g–i*. The auxiliary plates and angles are shop-welded to the columns and provide for bolted connections in the field before the permanent welds are made.

Figure 10.25*j* shows a type of welded construction used in light trusses in which the lower chord consists of a structural tee. Truss web members consisting of pairs of angles are welded to the stem of the tee chord. Welding may also be used for the connections for trusses using gusset plates, as shown in Fig. 10.17. In general, shop fabrication is most often achieved with welding; thus the various seats and bracing clips shown in Fig. 10.13 and 10.14 would most likely be attached to members in the shop by welding, even though field connections are achieved with bolts.

Some additional connection details are given in Fig. 10.26. The detail in Fig. 10.26*a* is an arrangement for framing a beam to a girder, in which welds are substituted for bolts or rivets. In this figure welds replace the fasteners that secure the connection angles to the web of the supported beam.

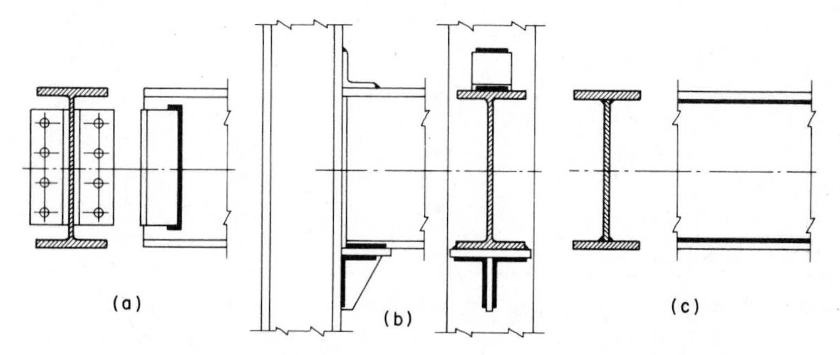

FIGURE 10.26 Additional welded framing connections.

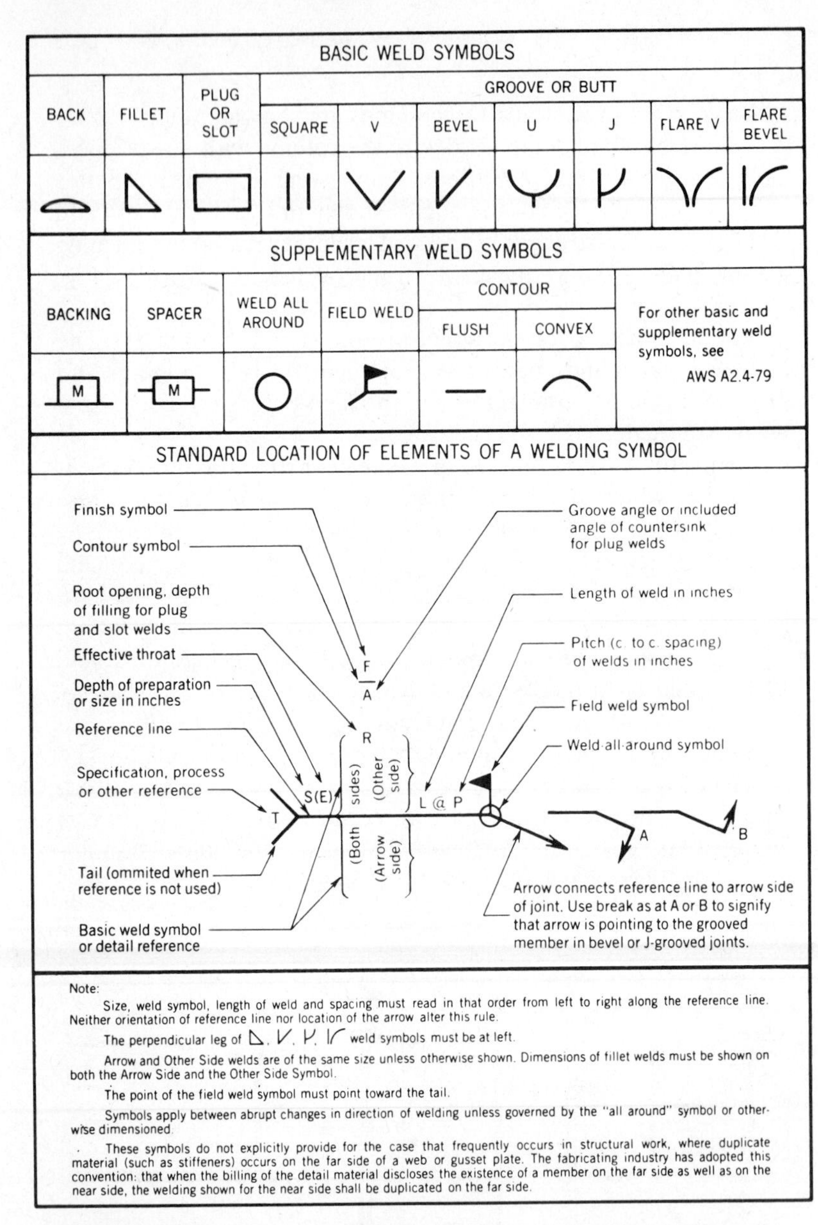

FIGURE 10.27 Standard weld symbols used on construction drawings. Reproduced from the *Manual of Steel Construction* (Ref. 1), with permission of the publishers, the American Institute of Steel Construction.

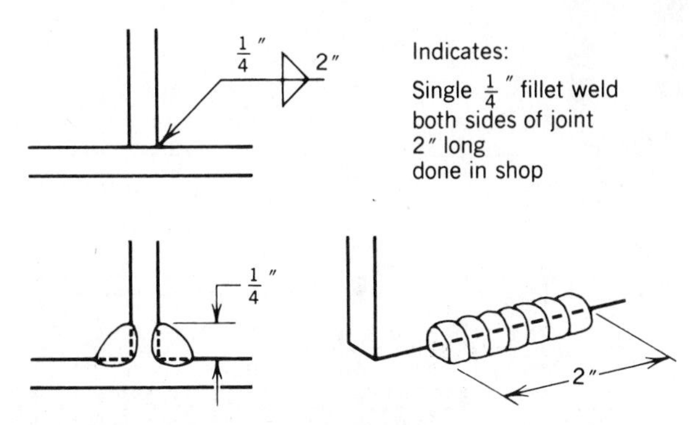

FIGURE 10.28 Use of standard weld symbols.

A welded connection for a stiffened seat beam connection to a column is shown in Fig. 10.26*b*. Figure 10.26*c* shows the simplicity of welding in connecting the upper and lower flanges of a plate girder to the web plate.

10.16 SYMBOLS FOR WELDS

Standard symbols are used in detail drawings of welded connections of structural elements. In addition to the type of weld, other information to be conveyed includes size, exact location, and finishes. Figure 10.27, reproduced from the AISC Manual, gives the standard symbols for welded joints. It will be noted that the symbol for a fillet weld is a triangle; this is drawn below the horizontal line if the weld is on the near side, above if it is on the far side; two triangles, one above and one below, are drawn for welds on both sides of the joint. The size of the weld is placed to the left of the vertical line of the triangle and the length to the right side of the hypotenuse. An example of the use of standard symbols for a simple fillet weld is shown in Fig. 10.28.

11

GENERAL CONCERNS FOR STRUCTURES

This chapter contains some discussions of general issues relating to design of building structures. These concerns have mostly not been addressed in the presentations in earlier chapters, but require some general consideration when dealing with whole building design situations. General application of these materials is illustrated in the design examples in Chapter 12.

11.1 INTRODUCTION

Materials, methods, and details of building construction vary considerably on a regional basis. There are many factors that affect this situation, including the real effects of response to climate and the availability of construction materials. Even in a single region, differences occur between individual buildings, based on individual styles of architectural design and personal techniques of builders. Nevertheless, at any given time there are usually a few predominant, popular methods of construction that are employed for most buildings of a given type and size. The construction methods and details shown here are reasonable, but in no way are they intended to illustrate a singular, superior style of building.

11.2 DEAD LOADS

Dead load consists of the weight of the materials of which the building is constructed such as walls, partitions, columns, framing, floors, roofs, and ceilings. In the design of a beam, the dead load must include an allowance for the weight of the beam itself. Table 11.1, which lists the weights of many construction materials, may be used in the computation of dead loads. Dead loads are due to gravity and they result in downward vertical forces.

Dead load is generally a permanent load, once the building construction is completed, unless frequent remodeling or rearrangement of the construction occurs. Because of this permanent, long-time, character, the dead load requires certain considerations in design, such as the following:

1. It is always included in design loading combinations, except for investigations of singular effects, such as deflections due to only live load.
2. Its long-time character has some special effects causing sag and requiring reduction of design stresses in wood structures, producing creep effects in concrete structures, and so on.
3. It contributes some unique responses, such as the stabilizing effects that resist uplift and overturn due to wind forces.

11.3 BUILDING CODE REQUIREMENTS

Structural design of buildings is most directly controlled by building codes, which are the general basis for the granting of building permits—the legal permission required for construction. Building

TABLE 11.1 Weights of Building Construction

	lb/ft²	kN/m²
Roofs		
3-ply ready roofing (roll, composition)	1	0.05
3-ply felt and gravel	5.5	0.26
5-ply felt and gravel	6.5	0.31
Shingles		
wood	2	0.10
asphalt	2–3	0.10–0.15
clay tile	9–12	0.43–0.58
concrete tile	8–12	0.38–0.58
slate, 1/4 in.	10	0.48
fiber glass	2–3	0.10–0.15
aluminum	1	0.05
steel	2	0.10
Insulation		
fiber glass batts	0.5	0.025
rigid foam plastic	1.5	0.075
foamed concrete, mineral aggregate	2.5/in.	0.0047/mm
Wood rafters		
2 × 6 at 24 in.	1.0	0.05
2 × 8 at 24 in.	1.4	0.07
2 × 10 at 24 in.	1.7	0.08
2 × 12 at 24 in.	2.1	0.10
Steel deck, painted		
22 ga	1.6	0.08
20 ga	2.0	0.10
18 ga	2.6	0.13
Skylight		
glass with steel frame	6–10	0.29–0.48
plastic with aluminum frame	3–6	0.15–0.29
Plywood or softwood board sheathing	3.0/in.	0.0057/mm
Ceilings		
Suspended steel channels	1	0.05
Lath		
steel mesh	0.5	0.025
gypsum board, 1/2 in.	2	0.10
Fiber tile	1	0.05
Dry wall, gypsum board, 1/2 in.	2.5	0.12
Plaster		
gypsum, acoustic	5	0.24
cement	8.5	0.41
Suspended lighting and air distribution systems, average	3	0.15

TABLE 11.1 (*Continued*)

Floors		
Hardwood, 1/2 in.	2.5	0.12
Vinyl tile, 1/8 in.	1.5	0.07
Asphalt mastic	12/in.	0.023/mm
Ceramic tile		
3/4 in.	10	0.48
thin set	5	0.24
Fiberboard underlay, 5/8 in.	3	0.15
Carpet and pad, average	3	0.15
Timber deck	2.5/in.	0.0047/mm
Steel deck, stone concrete fill, average	35–40	1.68–1.92
Concrete deck, stone aggregate	12.5/in.	0.024/mm
Wood joists		
2 × 8 at 16 in.	2.1	0.10
2 × 10 at 16 in.	2.6	0.13
2 × 12 at 16 in.	3.2	0.16
Lightweight concrete fill	8.0/in.	0.015/mm
Walls		
2 × 4 studs at 16 in., average	2	0.10
Steel studs at 16 in., average	4	0.20
Lath, plaster; see Ceilings		
Gypsum dry wall, 5/8 in. single	2.5	0.12
Stucco, 7/8 in., on wire and paper or felt	10	0.48
Windows, average, glazing + frame		
small pane, single glazing, wood or metal frame	5	0.24
large pane, single glazing, wood or metal frame	8	0.38
increase for double glazing	2–3	0.10–0.15
curtain walls, manufactured units	10–15	0.48–0.72
Brick veneer		
4 in., mortar joints	40	1.92
1/2 in., mastic	10	0.48
Concrete block		
lightweight, unreinforced—4 in.	20	0.96
6 in.	25	1.20
8 in.	30	1.44
heavy, reinforced, grouted—6 in.	45	2.15
8 in.	60	2.87
12 in.	85	4.07

codes (and the permit-granting process) are administered by some unit of government: city, county, or state. Most building codes, however, are based on some model code, of which there are three widely used in the United States:

1. The *Uniform Building Code* (Ref. 2), which is widely used in the West, as it has the most complete data for seismic design.
2. *The BOCA Basic National Building Code,* used widely in the East and Midwest.
3. *The Standard Building Code,* used in the Southeast.

These model codes are more similar than different, and are in turn largely derived from the same basic data and standard reference sources, including many industry standards. In the several model codes and many city, county, and state codes, however, there are some items that reflect particular regional concerns.

With respect to control of structures, all codes have materials (all essentially the same) that relate to the following issues:

1. *Minimum Required Live Loads.* This is addressed in Sec. 11.4; all codes have tables similar to those shown in Tables 11.2 and 11.3, which are reproduced from the *Uniform Building Code.*
2. *Wind Loads.* These are highly regional in character with respect to concern for local windstorm conditions. Model codes provide data with variability on the basis of geographic zones.
3. *Seismic (Earthquake) Effects.* These are also regional with predominant concerns in the western states. This data, including recommended investigations, is subject to quite frequent modification, as the area of study responds to ongoing research and experience.
4. *Load Duration.* Loads or design stresses are often modified on the basis of the time span of the load, varying from the life of the structure for dead load to a fraction of a second for a wind gust or a single major seismic shock.

Safety factors are frequently adjusted on this basis. Some applications are illustrated in the work in the design examples in this part.

5. *Load Combinations.* These were formerly mostly left to the discretion of designers, but are now quite commonly stipulated in codes, mostly because of the increasing use of ultimate strength design and the use of factored loads.
6. *Design Data for Types of Structures.* These deal with basic materials (wood, steel, concrete, masonry, etc.), specific structures (towers, balconies, pole structures, etc.), and special problems (foundations, retaining walls, stairs, etc.) Industry-wide standards and common practices are generally recognized, but local codes may reflect particular local experience or attitudes. Minimal structural safety is the general basis, and some specified limits may result in questionably adequate performances (bouncy floors, cracked plaster, etc.)
7. *Fire Resistance.* For the structure, there are two basic concerns, both of which produce limits for the construction. The first concern is for structural collapse or significant structural loss. The second concern is for containment of the fire to control its spread. These concerns produce limits on the choice of materials (e.g., combustible or noncombustible) and some details of the construction (cover on reinforcement in concrete, fire insulation for steel beams, etc.)

The work in the design examples in this part is based largely on criteria from the *Uniform Building Code* (Ref. 2). The choice of this model code reflects only the fact of the degree of familiarity of the author with specific codes in terms of his recent experience.

11.4 LIVE LOADS

Live loads technically include all the nonpermanent loadings that can occur, in addition to the dead loads. However, the term as commonly used usually refers only to the vertical gravity loadings on roof and floor surfaces. These loads occur in combination with the dead loads, but are generally random in character and must be dealt with as potential contributors to various loading combinations, as discussed in Sec. 11.3.

Roof Loads. In addition to the dead loads they support, roofs are designed for a uniformly distributed live load that includes snow accumulation and the general loadings that occur during construction and maintenance of the roof. Snow loads are based on local snowfalls and are specified by local building codes.

Table 11.2 gives the minimum roof live-load requirements specified by the 1988 edition of the *Uniform Building Code*. Note the adjustments for roof slope and for the total area of roof surface supported by a structural element. The latter accounts for the increase in probability of the lack of total surface loading as the size of the surface area increases.

Roof surfaces must also be designed for wind pressure, for which the magnitude and manner of application are specified by local building codes based on local wind histories. For very light roof construction, a critical problem is sometimes that of the upward (suction) effect of the wind, which may exceed the dead load and result in a net upward lifting force.

Although the term *flat roof* is often used, there is generally no such thing; all roofs must be designed for some water drainage. The minimum required pitch is usually $\frac{1}{4}$ in./ft, or a slope of approximately 1 : 50. With roof surfaces that are this close to flat, a potential problem is that of *ponding,* a phenomenon in which the weight of water on the surface causes deflection of the supporting structure, which in turn allows for more water accumulation (in a pond), causing more deflection, and so on, resulting in an accelerated collapse condition.

Floor Loads. The live load on a floor represents the probable effects created by the occupancy. It includes the weights of human occupants, furniture, equipment, stored materials, and so on. All building codes provide minimum live loads to be used in the design of buildings for various occupancies. Since there is a

TABLE 11.2 Minimum Roof Live Loads

	Minimum Uniformly Distributed Load (lb/ft^2)			Minimum Uniformly Distributed Load (kN/m^2)		
	Tributary Loaded Area for Structural Member (ft^2)			Tributary Loaded Area for Structural Member (m^2)		
Roof Slope Conditions	0–200	201–600	Over 600	0–18.6	18.7–55.7	Over 55.7
1. Flat or rise less than 4 in./ft (1 : 3). Arch or dome with rise less than 1/8 span.	20	16	12	0.96	0.77	0.575
2. Rise 4 in./ft (1 : 3) to less than 12 in./ft (1 : 1). Arch or dome with rise 1/8 of span to less than 3/8 of span.	16	14	12	0.77	0.67	0.575
3. Rise 12 in./ft (1 : 1) or greater. Arch or dome with rise 3/8 of span or greater.	12	12	12	0.575	0.575	0.575
4. Awnings, except cloth covered.	5	5	5	0.24	0.24	0.24
5. Greenhouses, lath houses, and agricultural buildings.	10	10	10	0.48	0.48	0.48

Source: Adapted from the *Uniform Building Code*, 1988 ed. (Ref. 2), copyright © 1988, with the permission of the publishers, the International Conference of Building Officials.

lack of uniformity among different codes in specifying live loads, the local code should always be used. Table 11.3 contains values for floor live loads as given by the 1988 edition of the *Uniform Building Code*.

Although expressed as uniform loads, code-required values are usually established large enough to account for ordinary concentrations that occur. For offices, parking garages, and some other occupancies, codes often require the consideration of a specified concentrated load as well as the distributed loading. Where buildings are to contain heavy machinery, stored materials, or other contents of unusual weight, these must be provided for individually in the design of the structure.

When structural framing members support large areas, most codes allow some reduction in the total live load to be used for design. These reductions, in the case of roof loads, are incorporated into the data in Table 11.2. The following is the method given in the 1988 edition of the *Uniform Building Code* for determining the reduction permitted for beams, trusses, or columns that support large floor areas.

Except for floors in places of assembly (theaters, etc.), and except for live loads greater than 100 psf [4.79 kN/m^2], the design live load on a member may be reduced in accordance with the formula

$$R = 0.08\,(A - 150)$$

$$[R = 0.86\,(A - 14)]$$

The reduction shall not exceed 40% for horizontal members or for vertical members receiving load from one level only, 60% for other vertical members, nor R as determined by the formula

$$R = 23.1\left(1 + \frac{D}{L}\right)$$

In these formulas

TABLE 11.3 Minimum Floor Loads

Use or Occupancy		Uniform Load		Concentrated Load	
Description	Description	(psf)	(kN/m²)	(lb)	(kN)
Armories		150	7.2		
Assembly areas and auditoriums and balconies therewith	Fixed seating areas	50	2.4		
	Movable seating and other areas	100	4.8		
	Stages and enclosed platforms	125	6.0		
Cornices, marquees, and residential balconies		60	2.9		
Exit facilities		100	4.8		
Garages	General storage, repair	100	4.8	*	
	Private pleasure car	50	2.4	*	
Hospitals	Wards and rooms	40	1.9	1000	4.5
Libraries	Reading rooms	60	2.9	1000	4.5
	Stack rooms	125	6.0	1500	6.7
Manufacturing	Light	75	3.6	2000	9.0
	Heavy	125	6.0	3000	13.3
Offices		50	2.4	2000	9.0
Printing plants	Press rooms	150	7.2	2500	11.1
	Composing rooms	100	4.8	2000	9.0
Residential		40	1.9		
Rest rooms		**			
Reviewing stands, grandstands, and bleachers		100	4.8		
Roof decks (occupied)	Same as area served				
Schools	Classrooms	40	1.9	1000	4.5
Sidewalks and driveways	Public access	250	12.0	*	
Storage	Light	125	6.0		
	Heavy	250	12.0		
Stores	Retail	75	3.6	2000	9.0
	Wholesale	100	4.8	3000	13.3

Source: Adapted from the *Uniform Building Code*, 1988 ed. (Ref. 2), copyright © 1988, with the permission of the publishers, the International Conference of Building Officials.

* Wheel loads related to size of vehicles that have access to the area.

** Same as the area served or minimum of 50 psf.

R = reduction, in percent,

A = area of floor supported by a member,

D = unit dead load/sq ft of supported area,

L = unit live load/sq ft of supported area.

In office buildings and certain other building types, partitions may not be permanently fixed in location but may be erected or moved from one position to another in accordance with the requirements of the occupants. In order to provide for this flexibility, it is customary to require an allowance of 15–20 psf [0.72–0.96 kN/m^2], which is usually added to other dead loads.

11.5 LATERAL LOADS

As used in building design, the term *lateral load* is usually applied to the effects of wind and earthquakes, as they induce horizontal forces on stationary structures. From experience and research, design criteria and methods in this area are continuously refined, with recommended practices being presented through the various model building codes, such as the *Uniform Building Code* (UBC) (Ref. 2).

Space limitations do not permit a complete discussion of the topic of lateral loads and design for their resistance. The following discussion summarizes some of the criteria for design in the latest edition of the UBC. Examples of application of these criteria are given in the chapters that follow containing examples of building structural design. For a more extensive discussion the reader is referred to *Simplified Building Design for Wind and Earthquake Forces* (Ref. 6).

Wind

Where wind is a major local problem, local codes are usually more extensive with regard to design requirements for wind.

However, many codes still contain relatively simple criteria for wind design. One of the most up-to-date and complex standards for wind design is contained in the *American National Standard Minimum Design Loads for Buildings and Other Structures*, ANSI A58.1-1982, published by the American National Standards Institute in 1982.

Complete design for wind effects on buildings includes a large number of both architectural and structural concerns. The following is a discussion of some of the requirements for wind as taken from the 1988 edition of the UBC (Ref. 2), which is in general conformance with the material presented in the ANSI Standard just mentioned.

Basic Wind Speed. This is the maximum wind speed (or velocity) to be used for specific locations. It is based on recorded wind histories and adjusted for some statistical likelihood of occurrence. For the continental United States the wind speeds are taken from UBC, Fig. No. 4. As a reference point, the speeds are those recorded at the standard measuring position of 10 m (approximately 33 ft) above the ground surface.

Exposure. This refers to the conditions of the terrain surrounding the building site. The ANSI Standard describes four conditions (A, B, C, and D), although the UBC uses only two (B and C). Condition C refers to sites surrounded for a distance of one-half mile or more by flat, open terrain. Condition B has buildings, forests, or ground-surface irregularities 20 ft or more in height covering at least 20% of the area for a distance of 1 mile or more around the site.

Wind Stagnation Pressure (q_s). This is the basic reference equivalent static pressure based on the critical local wind speed. It is given in UBC Table No. 23-F and is based on the following formula as given in the ANSI Standard:

$$q_s = 0.00256V^2$$

Example: For a wind speed of 100 mph

$$q_s = 0.00256V^2 = 0.00256(100)^2$$
$$= 25.6 \text{ psf } [1.23 \text{ kPa}]$$

which is rounded off to 26 psf in the UBC table.

Design Wind Pressure. This is the equivalent static pressure to be applied normal to the exterior surfaces of the building and is determined from the formula

$$p = C_e C_q q_s I$$

(UBC Formula 11-1, Section 2311), in which

p = design wind pressure in psf,
C_e = combined height, exposure, and gust factor coefficient as given in UBC Table No. 23-G,
C_q = pressure coefficient for the structure or portion of structure under consideration as given in UBC Table No. 23-H,
q_s = wind stagnation pressure at 30 ft given in UBC Table No. 23-F,
I = importance factor.

The importance factor is 1.15 for facilities considered to be essential for public health and safety (such as hospitals and government buildings) and buildings with 300 or more occupants. For all other buildings the factor is 1.0.

The design wind pressure may be positive (inward) or negative (outward, suction) on any given surface. Both the sign and the value for the pressure are given in the UBC table. Individual building surfaces, or parts thereof, must be designed for these pressures.

Design Methods. Two methods are described in the Code for the application of the design wind pressures in the design of structures. For design of individual elements particular values are given in UBC Table 23-H for the C_q coefficient to be used in determining p. For the primary bracing system the C_q values and their use is to be as follows:

Method 1 (Normal Force Method). In this method wind pressures are assumed to act simultaneously normal to all exterior surfaces. This method is required to be used for gabled rigid frames and may be used for any structure.

Method 2 (Projected Area Method). In this method the total wind effect on the building is considered to be a combination of a single inward (positive) horizontal pressure acting on a vertical surface consisting of the projected building profile and an outward (negative, upward) pressure acting on the full projected area of the building in plan. This method may be used for any structure less than 200 ft in height, except for gabled rigid frames. This is the method generally employed by building codes in the past.

Uplift. Uplift may occur as a general effect, involving the entire roof or even the whole building. It may also occur as a local phenomenon such as that generated by the overturning moment on a single shear wall. In general, use of either design method will account for uplift concerns.

Overturning Moment. Most codes require that the ratio of the dead load resisting moment (called the restoring moment, stabilizing moment, etc.) to the overturning moment be 1.5 or greater. When this is not the case, uplift effects must be resisted by anchorage capable of developing the excess overturning moment. Overturning may be a critical problem for the whole building, as in the case of relatively tall and slender tower structures. For buildings braced by individual shear walls, trussed bents, and rigid-frame bents, overturning is investigated for the individual bracing units. Method 2 is usually used for this investigation, except for very tall buildings and gabled rigid frames.

Drift. Drift refers to the horizontal deflection of the structure due to lateral loads. Code criteria for drift are usually limited to requirements for the drift of a single story (horizontal movement of one level with respect to the next above or below). The UBC does not provide limits for wind drift. Other standards give various recommendations, a common one being a limit of story drift to 0.005 times the story height (which is the UBC limit for seismic drift). For masonry structures wind drift is sometimes limited to 0.0025 times the story height. As in other situations involving structural deformations, effects on the building construction must be considered; thus the detailing of curtain walls or interior partitions may affect limits on drift.

Combined Loads. Although wind effects are investigated as isolated phenomena, the actions of the structure must be considered simultaneously with other phenomena. The requirements for load combinations are given by most codes, although common sense will indicate the critical combinations in most cases. With the increasing use of load factors the combinations are further modified by applying different factors for the various types of loading, thus permitting individual control based on the reliability of data and investigation procedures and the relative significance to safety of the different load sources and effects. Required load combinations are described in Sec. 2303 of the UBC.

Special Problems. The general design criteria given in most codes are applicable to ordinary buildings. More thorough investigation is recommended (and sometimes required) for special circumstances such as the following:

Tall Buildings. These are critical with regard to their height dimension as well as the overall size and number of occupants inferred. Local wind speeds and unusual wind phenomena at upper elevations must be considered.

Flexible Structures. These may be affected in a variety of ways, including vibration or flutter as well as the simple magnitude of movements.

Unusual Shapes. Open structures, structures with large overhangs or other projections, and any building with a complex shape should be carefully studied for the special wind effects that may occur. Wind-tunnel testing may be advised or even required by some codes.

Use of code criteria for various ordinary buildings is illustrated in the design examples in the following chapter.

Earthquakes

During an earthquake a building is shaken up and down and back and forth. The back-and-forth (horizontal) movements are typically more violent and tend to produce major unstabilizing effects on buildings; thus structural design for earthquakes is mostly done in terms of considerations for horizontal (called lateral) forces. The lateral forces are actually generated by the weight of the building—or, more specifically, by the mass of the building that represents both an inertial resistance to movement and the source for kinetic energy once the building is actually in motion. In the simplified procedures of the equivalent static force method, the building structure is considered to be loaded by a set of horizontal forces consisting of some fraction of the building weight. An analogy would be to visualize the building as being rotated vertically 90° to form a cantilever beam, with the ground as the fixed end and with a load consisting of the building weight.

In general, design for the horizontal force effects of earthquakes is quite similar to design for the horizontal force effects of wind. Indeed, the same basic types of lateral bracing (shear walls, trussed bents, rigid frames, etc.) are used to resist both force effects. There are indeed some significant differences, but in the main a system of bracing that is developed for wind bracing will most likely serve reasonably well for earthquake resistance as well.

Because of its considerably more complex criteria and procedures, we have chosen not to illustrate the design for earthquake effects in the examples in this part. Nevertheless, the development of elements and systems for the lateral bracing of the buildings in the design examples here is quite applicable in general to situations where earthquakes are a predominant concern. For structural investigation, the principal difference is in the determination of the loads and their distribution in the building. Another major difference is in the true dynamic effects, critical wind force being usually represented by a single, major, one-direction punch from a gust, while earthquakes represent rapid back-and-forth, reversing-direction actions. However, once the dynamic effects are translated into equivalent static forces, design concerns for the bracing systems are very similar, involving considerations for shear, overturning, horizontal sliding, and so on.

For a detailed explanation of earthquake effects and illustrations of the investigation by the equivalent static force method the reader is referred to *Simplified Building Design for Wind and Earthquake Forces* (Ref. 6).

12

BUILDING STRUCTURES: DESIGN EXAMPLES

This part contains examples of the design of complete structural systems for buildings. The buildings selected for design are not intended as examples of good architectural design, but rather have been made to create a range of situations in order to be able to demonstrate the use of various structural components. Design of individual elements of the structural systems is largely based on the materials presented in the earlier chapters. The principal purpose here is to show the broader context of design work by dealing with whole structures and the building in general.

12.1 BUILDING ONE: GENERAL DESIGN CONSIDERATIONS

Figure 12.1 shows a one-story, box-shaped building intended for commercial occupancy. As the section indicates, the exterior walls are principally of reinforced masonry with concrete blocks (concrete masonry units, or CMUs). We will consider some alter-

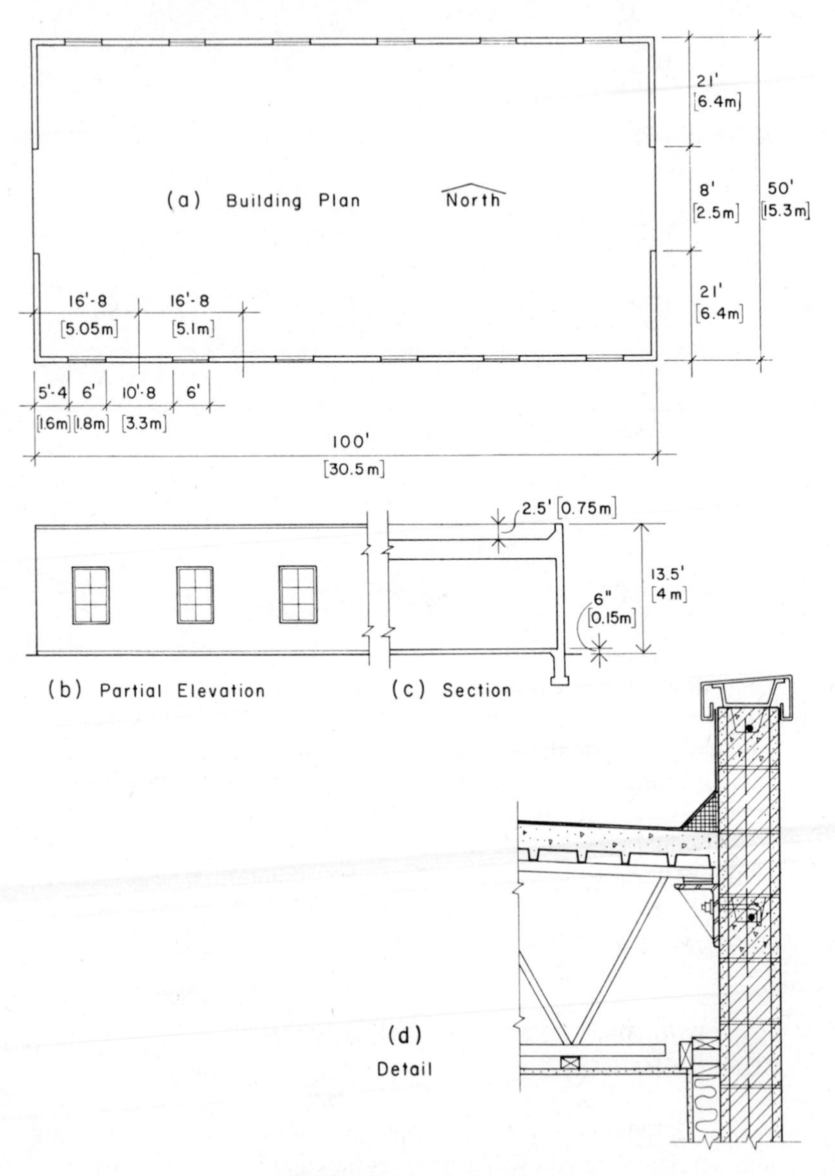

FIGURE 12.1 Building One.

natives for the design of the roof structure, using steel components.

We assume the following data for design:

$$\text{roof live load} = 20 \text{ psf (reducible)}$$

$$\text{assumed total roof dead load} = 15 \text{ psf (without structure)}$$

The section in Fig. 12.1*d* indicates that the wall continues above the roof level to form a parapet and that the clear span steel joists are supported at the wall face. The span of the joists is thus approximately 48 ft. The general building section in Fig. 12.1*c* indicates a flat roof. We will assume the roof deck to be placed directly on the top chords of the trusses and the ceiling to be directly attached to the bottom chords. For a minor roof slope of $\frac{1}{4}$ in. per ft, the top chords may be made to follow the profile of the slope, although we will assume a constant depth of the trusses for the design.

12.2 BUILDING ONE: ROOF STRUCTURE

Spacing of the open-web joists must be coordinated with the selection of the roof deck and the details for construction of the ceiling. For a trial design, we assume a joist spacing of 4 ft. From Table 5.5, with deck units typically achieving three spans or more, the lightest deck in the table (22 gage) may be used. Choice of the deck configuration (rib width) will depend on the type of insulation material on top of the deck and the means used for attaching the deck units to the joists.

Adding the weight of the deck to the roof dead load determined previously, we obtain approximately 17 psf dead load for the design of the joists. As illustrated in Sec. 5.12, we proceed to design for a K-series joist (Table 5.2) as follows:

$$\text{joist live load} = 4(20) = 80 \text{ lb/ft}$$

$$\text{total load} = 4(37) = 148 \text{ lb/ft} + \text{joist weight}$$

For the 48 ft span, we find the following alternative choices from Table 5.2:

$$24K9 \text{ at } 12.0 \text{ lb/ft, total load} = 148 + 12 = 160$$
$$< \text{table value of } 211 \text{ lb/ft}$$

$$28K6 \text{ at } 11.4 \text{ lb/ft, } 148 + 11.4 = 159.4 < \text{table value of } 184 \text{ lb/ft}$$

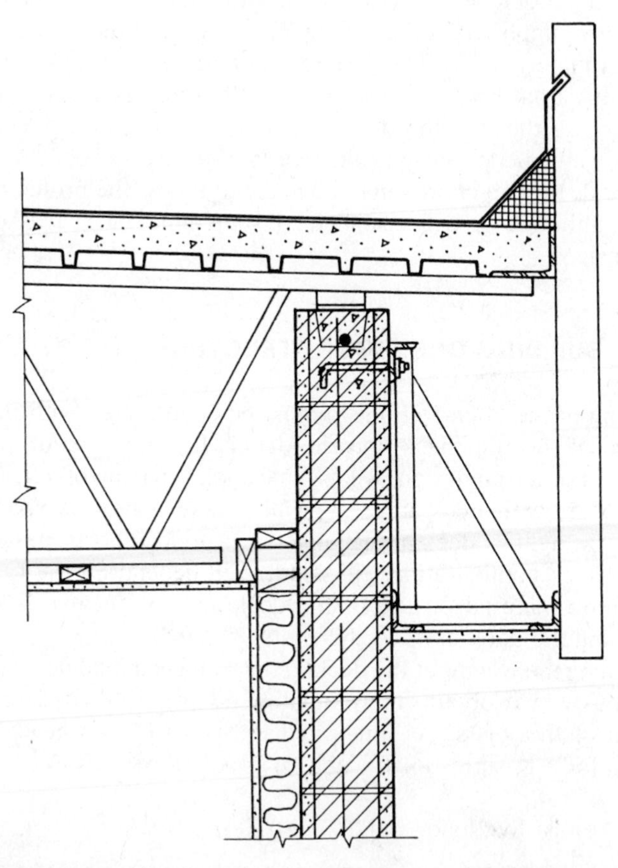

FIGURE 12.2 Building One: variation of the roof-to-wall detail. For comparison, see Fig. 12.1*d*.

While the 28K6 is the indicated choice for lightest weight of steel, there may be other compelling reasons for choosing a deeper joist. Creating a deeper space between the roof and ceiling to allow the incorporation of ducts, wiring, and piping may be one such reason. Deflection may also be a concern; the table allowable of 1/360 of the span means an allowable dimension of $(48 \times 12)/360 = 1.6$ in. for the joists. This may not be critical for the roof surface, but could present a problem for the ceiling sag on the interior. From Table 5.2, a 30K7 at 12.3 lb/ft indicates considerably greater load capacity to achieve the limiting deflection.

It should be noted that Table 5.2 is abridged from a larger table in the industry reference. There are therefore many more choices of joist sizes. Our example here is meant only to indicate the process for use of such references.

Specifications for open-web joists give requirements for end support details and bridging (see Ref. 9). If the 30K7 is used for the 48-ft span, for example, four rows of bridging are required. If evenly spaced, the bridging would be approximately 10 ft on center.

Although we will not design the masonry walls for this example, it should be noted that the support indicated for the joists in Fig. 12.1*d* would result in an eccentric compression on the wall. This will induce bending in the wall, which may be objectionable. An alternative detail for the roof-to-wall joint is shown in Fig. 12.2, in which the joist sits directly on the wall with the top chord projecting to form a short cantilever. This is a common detail, but the cantilever dimension and loading is quite limited.

12.3 BUILDING ONE: ALTERNATIVE ROOF STRUCTURE

If a clear spanning roof structure is not required for this building, it may be possible to use some interior columns and a framing system for the roof with quite modest spans. Figure 12.3*a* shows a framing plan for a system that uses columns at 16 ft 8 in. on center in each direction. While short joists may be used with this system, it would also be possible to use a longer span deck, as indicated on the plan. This span exceeds the span capability of

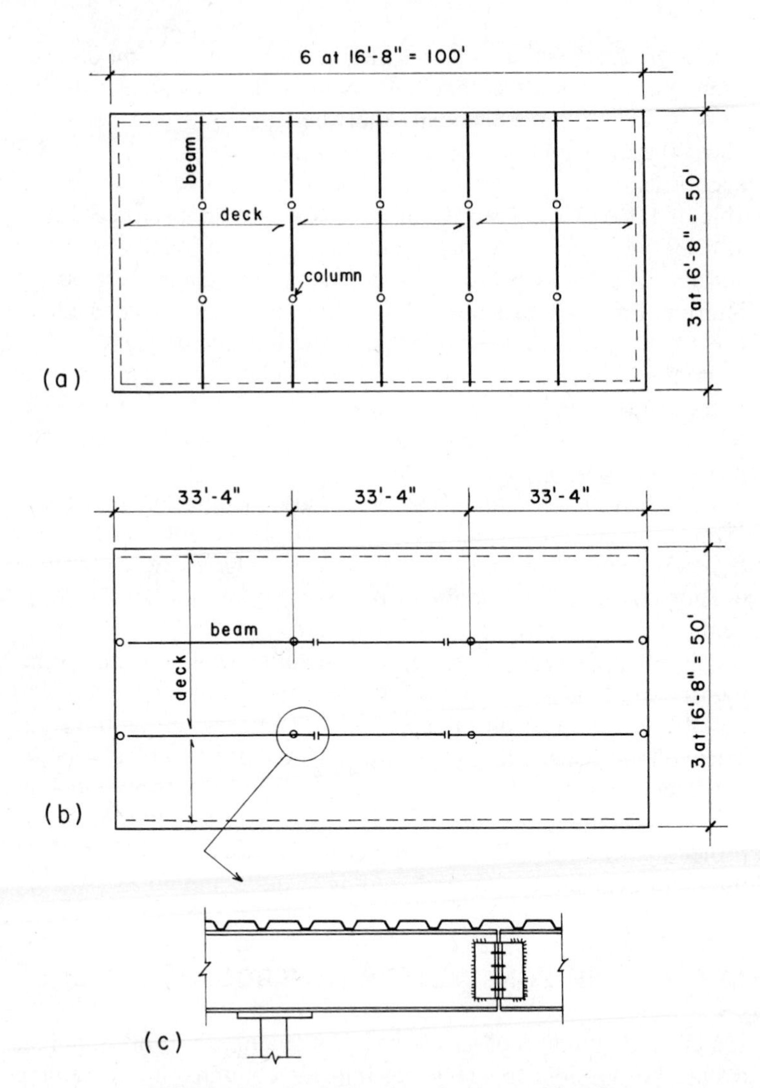

FIGURE 12.3 Considerations for the roof framing.

the 1.5-in.-high deck ribs shown for the decks in Table 5.5, but decks with deeper ribs are available, permitting spans of 30 ft or more.

A second possible framing arrangement is shown in Fig. 12.3*b*, in which the deck spans the other direction and only two rows of beams are used. This plan also allows for wider column spacing; while that increases the beam spans, a major cost savings is represented by the elimination of 60% of the interior columns and the column footings.

Beams in continuous rows can sometimes be made to simulate a continuous beam action without the need for moment connections. Use of beam joints off the columns, as shown in Fig. 12.3*c*, allows for simple framing but some advantages of the continuous beam, a principal one being reduction of deflections.

Let us assume the use of the framing scheme in Fig. 12.3*b* and consider the design of the supporting beams and columns. Based on use of a slightly heavier deck, we will assume the total roof and ceiling dead load to be 20 psf. A single beam span supports a load area of $33.3 \times 16.67 = 555 \text{ ft}^2$, which qualifies for a reduced roof live load of 16 psf (see Table 11.2). The uniformly distributed loading on the beam is thus

$$(16.67)(20 + 16) = 600 \text{ lb/ft} + \text{the beam weight, say } 640 \text{ lb/ft}$$

and for a simple span beam the maximum moment is

$$M = \frac{wL^2}{8} = \frac{(0.640)(33.3)^2}{8} = 88.9 \text{ kip-ft}$$

From Appendix Table B.1 the lightest weight W-shape beam is a W 16 × 31. Checking with the graph in Fig. 5.3, it may be observed that the total load deflection will be approximately L/240, which is not usually critical for roof structures. Furthermore, the live-load deflection will be less than one-half this amount, which is quite a modest value.

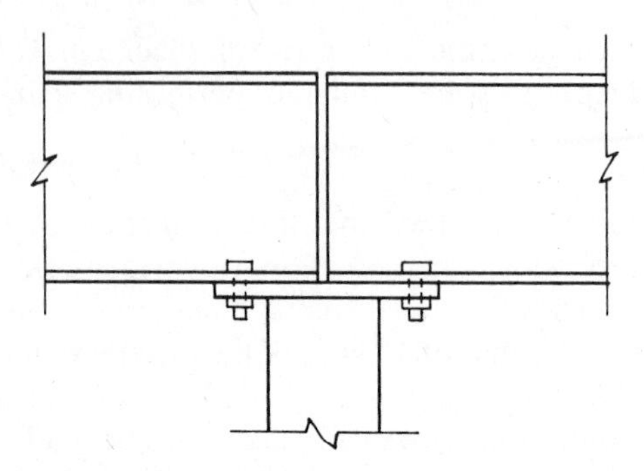

FIGURE 12.4 Framing detail at top of column; simple beam condition. For comparison, see detail in Fig. 12.3*c*.

If each three-span beam is constructed with three, simple beam segments, the detail at the top of the interior columns will be as shown in Fig. 12.4. Although this may be possible, the detail shown in Fig. 12.3, with the beam-to-beam connection off the column, is a better framing detail. The investigation shown in Fig. 12.5 is made with a beam splice joint made 4 ft from the column. The center portion of the three-span beam thus becomes a simple span of 25 ft 4 in. The end reactions for the center portion become loads on the ends of the extensions of the outer span portions, which are able to be analyzed by statics. The resulting reactions, shears, and moments are as shown in the illustration.

The maximum bending moment determined in Fig. 12.5 is 71.09 kip-ft, or approximately 80% of that for the simple span beam as computed previously. Reference to Appendix Table B.1 indicates the possibility of reducing the shape to a W 16 × 26 on the basis of the reduction in bending moment. The maximum bending moment could be reduced further by extending the beam splice an additional distance from the column. However, the W 16 × 26 is the lightest 16-in.-deep wide-flange beam, and change to a shallower depth is probably not advisable for deflection considerations.

The support reaction force of 22.458 kips at the interior of the beam in Fig. 12.5 is the total load for the column. Assuming a

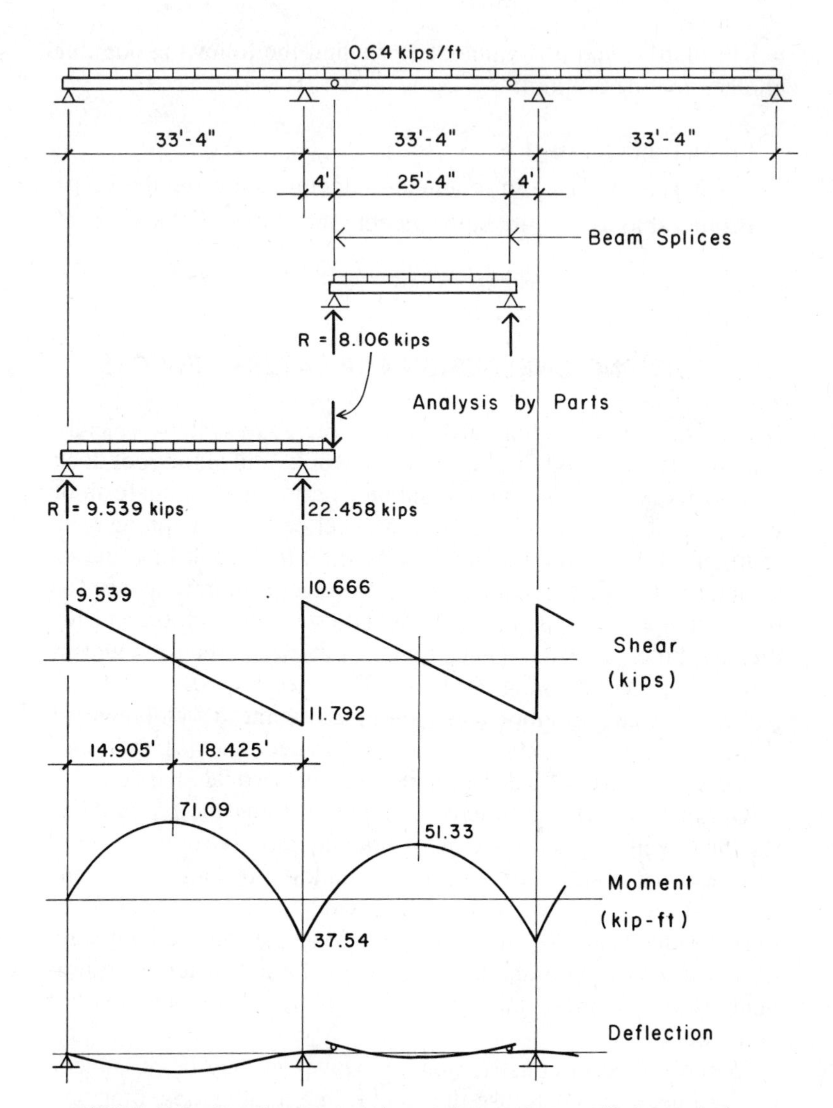

FIGURE 12.5 Development of the continuous beam with internal pins.

height of 10 ft and a K value of 1, we find the following possible choices for the column:

From Table 6.2 W 4 × 13

From Table 6.3, a 3-in.-nominal-diameter round standard pipe

From Table 6.4, a 3-in. square steel tube with wall thickness of $\frac{3}{16}$ in.

12.4 BUILDING ONE: DESIGN FOR LATERAL FORCES

For resistance of wind or earthquake forces, the system typically used for this size and form of building would utilize the roof deck as a horizontal diaphragm in combined action with some form of vertical bracing elements. Design of steel decks as diaphragms is routinely done with data supplied by the Steel Deck Institute or by individual deck manufacturers. Major considerations are for the code-approved, rated load capacities of deck units, and for the attachments between deck units and between the deck and its supporting steel framing. In addition, as for any structural diaphragm design, attention must be given to the development of details of the construction to achieve the necessary chords, collectors, drag struts, and the anchorages to vertical elements.

Options for vertical bracing elements include the possibilities for shear walls, rigid frames (with moment-resistive connections), or braced (trussed) frames. For this building the masonry walls would surely be used for lateral, as well as vertical, load resistance. Indeed, if another system were used, using the light steel frame, it would be a challenge to detail the construction to prevent the exceedingly stiff masonry walls from absorbing the lateral forces.

With other wall constructions, however, it may be possible—or even necessary—to use the steel frame to develop a moment-resistive or trussed system. Discussion of these options is given more extensively in the illustrations for development of the lateral load resistive systems for Building Three.

12.5 BUILDING TWO: GENERAL DESIGN CONSIDERATIONS

Figure 12.6 shows a partial framing plan for the roof structure of a one-story industrial building. It is assumed that the system as shown, with 48-ft square bays, is repeated a number times in each direction. We will assume a general roof construction with the live-load and dead-load criteria as used for Building One (Sec. 12.1).

The plan in Fig. 12.6 indicates a series of girders supported by columns, a series of joists perpendicular to the girders, and a roof deck supported directly by the joists. A critical planning decision for this system is the spacing of the joists, which affects the following:

1. *The Load on an Individual Joist.* Based on the span, there is some feasible range for the spacing, depending on the type of joist to be used.

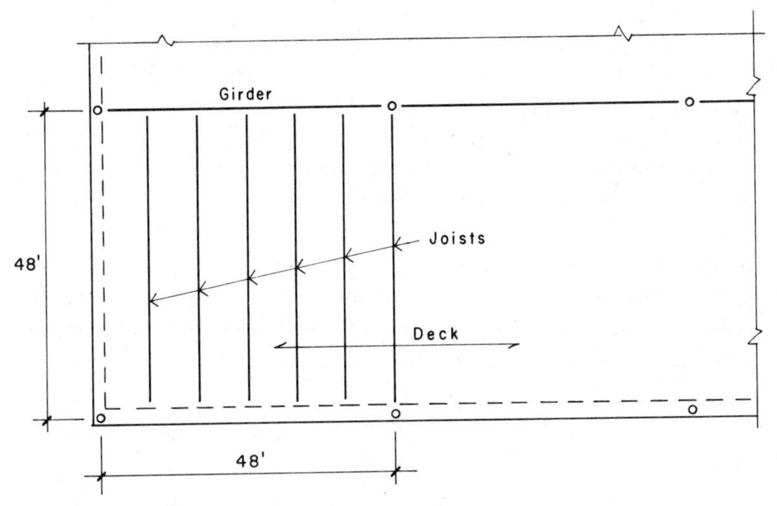

FIGURE 12.6 Building Two: partial roof framing plan.

2. *The Deck Required.* If the joist spacing is considerable, a very deep or heavy deck may be required.
3. *The Load Pattern and Magnitude of Individual Loads on the Girders.* This is of special concern if the girder is actually a truss, for which the joist support points must coincide with truss panel points.

A very wide range of possible combinations exists for the choice of the deck, joists, girders, and columns for this structure. Local building codes, regional market considerations, and availability of products all tend to make for some local preferences. We will illustrate two possible solutions here, but many others are also possible—and possibly much preferred in special circumstances.

12.6 BUILDING TWO: ALTERNATIVE ROOF SYSTEM ONE

For this system we assume joists at 4 ft on center supported by W-shape girders. For the span and spacing the choices for the deck and joists will be approximately the same as for Building One. Let us then consider the design of the 48-ft-span girders.

Because of the large area supported by a single girder, the roof live load may be reduced to 12 psf. We will assume a dead load of 20 psf, including the weights of the deck, joists, bridging, and joist connection elements. The total design load for the girders is thus 32 psf, and the uniformly distributed load on a girder is (32)(48) = 1540 lb/ft, or 1.54 kips/ft. The joist loads actually constitute point loads on the girder, but with the 4-ft spacing on the 48-ft span girder, this is very much like a uniformly distributed loading. Adding some weight for the girder, we will design for a load of 1.6 kips/ft. For a simple beam, the maximum moment is

$$M = \frac{wL^2}{8} = \frac{(1.6)(48)^2}{8} = 518.4 \text{ kip-ft}$$

If the simple beam is used, Appendix B indicates the least weight shape to be a W 30 × 99.

It is possible here to use the girder splice joint off of the column, as was done in Sec. 12.3. If a similar reduction in the maximum bending moment of 20% can be achieved, the design moment here drops to (0.80)(518.4) = 415 kip-ft, and the lightest shape from Appendix Table B.1 is a W 27 × 84.

For the column, we determine a total load as follows:

$$(34 \text{ psf})(48 \times 48) = 78{,}336 \text{ lb or approximately } 78.3 \text{ kips}$$

Note that the additional 2 psf of dead load accounts for the weight of the girders. Clear height may be somewhat greater for this type of building; we will assume a height of 20 ft and a K value of 1. Thus some choices for the interior column are:

From Table 6.2, W 8 × 31

From Table 6.3, 8-in.-nominal-diameter standard pipe

From Table 6.4, no option (*Note:* Larger sizes are available, up to 16 in. square; the AISC Manual yields a 6 × 6 tube with $\frac{5}{16}$-in. wall thickness for this example.)

12.7 BUILDING TWO: ALTERNATIVE ROOF SYSTEM TWO

A variation on the system developed in the preceding section is one that uses trusses for the girders. One option for this is to use the joist girder system in combination with the open web joists, as described in Sec. 5.12. As usual, the joist spacing must be worked out with concerns for the deck and girders. In this case the spacing of the joists determines the layout of the joist girders, as shown in Fig. 5.11. The girder depth and loading interval must be maintained in some limited ratio at each other, so as not to create the undesirable extremes shown in Fig. 12.7.

A general rule of thumb (actually mentioned in a design example in Ref. 9) is that the girder depth in inches should approximate the girder span in feet. Thus, for our case, a depth of 48 in. would be appropriate. Tables in Ref. 9 indicate a preferred range for the number of joist spacings in the girder span of 48 ft as between 7

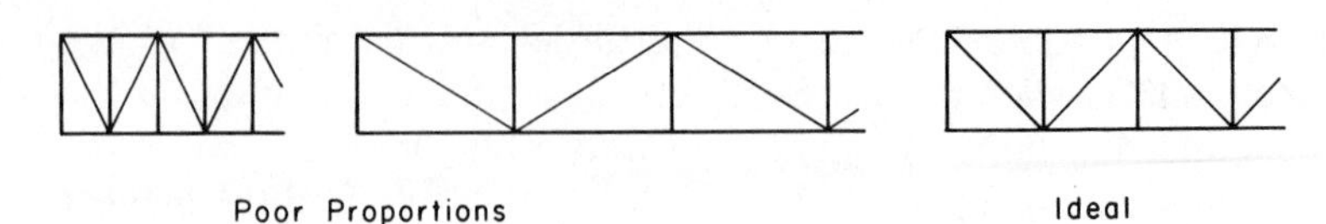

FIGURE 12.7 Considerations for the girder form.

and 9. If eight spacings are used, the joist spacing is 48/8 = 6 ft. The joist load, with a live load of 16 psf and dead load of 20 psf, is

$$\text{live load} = (6)(16) = 96 \text{ lb/ft}$$

$$\text{total load} = (6)(36) = 216 \text{ lb/ft (plus joist weight/ft)}$$

For this loading, Table 5.2 yields the lightest joist as a 30K8 at 13.2 lb/ft.

For the joist girders, the live load may be reduced to 12 psf (see Table 11.2). The panel point load on the girder thus becomes

$$[(6)(12 + 20) + 13.2](48) = 9850 \text{ lb or } 9.85 \text{ kips}$$

As described in Sec. 5.12, the designation for the joist girder is thus a 48G8N9.85K, indicating a 48-in.-deep girder with eight spaces for supported joists and a joist load of 9.85 kips (see Fig. 5.11).

Options for the columns for this structure would generally be the same as those determined in the preceding section.

12.8 BUILDING TWO: DESIGN FOR LATERAL FORCES

In general, the options for lateral bracing for wind or earthquake forces are the same as those discussed for Building One in Sec. 12.4. Because of the larger size, and possibly more complex plan, for such a building—compared to the relatively small, simple rectangular plan building—there may be some additional concerns or different conditions. Some of these considerations are as follows:

1. The size of the building in plan may require that some interior bracing be used to define multiple units for the roof diaphragm, whereas the size and form of the plan in Building One would permit the use of a single roof diaphragm with vertical bracing only in the exterior walls (called a *perimeter bracing system*).
2. If the plan is other than a simple, single rectangle, discontinuities in the plan may require the use of structural separation joints for development of the lateral bracing as well as control of thermal movements.
3. Use of a very light steel deck—as may be economical with closely spaced joists—may result in low diaphragm strength in proportion to the building size. This may be less of a problem if vertical bracing defines relatively small units of the diaphragm. Otherwise, it may be necessary to develop a horizontal trussing system in the roof plane to replace the usual roof diaphragm.

For this case it may be possible to develop trussed bents on the column lines as the vertical bracing elements. This would, however, require the design of special bents instead of the use of the usual manufactured joist girders.

More extensive discussion of lateral bracing in general is given for Building Three.

12.9 BUILDING THREE: GENERAL DESIGN CONSIDERATIONS

Building Three is a three-story office building. We will design a structure for this building using an all-steel frame of W shape columns, girders, and beams, with roof and floor decks of formed sheet steel. Figure 12.8 presents a plan of the upper floor and a full building profile section. We assume that a fundamental requirement for the building is the provision of a significant amount of exterior window surface and the avoidance of long expanses of unbroken solid wall surface. Another assumption is that the build-

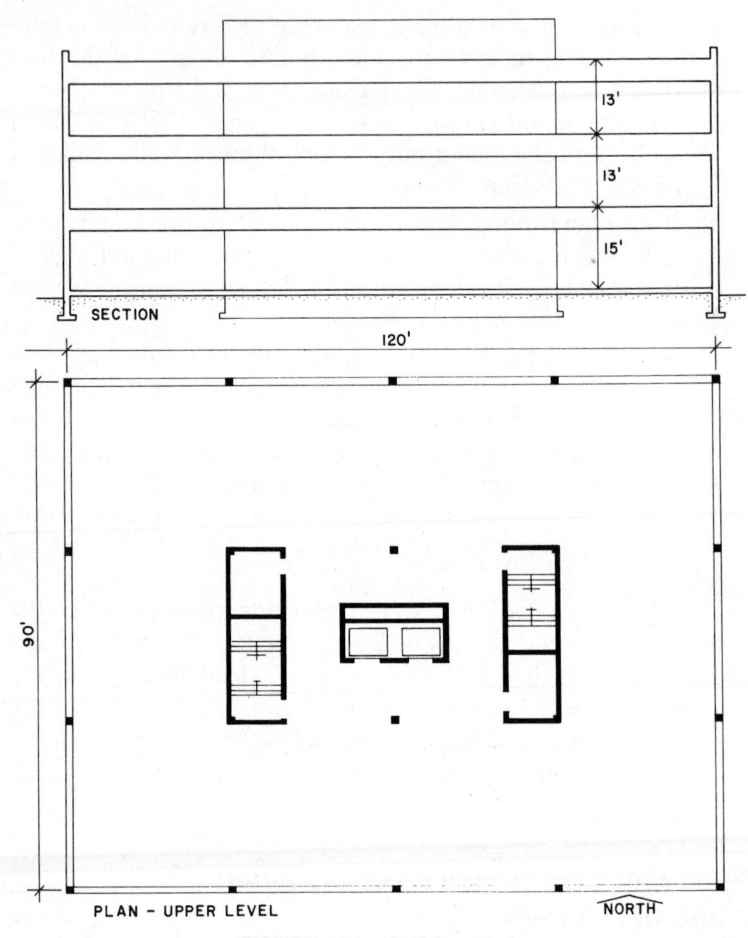

FIGURE 12.8 Building Three.

ing is freestanding on the site, and all sides having a clear view.

The following will be assumed as criteria for the design work:

Building code: 1988 edition of Uniform Building Code (Ref. 2).

Live Loads:

Roof: Table 11.2 (UBC Table 23-C).

Floors: Table 11.3 (UBC Table 23-A).

Office areas: 50 psf [2.39 kPa].
Corridor and lobby: 100 psf [4.79 kPa].
Partitions: 20 psf (UBC minimum per Sec. 2304) [0.96 kPa].

Wind: map speed, 80 mph [129 km/h]; exposure B.

Assumed construction loads:
Floor finish: 5 psf [0.24 kPa].
Ceilings, lights, ducts: 15 psf [0.72 kPa].
Walls (average surface weight):
Interior, permanent: 10 psf [0.48 kPa].
Exterior curtain wall: 15 psf [0.72 kPa].

12.10 BUILDING THREE: STRUCTURAL ALTERNATIVES

The plan as shown, with 30-ft square bays and a general open interior, is an ideal arrangement for a beam and column system in either steel or reinforced concrete. Other types of systems may be made more effective if some modifications of the basic plans are made. These changes may affect the planning of the building core, the plan dimensions for the column locations, the articulation of the exterior wall, or the vertical distances between the levels of the building.

The general form and basic type of the structural system must relate to both the gravity and lateral force problems. Considerations for gravity require the development of the horizontal spanning systems for the roof and floors and the arrangement of the vertical elements (walls and columns) that provide support for the spanning structure. Vertical elements should be stacked, thus requiring coordinating the plans of the various levels.

The most common choices for the lateral bracing system would be the following (see Fig. 12.9):

1. *Core Shear Wall System* (Fig. 12.9*a*). This consists of using solid walls to produce a very rigid central core. The rest of the structure leans on this rigid interior portion, and the roof and floor constructions outside the core, as well as the

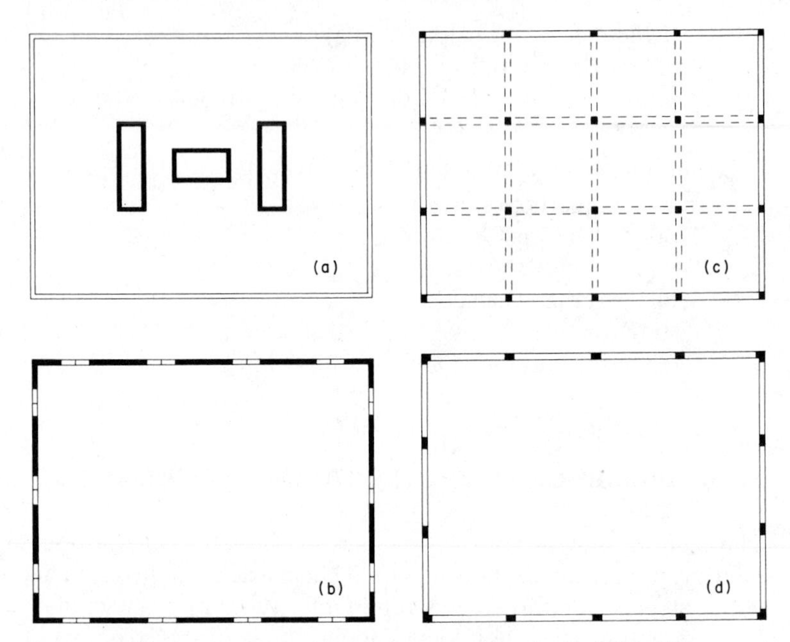

FIGURE 12.9 Options for the lateral bracing for Building Three.

exterior walls, are free for concerns for lateral forces as far as the structure as a whole is concerned.

2. *Truss-Braced Core*. This is similar in nature to the shear-wall-braced core, and the planning considerations would be essentially similar. The solid walls would be replaced by bays of trussed framing (in vertical bents) using various possible patterns for the truss elements.
3. *Peripheral Shear Walls* (Fig. 12.9*b*). This in essence makes the building into a tubelike structure. Because doors and windows must pierce the exterior, the peripheral shear walls usually consist of linked sets of individual walls (sometimes called piers).
4. *Mixed Exterior and Interior Shear Walls*. This is essentially a combination of the core and peripheral systems.
5. *Full Rigid-Frame System* (Fig. 12.9*c*). This is produced by using the vertical planes of columns and beams in each

direction as a series of rigid bents. For this building there would thus be four bents for bracing in one direction and five for bracing in the other direction. This requires that the beam-to-column connections be moment resistive.

6. *Peripheral Rigid-Frame System* (Fig. 12.9*d*). This consists of using only the columns and beams in the exterior walls, resulting in only two bracing bents in each direction.

In the right circumstances any of these systems may be acceptable. Each has advantages and disadvantages from both structural design and architectural planning points of view. The core-braced schemes were popular in the past, especially for buildings in which wind was the major concern. The core system allows for the greatest freedom in planning the exterior walls, which are obviously of major concern to the architect. The peripheral system, however, produces the most torsionally stiff building—an advantage for seismic resistance.

The rigid-frame schemes permit the free planning of the interior and the greatest openness in the wall planes. The integrity of the bents must be maintained, however, which restricts column locations and planning of stairs, elevators, and duct shafts so as not to interrupt any of the column-line beams. If designed for lateral forces, columns are likely to be large, and thus offer more intrusion in the building plan.

12.11 BUILDING THREE: DESIGN OF THE FLOOR STRUCTURE

There are a number of options for this system, involving the plan layout and choices for the structural deck and its supporting framing. The spans, the plan layout, the fire code requirements, the anticipated surfacing of floors and ceilings, and the need for incorporating elements for wiring, piping, heat and cooling, ventilation, fire sprinklers, and lighting are all influences on the choices of both materials and details of the construction. For this example we assume that the various considerations can be met with a system consisting of a formed steel deck with concrete fill, a series of closely spaced beams, and girders on the column lines.

A partial framing system for the floor is shown in Fig. 12.10. The general pattern of the evenly spaced beams that support the deck is interrupted to frame the holes for stairs, elevators, and vertical ducts, but the typical spans for the deck are maintained as much as possible.

The Deck

The floor deck is used both for spanning between beams to resist gravity loads and as a horizontal diaphragm for distribution of

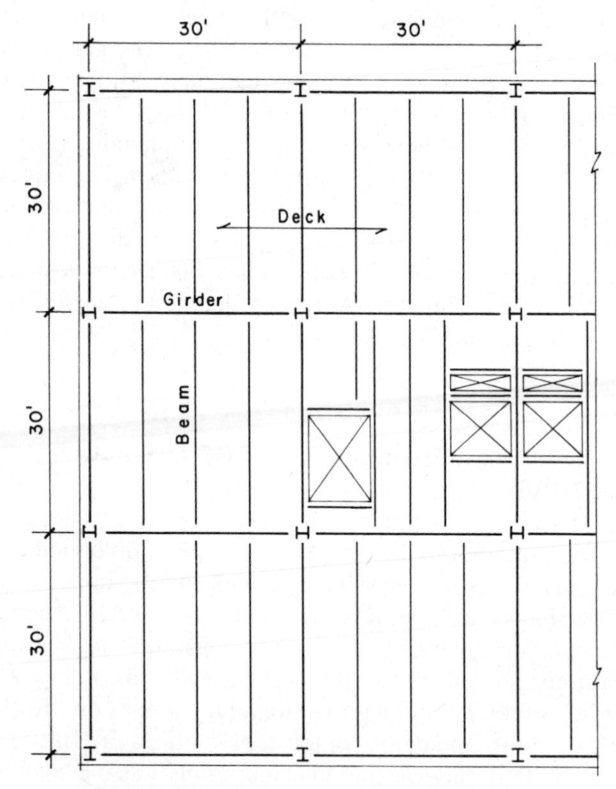

FIGURE 12.10 Floor framing for Building Three.

wind or seismic forces to the vertical bracing system. Choice of the deck materials, attachment of the deck to its supports, and various detailing of the deck and framing systems must respond to both of these functions. It is also necessary to anticipate the type of floor surfacing and the materials and required support system for ceilings.

We assume that all requirements can be met in this case by use of a formed steel deck with 1.5-in.-high ribs and a 3-in. structural-grade concrete fill on top of the deck, providing a total deck thickness of 4.5 in. If a composite-type deck (see Sec. 8.3) is used, the 7.5-ft span should be easily achieved with such a deck.

The type of deck used may be the WR units shown in Table 5.5, although the deck surface would be modified to develop the necessary bonding with the concrete fill to achieve the composite action, as discussed in Sec. 8.3. This deck system may also incorporate some of the wiring elements for both power and communication systems. Some of the deck folds may be closed with a bottom flat sheet of steel to form conduits in one direction that integrate with some pattern of perpendicular conduits incorporated in the concrete fill to provide an accessible, two-way system for wiring.

Attachment of the deck to beams may be by welding. The relatively thin deck may be too thin for this, but thicker elements are placed in the bottom of some ribs and welded to the beams by burning through the deck. If a composite concrete slab and steel beam system is used, attachment is essentially achieved by the shear developers (see Sec. 8.3).

Although the Steel Deck Institute provides useful information for the industry as a whole, specific information about decking products should be obtained directly from the product manufacturers.

The Beams

The beams are simply supported, single span, uniformly loaded elements. With the span established and the loading determined, they can be selected as illustrated in Sec. 5.2, with primary structural concerns generally limited to bending and deflection. For the loading we assume the following:

Live Load:

Assume a design load of 100 psf to allow for any location of a corridor; reduce as follows (see the discussion in Sec. 11.4).

$$R = 0.08(A - 150) = 0.08[(7.5 \times 30) - 150] = 6\%$$

Then, for the beam, LL = 7.5(0.94)(100) = 705 lb/ft.

Dead Load:

Carpet + pad at 5 psf
Ceiling, lights, ducts, etc. at 15 psf
Deck (assumed) at 35 psf
Total unit dead load = 55 psf
For the beam, DL = 7.5(55) = 412.5 lb/ft (applied)
Add for estimate of beam weight: 30 lb/ft

Total beam load = 705 + 412.5 + 30 = 1147.5 lb/ft, say 1.15 kips/ft

Or, for use of Appendix Table D.1, W = 30(1.15) = 34.5 kips. From Appendix Table B.1: W 16 × 40 is lightest, table load = 34.5 kips.

Since the table value is exactly that computed as required, and the beam weight actually slightly exceeds the estimate, the assumptions made for dead loads should be carefully reviewed. If they are determined to be conservative, the choice is probably all right. In this situation, there is a deeper beam (W 18 × 40) with the same weight; which may be a better choice if its depth does not present a problem.

For deflection, the usual criteria is for a maximum total load deflection of less than $L/240$ and a maximum live-load deflection of less than $L/360$. Inspection of Fig. 5.3 should assure that the 16-in. beam will satisfy these criteria; and, of course, the 18-in. beam will have less deflection.

The Interior Girder

The beam reactions deliver concentrated loads from each side to the interior girder, forming a total load equal to the full load on a single beam. However, the girder carries more load area, which allows for more reduction of the live load. We will therefore recompute the beam live load for the girder. Each girder supports three beams for a total load area of (3)(7.5)(30) = 675 ft². Using the 100-psf live load, the design load for the girder is determined as follows:

$$R = 0.08(A - 150) = 0.08(675 - 150) = 42\%$$

But the maximum reduction is 40%, or

$$R = 23.1\left(1 + \frac{D}{L}\right) = 23.1\left(1 + \frac{100}{55}\right) = 65.1\%$$

Thus we use a 40% reduction, or a unit live load of 60 psf. For the beam load,

$$W = (7.5)(30)(60 + 55) = 28{,}875 \text{ lb, say 26 kips}$$

The girder also carries its own dead weight as a uniformly distributed loading, plus part of the floor loading that is directly on top of the beam as a uniformly distributed loading. As we have included all of the floor load in the concentrated load, we will ignore the girder weight for design.

From Appendix Fig. E.6, the maximum moment for this load arrangement is $M = PL/2$. For the girder, we thus find

$$M = \frac{PL}{2} = \frac{(26)(30)}{2} = 390 \text{ kip-ft}$$

From Appendix Table B.1, the lightest section is a W 24 × 84.

It should be noted in Appendix Table B.1 that the L_c value for this shape is 9.5 ft. As this is greater than the beam spacing in this case, lateral support is not a problem. It should also be noted that there is a 27-in.-deep member with the same weight, and if its depth is not a problem, the increased strength and lower deflection may be advantageous gains with no additional cost.

There is a considerable shear force on the girders, but the stress in the web will be found to be less than allowable in this case. Consideration should be given to any substantially reduced web cross section at the end connection, however.

There are several other beams to be designed for the full floor framing system, including the spandrel beams and girders and the special beams at floor openings. Most of these directly support walls and should be designed for the specific dead load of the construction with very conservative consideration for deflections that may cause undesirable distortions of the wall construction.

12.12 BUILDING THREE: DESIGN OF THE COLUMNS

There are several different cases for the columns, due to the framing arrangements and column locations. For a complete design it would be necessary to tabulate the loading for each different case, although simplification of the framing would probably result in some grouping of actual column sizes to produce the fewest different sections. The five likely typical columns are the following:

The typical interior column.
The typical exterior corner column.
The intermediate column on the north and south sides.
The intermediate column on the east and west sides.
The four special columns at the building core, to be used to develop the lateral bracing system.

For an example design we will illustrate the computations for the intermediate exterior columns on the east and west sides. For gravity loading, Table 12.1 summarizes the design data and design load computations for this column. With the design loads for

TABLE 12.1 Column Load and Design[a]

Level	Story	Unbraced Height (ft)	Loads (kips)				Section Choices[b]	
			Source	DL	LL	Total	8 in.	10 in.
Roof								
			Roof	25	9			
			Wall	7				
	Third	13	Column	2			W 8 × 24	W 10 × 33
			Total/story	34	9			
			LL reduction		0			
			Design/story	34	9	43		
Third floor								
			Floor	25	32			
			Wall	6				
	Second	13	Column	2			W 8 × 24	W 10 × 33
			Total/story	67	41			
			LL reduction		60%			
			Design/story	67	16	83		
Second floor								
			Floor	25	32			
			Wall	7				
	First	15	Column	2			W 8 × 31	W 10 × 33
			Total/story	101	73			
			LL reduction		60%			
			Design/story	101	29	130		
First floor								

[a] Assumptions: $K = 1$, roof and floor DL = 55 psf, wall DL = 15 psf, roof LL = 20 psf, floor LL = 70 psf (50 LL + 20 partition); LL reduction by UBC formulas.
[b] Lightest sections from Table 6.2.

the three-story column and the unbraced heights for each story, choices for the column sections can be found from Table 6.2. Although the loads are quite modest, it is likely that the selection would be made from available 10-in.-nominal-depth wide-flange sections, as this provides for better access for beam framing connections. At this size, it is conceivable that the column might be made of a single piece—from footing to roof. This would make it redundantly strong in the upper stories, but the cost of splicing may likely be greater than cost of the extra weight of steel for the upper columns. We show the possibility for three column sizes in Table 12.1, but it is unlikely that more than a single size change—if any—would be made.

For the load tabulations in Table 12.1 we have assumed a dead weight for the roof construction approximately equal to that for the floor construction.

12.13 BUILDING THREE: DESIGN FOR WIND

Referring to Fig. 12.10, it may be noted that there are some extra columns in the framing plan at the location of the stair, rest room, and elevator walls. These columns will be used in conjunction with the regular columns and some of the horizontal framing to define vertical planes of framing for the development of the truss-bracing system shown in Fig. 12.11. These frames will be braced for resistance to lateral loads by the addition of diagonal X-braces. For a simplified design, we will consider the diagonal

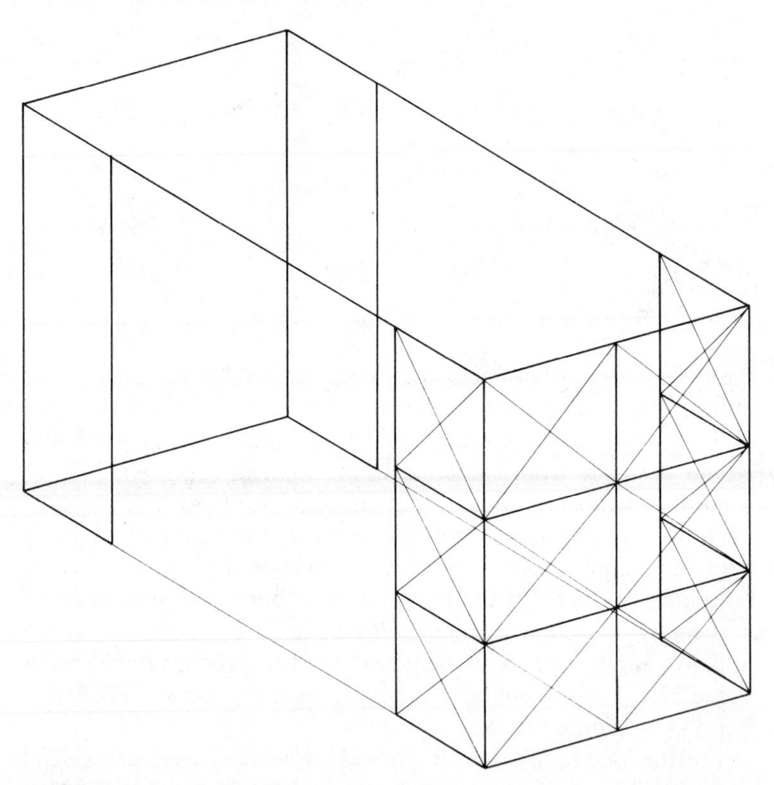

FIGURE 12.11 Development of the core-bracing system for lateral load resistance for Building Three.

members to function only in tension, making the vertical frames consist of statically determinate trusses. There are thus four vertically cantilevered trusses in each direction, placed symmetrically at the building core.

With the symmetrical building exterior form and the symmetrically placed core bracing, this is a reasonable system to use in conjunction with the horizontal roof and floor structures to develop resistance to horizontal forces due to wind or seismic actions. We will illustrate the design of the trussed bents for wind.

The 1988 edition of the *Uniform Building Code* (Ref. 2) provides for the use of the projected profile method for wind using a pressure on vertical surfaces defined as

$$p = C_e C_q q_s I$$

where C_e is a combined factor including concerns for the height above grade, exposure conditions, and gust effects. From UBC Table 23-G, assuming exposure condition B,

$$\begin{aligned} C_e &= 0.7 \text{ for surfaces from 0 to 20 ft above grade} \\ &= 0.8 \text{ from 20 to 40 ft above grade} \\ &= 1.0 \text{ from 40 to 60 ft above grade} \end{aligned}$$

and C_q is the pressure coefficient, which the UBC defines as follows:

$$\begin{aligned} C_q &= 1.3 \text{ for surfaces up to 40 ft above grade} \\ &= 1.4 \text{ from 40 ft up} \end{aligned}$$

The symbol q_s stands for the wind stagnation pressure as related to wind speed and measured at the standard height above ground of 10 m (approximately 30 ft). For the wind speed of 80 mph assumed earlier, the UBC yields a value for q_s of 17 psf.

Table 12.2 summarizes the forgoing data for the determination of the wind pressures at the various height zones on Building

TABLE 12.2 Design Wind Pressures for Building Three

Height Above Average Level of Adjoining Ground (ft)	C_e	C_q	Pressure[a], p (psf)
0–20	0.7	1.3	15.47
20–40	0.8	1.3	17.68
40–60	1.0	1.4	23.80

[a] Horizontally directed pressure on vertical projected area: $p = C_e \times C_q \times 17$ psf.

Three. For investigation of the wind effects on the lateral bracing system, the wind pressures on the exterior wall are translated into edge loadings for the horizontal roof and floor diaphragms, as shown in Fig. 12.12. Note that we have rounded off the wind pressures from Table 12.2 for use in Fig. 12.12.

The accumulated forces noted as H_1, H_2, and H_3 in Fig. 12.12 are shown applied to one of the vertical trussed bents in Fig. 12.13*a*. For the east–west bents, the loads will be as shown in Fig. 12.13*b*. These loads are determined by multiplying the edge

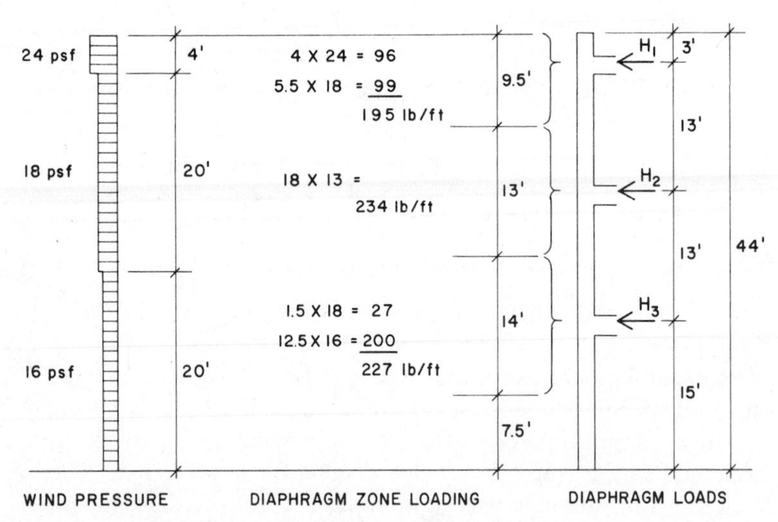

FIGURE 12.12 Wind loading for Building Three.

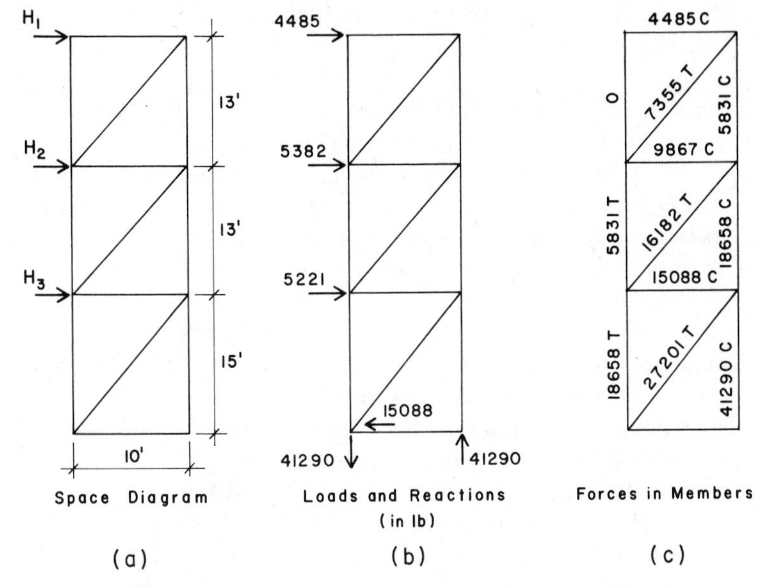

FIGURE 12.13 Investigation of the east–west trussed bents.

loadings for the diaphragms as shown in Fig. 12.12 by the 92-ft overall width of the building on the east and west sides. For a single bent, this total force is divided by 4. Thus for H_1 the bent load is determined as

$$H_1 = 195 \times 92 \div 4 = 4485 \text{ lb}$$

Analyzed as a truss, ignoring the compression diagonals, the resulting internal forces in the bent are as shown in Fig. 12.13*c*. The forces in the diagonals may be used to design tension members, using the usual one-third increase in allowable stress. The forces in the vertical columns may be added to the gravity loads and checked for possible critical conditions for the columns previously designed for gravity load only. The anchorage force in tension (uplift) may require tension-resistive anchor bolts and some special considerations for the foundations.

12.14 BUILDING THREE: ALTERNATIVE LATERAL BRACING

The horizontal decks of the roof and upper floors would certainly be utilized in the development of any option for the general resistance of lateral forces on this building. The vertical elements of such a system, however, may include several alternatives. Shear walls may be utilized if architectural planning permits. The height of the building requires that shear walls have some reasonable length in plan to avoid serious overturning problems. Reinforced concrete, masonry, or heavily nailed plywood would all be possible choices for shear wall materials. If the relatively complete steel framing system is retained, a popular system for high-risk seismic zones is the *dual bracing system,* consisting of interactive shear walls with a rigid frame or braced frame.

Various forms of eccentric bracing (Sec. 7.4) may be used with core bracing as developed in Sec. 12.13, or with a *perimeter bracing system* located in exterior walls. As discussed in Chapter 7, the eccentric bracing offers some of the advantages of both ordinary, triangulated bracing and rigid-frame bracing; having the relative overall lateral stiffness of trussed bracing and the enhanced energy absorption of rigid-frame bracing.

For architectural planning, one of the most popular vertical bracing systems is the rigid-frame system. This system makes the least intrusion in the architectural plan and leaves the most open space in vertical planes. For this, and possibly other reasons, it is highly favored when the general structure is of steel or reinforced concrete.

The general nature of rigid frames is discussed in Secs. 7.1 and 7.2. A critical concern for multistory, multiple-bay frames is the lateral strength and stiffness of the columns. As frames must be developed to resist lateral loads in all directions, it becomes necessary in many cases to consider the shear and bending resistance of the columns in two directions (north–south and east–west, for example). This presents a problem for W-shape columns, as they have considerably greater resistance on their major (x–x) axis versus their minor (y–y) axis. Orientation of W-shape columns thus sometimes becomes a major consideration in structural planning.

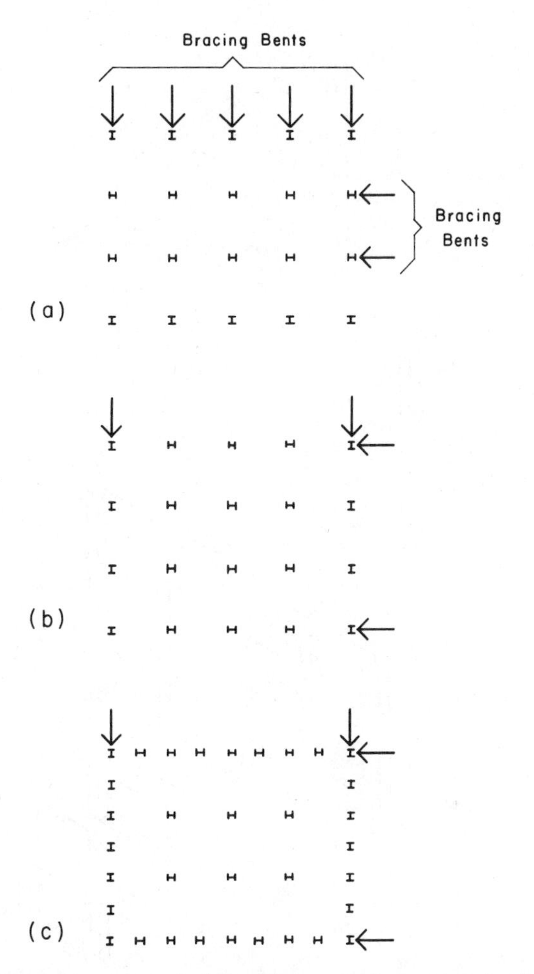

FIGURE 12.14 Arrangement of the W-shape columns for development of rigid frame bents.

Figure 12.14*a* shows a possible plan arrangement for column orientation for Building Three, relating to the development of two major bracing bents in the east–west direction and five shorter and less stiff bents in the north–south direction. The two stiff bents may well be approximately equal in resistance to the five shorter bents, giving the building a reasonably symmetrical response in the two directions.

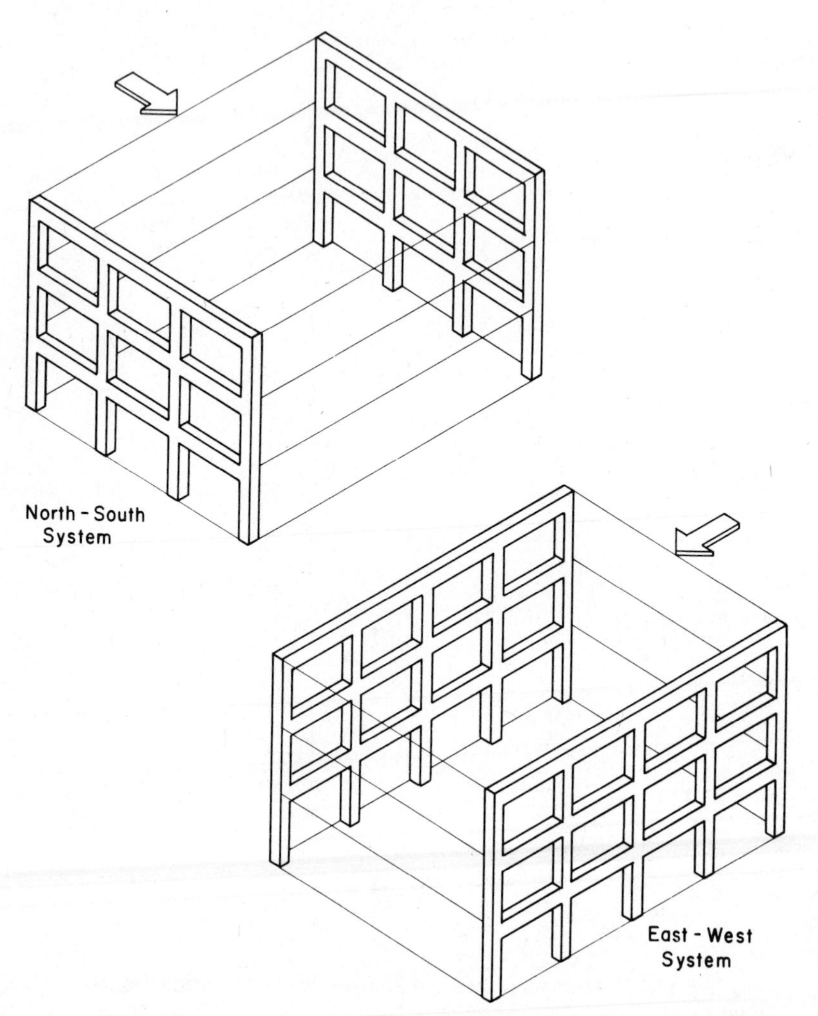

FIGURE 12.15 The perimeter bent system (see Fig. 12.14*c*).

Figure 12.14*b* shows a plan arrangement for columns designed to produce approximately symmetrical bents on the building perimeter. The form of such perimeter bracing is shown in Fig. 12.15.

One advantage of perimeter bracing is the potential for using deeper (and thus stiffer) beams, as the restriction on depth of the spandrel beams is usually less than that on interior framing, which must allow for clearance of ducts, piping, wiring, and recessed lighting fixtures. If this beam stiffness is fully utilized, it may also be possible to increase the number of columns at the perimeter, where intrusion in the plan is not a problem. The result can be a very stiff and strong bent—not likely a critical concern for the three-story building, but important for very tall buildings.

13

PLASTIC BEHAVIOR AND STRENGTH DESIGN

The working stress method of design, which visualizes the development of stresses and deformations at service load levels, has been used primarily in the preceding work in this book. The strength method of design, developed originally mostly for design of reinforced concrete, visualizes conditions at failure. The structure is in fact designed for failure at a load level above that of the service load. For the steel framed structure, failure is usually induced by one of two primary developments: buckling or ductile yielding. In this chapter we consider briefly the case of plastic, yielding-level stress behavior and the use of the plastic hinge for design by the strength method. For a full explanation of the strength method and investigation of plastic behavior the reader is referred to any textbook used for a steel design course in a collegiate program in civil engineering, such as *Steel Structures: Design and Behavior*, by Salmon and Johnson (Ref. 3).

13.1 PLASTIC VERSUS ELASTIC BEHAVIOR

The discussions up to this point of the design of members in bending have been based on bending stresses well within the yield point stress. In general, allowable stresses are based on the *theory of elastic behavior*. However, it has been found by tests that steel members can carry loads much higher than anticipated, even when the yield point stress is reached at sections of maximum bending moment. This is particularly evident in continuous beams and in structures with rigid connections. An inherent property of structural steel is its ability to resist large deformations without failure. These large deformations occur chiefly in the *plastic range*, with no increase in the magnitude of the bending stress. Because of this phenomenon, the *plastic design theory*, sometimes called the *ultimate strength design theory* (or more recently *strength design*), has been developed.

Figure 13.1 represents the typical form of a load-test response for a specimen of ductile steel. The graph shows that up to a stress f_y, the yield point, the deformations are directly proportional to the applied stresses and that beyond the yield point there is a deformation without an increase in stress. For A36 steel this additional deformation, called the *plastic range*, is approximately

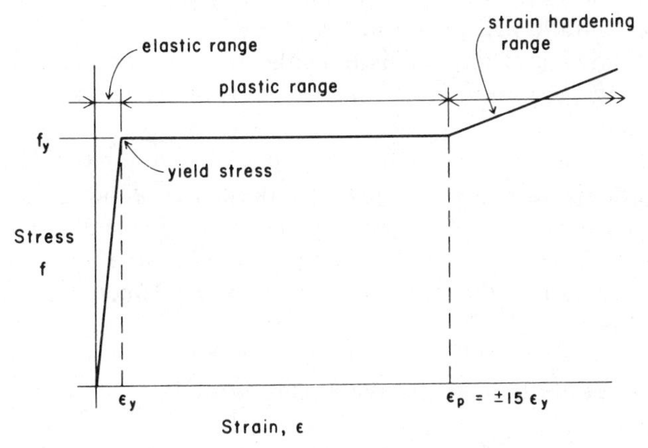

FIGURE 13.1 Idealized form of the stress-strain response for ductile steel.

15 times that produced elastically. This magnitude of the plastic range is the basis for qualification of a metal as *ductile*. Note that beyond this range *strain hardening* (loss of ductility) begins, when further deformation can occur only with an increase in stress.

For plastic behavior to be significant, the extent of the plastic range of deformation must be several times the elastic deformation. As the yield point is increased in magnitude, this ratio of deformations decreases, which is to say that higher-strength steels tend to be less ductile. At present, the theory of plastic design is generally limited to steels with a yield point of not more than 65 ksi [450 MPa].

13.2 PLASTIC MOMENT AND THE PLASTIC HINGE

In Sec. 5.2 we explain the design of members in bending in accordance with the theory of elasticity. When the extreme fiber stress does not exceed the elastic limit, the bending stresses in the cross section of a beam are directly proportional to their distances from the neutral surface. In addition, the strains (deformations) in these fibers are also proportional to their distances from the neutral surface. Both stresses and strains are zero at the neutral surface, and both increase to maximum magnitudes at the fibers farthest from the neutral surface.

The following example illustrates the analysis of a steel beam for bending, According to the theory of elastic behavior.

Example. A simple steel beam has a span of 16 ft [4.88 m] with a concentrated load of 18 kips [80 kN] at the center of the span. The section used is a W 12 × 30, the beam is adequately braced throughout its length, and the beam weight is ignored in the computations. Let us compute the maximum extreme fiber stress (see Fig. 13.2*d*).

Solution. To do this we use the flexure formula

$$f = \frac{M}{S} \qquad \text{(Sec. E.9)}$$

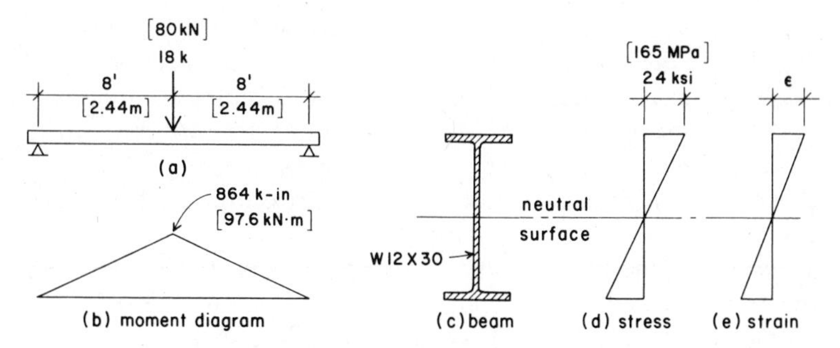

FIGURE 13.2 Elastic behavior of the beam.

Then

$$M = \frac{PL}{4} = \frac{18 \times 16}{4} = 72 \text{ kip-ft } [98 \text{ kN-m}]$$

which is the maximum bending moment. In Table A.1 we find $S =$ 38.6 in.^{3} [632×10^3 mm^3]. Thus

$$f = \frac{M}{S} = \frac{72 \times 12}{38.6} = 22.4 \text{ ksi } [154 \text{ MPa}]$$

which is the stress on the fiber farthest from the neutral surface (see Fig. 13.2*d*).

Note that this stress occurs only at the beam section at the center of the span, where the bending moment has its maximum value. Figure 13.2*e* shows the deformations that accompany the stresses shown in Fig. 13.2*d*. Note that both stresses and deformations are directly proportional to their distances from the neutral surface in elastic analysis.

When a steel beam is loaded to produce an extreme fiber stress in excess of the yield point the property of the material's ductility affects the distribution of the stresses in the beam cross section. Elastic analysis does not suffice to explain this phenomenon because the beam will experience some plastic deformation.

Assume that the bending moment on a beam is of such magnitude that the extreme fiber stress is f_y, the yield stress. Then, if M_y is the elastic bending moment at the yield stress, $M = M_y$, and the distribution of the stresses in the cross section is as shown in Fig. 13.3*a*; the maximum bending stress f_y is at the extreme fiber.

Next consider that the loading and the resulting bending moment have been increased; M is now greater than M_y. The stress on the extreme fiber is still f_y, but *the material has yielded* and a greater area of the cross section is also stressed to f_y. The stress distribution is shown in Fig. 13.3*b*.

Now imagine that the load is further increased. The stress on the extreme fiber is still f_y and, theoretically, *all fibers in the cross section are stressed to* f_y. This idealized plastic stress distribution is shown in Fig. 13.3*d*. The bending moment that produces this condition is M_p, the plastic bending moment. In reality about 10% of the central portion of the cross section continues to resist in an elastic manner, as indicated in Fig. 13.3*c*. This small resistance is quite negligible, and we assume that the stresses on all fibers of the cross section are f_y, as shown in Fig. 13.3*d*. The section is now said to be fully plastic and any further increase in load will result in large deformations; the beam acts as if it were hinged at this section. We call this a plastic hinge at which free rotation is permitted only after M_p has been attained (see Fig. 13.4). At sections of a beam in which this condition prevails the bending resistance of the cross section has been exhausted.

13.3 PLASTIC SECTION MODULUS

In elastic design the moment that produces the maximum allowable resisting moment may be found by the flexure formula

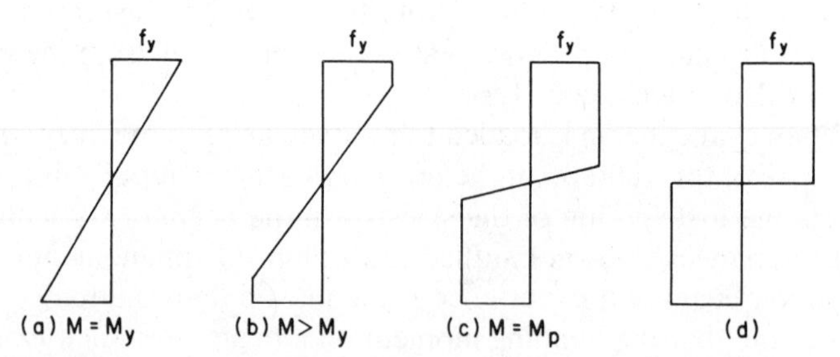

FIGURE 13.3 Progression of development of flexural stress—elastic to plastic.

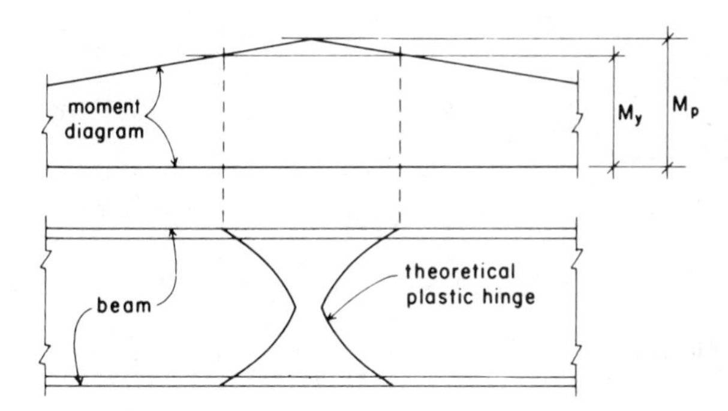

FIGURE 13.4 Development of the plastic hinge.

$$M = f \times S$$

where M = maximum allowable bending moment, in inch-pounds.

f = maximum allowable bending stress, in pounds per square inch,

S = section modulus, in inches to the third power.

If the extreme fiber is stressed to the yield stress

$$M_y = f_y \times S$$

where M_y = elastic bending moment at yield stress,

f_y = yield stress, in pounds per square inch,

S = section modulus, in inches to the third power.

Now let us find a similar relation between the plastic moment and its plastic resisting moment. Refer to Fig. 13.5, which shows the cross section of a W or S section in which the bending stress f_y, the yield stress, is constant over the cross section. In the figure

A_u = upper area of the cross section above the neutral axis, in square inches,

y_u = distance of the centroid of A_u from the neutral axis,

A_l = lower area of the cross section below the neutral axis, in square inches,

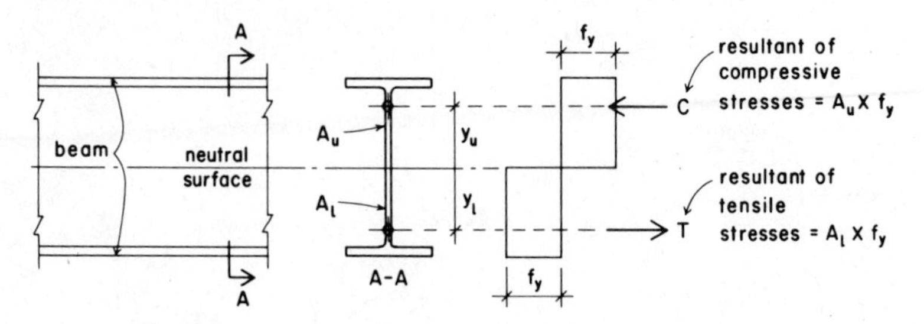

FIGURE 13.5 Development of the plastic resisting moment.

y_l = distance of the centroid of A_l from the neutral axis.

For equilibrium the algebraic sum of the horizontal forces must be zero. Then

$$\sum H = 0$$

or

$$[A_u \times (+f_y)] + [A_l \times (-f_y)] = 0$$

and

$$A_u = A_l$$

This shows that the neutral axis divides the cross section into equal areas, which is apparent in symmetrical sections, but it applies to unsymmetrical sections as well. Also the bending moment equals the sum of the moments of the stresses in the section. Thus for M_p, the plastic moment,

$$M_p = (A_u \times f_y \times y_u) + (A_l \times f_y \times y_l)$$

or

$$M_p = f_y[(A_u \times y_u) + A_l \times y_l)]$$

and

$$M_p = f_y \times Z$$

The quantity $(A_u y_u + A_l y_l)$ is called the *plastic section modulus* of the cross section and is designated by the letter Z; because it is an area multiplied by a distance, it is in units to the third power. If the area is in units of square inches and the distance is in linear inches, Z, the section modulus, is in units of inches to the third power.

The plastic section modulus is always larger than the elastic section modulus.

It is important to note that in plastic design the neutral axis for unsymmetrical cross sections does not pass through the centroid of the section. In plastic design the neutral axis divides the cross section into *equal* areas.

13.4 COMPUTATION OF THE PLASTIC SECTION MODULUS

The notation used in Sec. 13.3 is appropriate for both symmetrical and unsymmetrical sections. Consider now a symmetrical section such as a W or S shape, as shown in Fig. 13.5. $A_u = A_l$, $y_u = y_l$ and $A_u + A_l = A$, the total area of the cross section. Then

$$M_p = (A_u \times f_y \times y_u) + (A_l \times f_y \times y_l)$$

and

$$M_p = f_y \times A \times y \quad \text{or} \quad M_p = f_y \times Z$$

where f_y = yield stress,
A = total area of the cross section,
y = distance from the neutral axis to the centroid of the portion of the area on either side of the neutral axis,
Z = plastic modulus of the section, in in.^{3} or mm^3.

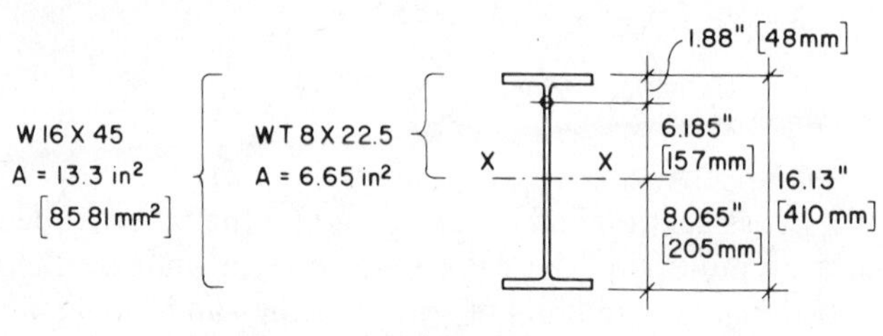

FIGURE 13.6

Now because $Z = A \times y$ we can readily compute the value of the plastic section modulus of a given cross section.

Consider a W 16 × 45. In Appendix Table A.1 we find that its total depth is 16.13 in. [410 mm] and its cross-sectional area is 13.3 in.^{2} [8581 mm^2]. In Appendix Table A.4 we find that a WT 8 × 22.5 (which is one-half a W 16 × 45) has its centroid located 1.88 in. [48 mm] from the outside of the flange. Therefore the distance from the centroid of either half of the W shape to the neutral axis is one half the beam depth less the distance obtained for the tee. Thus the distance y is (see Fig. 13.6)

$$y = (16.13/2) - 1.88 = 6.185 \text{ in. } [157 \text{ mm}]$$

Then the plastic modulus of the W 16 × 45 is

$$Z = A \times y = 13.3 \times 6.185 = 82.26 \text{ in.}^3 \ [1347 \times 10^3]$$

Use of the full value of the plastic hinge moment requires the shape to have limited values for the width/thickness ratio of the flanges and the depth/thickness ratio of the web. These requirements are given in Sec. 2.7 of the AISC Specification.

13.5 SHAPE FACTOR

Consider a beam subjected to a bending moment that produces bending stresses for which the extreme fiber stress is f_y.

By elastic design

$$M_y = f_y \times S$$

By plastic design

$$M_p = f_y \times Z$$

Then

$$\frac{\text{plastic design}}{\text{elastic design}} = \frac{M_p}{M_y} = \frac{f_y \times Z}{f_y \times S} = \frac{Z}{S}$$

The relation between the two moments, M_p/M_y is called the *shape factor*. It is represented by the letter u. Thus

$$u = \frac{Z}{S}$$

Let us compute the value of u, the shape factor for a rectangle whose width is b and whose depth is d (see Fig. 13.7).

The elastic section modulus of this rectangle about the neutral axis parallel to the base is

$$S = \frac{bd^2}{6} \qquad \text{(Fig. E.24)}$$

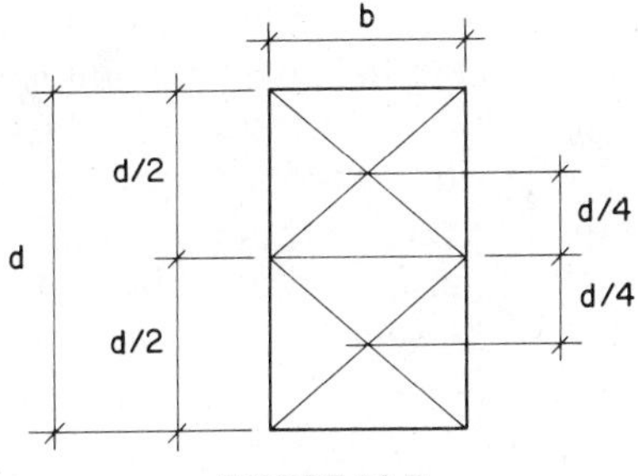

FIGURE 13.7

For the plastic section modulus of the rectangle

$$Z = \left(b \times \frac{d}{2} \times \frac{d}{4}\right) + \left(b \times \frac{d}{2} \times \frac{d}{4}\right)$$

or

$$Z = \frac{bd^2}{8} + \frac{bd^2}{8} = \frac{bd^2}{4}$$

Then

$$\frac{Z}{S} = \frac{bd^2}{4} \div \frac{bd^2}{6} = \frac{bd^2}{4} \times \frac{6}{bd^2} = \frac{3}{2} = 1.5$$

which is the shape factor for rectangles. Thus M_p, the plastic moment for a rectangular cross section, is 50% greater than M_y, the elastic moment at yield stress.

For the commonly used steel sections, such as W and S, the shape factor is approximately 1.12 for bending about the strong axis of the section. This means that these sections can support approximately 12% more than we expect them to carry on the basis of elastic design.

13.6 RESTRAINED BEAMS

Figure 13.8 shows a uniformly distributed load of w lb per lin ft on a beam that is fixed (restrained from rotation) at both ends. The maximum bending moment for this condition is $wl^2/12$; it occurs at the supports and is a negative quantity. At the center of the span the moment is positive and its magnitude is $wl^2/24$.

Considering now that the load is increased from w to w_y, thus producing a stress of f_y in the extreme fibers of the cross section over the supports. At the supports

$$M_y = -\frac{w_y l^2}{12}$$

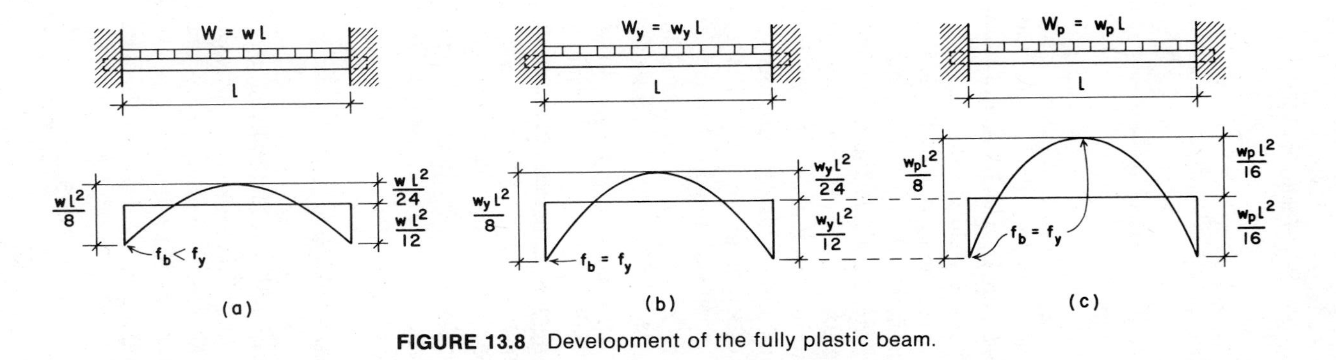

FIGURE 13.8 Development of the fully plastic beam.

and at the center of the span

$$M_y = +\frac{w_y l^2}{24}$$

as shown in Fig. 13.8*b*. This is the limit of the elastic behavior; the moments at the supports are still twice the magnitude of the moment at the midspan, at which none of the fibers of the cross section has reached the yield stress.

Now let us increase the load to w_p lb per lin ft (see Fig. 13.8*c*). The stresses in the fibers of the cross section *over the supports* will now begin to yield until *all* the fibers are stressed to f_y and plastic hinges are formed at the ends. The beam can no longer resist further rotation at its ends and the increase in load must be resisted by sections of the beam that are less stressed. The critical section lies at the center of the span. If the load w_p has been increased so that a plastic hinge is formed at the midspan, the beam will become a *mechanism*. A mechanism exists when sufficient plastic hinges have been formed and no further loading may be supported by the member. Note that the beam when loaded produces three plastic hinges, one at the midspan and two at the supports. The moments at the supports and at midspan are now equal and have magnitudes of $w_p l^2/16$, which means that the full strength of the beam is used at three sections instead of two. A load whose magnitude is greater than w_p lb per lin ft would result in a permanent deformation and consequently a practical failure.

13.7 FACTORED LOADS

Consider a beam of A36 steel laterally supported throughout its length. Its span is 24 ft [7.315 m] and it carries a concentrated load of 42 kips [186.8 kN] at the center. Let us determine the size of the beam in accordance with the theory of elastic behavior.

The maximum bending moment for this beam is

$$M = \frac{PL}{4} = \frac{42 \times 24}{4} = 252 \text{ kip-ft} \;\; [342 \text{ kN-m}]$$

and the required section modulus is

$$S = \frac{M}{f_b} = \frac{252 \times 12}{24} = 126 \text{ in.}^3 \ [2070 \times 10^3 \text{ mm}^3]$$

Appendix Table A.1 shows that a W 21 × 62 has a section modulus of 127 in.^{3} and is acceptable.

Now let us compute the magnitude of the concentrated load at the center of the span that would produce a bending moment equal to the plastic resisting moment. In Appendix Table A.1 the plastic modulus for the W 21 × 62 is 144 in.^{3} Thus the plastic moment is

$$M_p = F_y \times Z_x = 36 \times 144 = 5184 \text{ kip-in., or } 432 \text{ kip-ft} \quad [585.3 \text{ kN-m}]$$

Then the load that corresponds to this moment is

$$M_p = 432 = \frac{PL}{4} = \frac{P \times 24}{4}$$

and

$$P = \frac{4 \times 432}{24} = 72 \text{ kips} \ [320 \text{ kN}]$$

This load would produce a plastic hinge at the center of the span and a slight increase in load would result in failure.

The term *load factor* is given to the ratio of the ultimate load to the design load. In this example it is 72/42 = 1.714.

Referring to Fig. 13.8*c*, we read in Sec. 13.7 that a load greater than w_p lb/ft would result in a permanent deformation and a practical failure. In elastic design the allowable bending stress F_b is decreased to a fraction of F_y, the yield stress. For compact sections $F_b = 0.66F_y$. Therefore, the implied factor of safety against yielding is 1/0.66, or 1.5. This factor is higher, of course, if the beam's limiting capacity is taken as the plastic moment.

In plastic design the concept of allowable stress is not used and computations are based strictly on the limit of the yield stress. Safety is produced by the use of the load factor, by which the beam is literally designed to fail, but at a load larger than that it actually must sustain. For simple and continuous beams the load factor is specified as 1.7; for rigid frames it is 1.85.

13.8 DESIGN OF A SIMPLE BEAM

The design of simple beams by the elastic or plastic theory will usually result in the same size beam, as illustrated in the following examples:

Example 1. A simple beam of A36 steel has a span of 20 ft [6.1 m] and supports a uniformly distributed load of 4.8 kips/ft [70 kN/m], including its own weight. Design this beam in accordance with the elastic theory, assuming that it is laterally supported throughout its length.

Solution. The maximum bending moment is $wL^2/8$. Then

$$M = \frac{4.8 \times (20)^2}{8} = 240 \text{ kip-ft} \quad [326 \text{ kN-m}]$$

By using the allowable bending stress of 24 ksi [165 MPa] for a compact section the required section modulus is

$$S = \frac{M}{F_b} = \frac{240 \times 12}{24} = 120 \text{ in.}^3 \quad [1973 \times 10^3 \text{ mm}^3]$$

In Table A.1 we find a W 21 × 62 with $S_x = 127$ in.^{3}

Example 2. Design the same beam in accordance with the plastic theory.

Solution. We adjust the load with the load factor, which is given in Sec. 13.7 as 1.7. Thus

$$w_p = 4.8 \times 1.7 = 8.16 \text{ kips/ft} \ [119 \text{ kN-m}]$$

and the maximum bending moment is

$$M_p = \frac{w_p L^2}{8} = \frac{8.16 \times (20)^2}{8} = 408 \text{ kip-ft} \ [544 \text{ kN-m}]$$

and

$$M_p = F_y \times Z, \ Z = \frac{M_p}{F_y} = \frac{408 \times 12}{36} = 136 \text{ in.}^3 \ [2232 \times 10^3 \text{ mm}^3]$$

which is the minimum plastic section modulus. From Appendix Table A.1 select a W 21 × 62 for which $Z_x = 144$ in.^{3} Note that the elastic and plastic theories have yielded the same result in this example, which is common for simple beams.

13.9 DESIGN OF A BEAM WITH FIXED ENDS

A beam that has both ends fixed or restrained is similar to an interior span of a fully continuous beam; the magnitudes of the maximum bending moments are the same. In structural steel it is economical to design these beams in accordance with the plastic theory because there is a saving of material.

Example 1. A beam of A36 steel with both ends fixed has a clear span of 20 ft [6.1 m] and carries a uniformly distributed load of 4.8 kips/ft [70 kN/m], including its own weight. (This load and span are the same as those for the simple span beam in the example in Sec. 13.8.) Design the beam in accordance with elastic theory, assuming that full lateral support is provided.

Solution. Referring to Fig. 13.8*a* we see that the maximum bending moment is at the supports; a negative moment of $wL^2/12$. Then

$$M = \frac{4.8 \times (20)^2}{12} = 160 \text{ kip-ft} \ [217 \text{ kN-m}]$$

Assuming a compact section so that F_b = 24 ksi [165 MPa],

$$S = \frac{M}{F_b} = \frac{160 \times 12}{24} = 80 \text{ in.}^3 \ [1315 \times 10^3 \text{ mm}^3]$$

which is the required section modulus. Referring to Appendix Table A.1, we find that a W 18 × 50 has S_x = 88.9 in.^{3} and is acceptable.

Example 2. Design the same beam in accordance with the plastic theory.

Solution. Using the load factor of 1.7 for a continuous beam

$$w_p = 4.8 \times 1.7 = 8.16 \text{ kips/ft} \ [119 \text{ kN-m}]$$

Assuming plastic hinging at both the midspan and supports, as shown in Fig. 13.8*c*, we determine that the critical maximum moment is

$$M_p = \frac{wL^2}{16} = \frac{8.16 \times (20)^2}{16} = 204 \text{ kip-ft}$$

Then

$$Z = \frac{M_p}{F_y} = \frac{204 \times 12}{36} = 68 \text{ in.}^3 \ [1116 \times 10^3 \text{ mm}^3]$$

which is the required plastic section modulus. In Appendix Table A.1 we see that a W 18 × 40 has a plastic section modulus of 78.4 in.^{3} and is acceptable.

Note that a savings of 10 lb/ft in the weight of steel was achieved in this case by the use of the plastic theory. This represents a 20% reduction in material used.

Problem 13.9.A. A beam of A36 steel has a span of 36 ft [10.97 m] and is fixed at both ends. The load is 30 kips/ft [133.4 kN/m], which includes the beam weight, and full lateral support is provided. Using the elastic theory, determine the lightest weight W shape that will support the load.

Problem 13.9.B. Using the data given for the beam of Problem 13.9.A, determine the lightest shape in accordance with plastic theory.

13.10 SCOPE OF PLASTIC DESIGN

The purpose of this brief chapter is to explain how the theory of plastic design in steel has evolved. The illustrative problems relate only to beams that are fixed at both ends; they show how the application of plastic design theory may result in economy of material. The theory, however, has applications far beyond fixed beams, particularly in the design of rigid and eccentrically braced frames. Because the design is based on very high stresses, strict attention must be given to possibilities for local buckling and care must be taken to prevent excessive deflections that might occur with the selection of lighter shapes.

The reader who wishes to pursue the subject of plastic design is referred to publications on the topic that are available from the AISC; particularly the 1986 edition of the AISC Manual, subtitled *Load and Resistance Factor Design*. (See discussion in Sec. 1.2.)

REFERENCES

1. *Manual of Steel Construction,* 8th ed., American Institute of Steel Construction, Chicago, 1980. Called simply the AISC Manual. See discussion in Sec. 1.2 regarding 9th edition.
2. *Uniform Building Code,* 1988 ed., International Conference of Building Officials, Whittier, CA. Called simply the UBC.
3. C. G. Salmon and J. E. Johnson, *Steel Structures: Design and Behavior,* 2nd ed., Harper & Row, New York, 1980.
4. S. W. Crawley and R. M. Dillon, *Steel Buildings: Analysis and Design,* 3rd ed., Wiley, New York, 1984.
5. H. Parker and J. Ambrose, *Simplified Engineering for Architects and Builders,* 7th ed., Wiley, New York, 1989.
6. J. Ambrose and D. Vergun, *Simplified Building Design for Wind and Earthquake Forces,* 2nd ed., Wiley, New York, 1990.
7. J. C. McCormac, *Structural Analysis,* 4th ed., Harper & Row, New York, 1984.
8. E. Allen, *Fundamentals of Building Construction: Materials and Methods,* Wiley, New York, 1985.
9. *Standard Specifications, Load Tables, and Weight Tables for Steel Joists and Joist Girders,* Steel Joist Institute, Myrtle Beach, SC, 1988.
10. *Steel Deck Institute Design Manual for Composite Decks, Form Decks, and Roof Decks,* Steel Deck Institute, St. Louis, 1981–1982.

APPENDIX A

PROPERTIES OF ROLLED STRUCTURAL SHAPES

The following tables contain data for standard rolled structural shapes. This material has been reproduced and adapted from the *Manual of Steel Construction*, 8th ed. (Ref. 1) with the permission of the publishers, the American Institute of Steel Construction. Reference may be made to Tables 2.1 and 2.2 to determine the availability of shapes in the various grades of steel. It should be noted that Table A.1 does not contain data for the deeper shapes (40 and 44 in.) or the "jumbo shapes" which are contained in the tables in the 9th edition of the manual. See the discussion in Sec. 1.2.

To convert table units of:	in.	in.^{2}	in.^{3}	in.^{4}
to SI units of:	mm	mm^2	mm^3	mm^4
multiply by:	25.4	645.2	16.39(10)^3	0.4162(10)^6

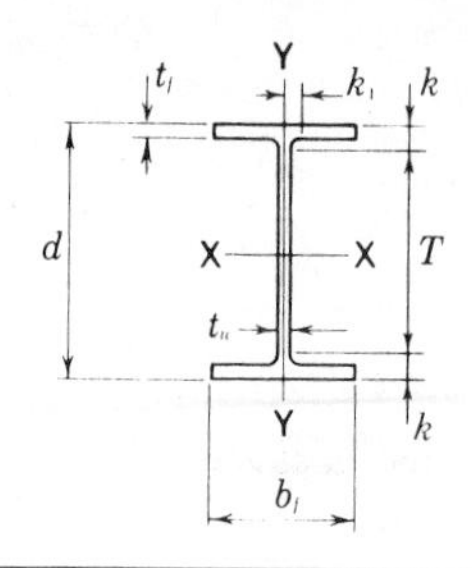

TABLE A.1 Properties of Wide Flange (W) Shapes

Designation	Area A	Depth d		Web Thickness t_w		Web $\frac{t_w}{2}$	Flange Width b_f		Flange Thickness t_f		Distance T	Distance k	Distance k_1
	In.^{2}	In.		In.		In.	In.		In.		In.	In.	In.
W 36x300	88.3	36.74	36 3/4	0.945	15/16	1/2	16.655	16 5/8	1.680	1 11/16	31 1/8	2 13/16	1 1/2
x280	82.4	36.52	36 1/2	0.885	7/8	7/16	16.595	16 5/8	1.570	1 9/16	31 1/8	2 11/16	1 1/2
x260	76.5	36.26	36 1/4	0.840	13/16	7/16	16.550	16 1/2	1.440	1 7/16	31 1/8	2 9/16	1 1/2
x245	72.1	36.08	36 1/8	0.800	13/16	7/16	16.510	16 1/2	1.350	1 3/8	31 1/8	2 1/2	1 7/16
x230	67.6	35.90	35 7/8	0.760	3/4	3/8	16.470	16 1/2	1.260	1 1/4	31 1/8	2 3/8	1 7/16
W 36x210	61.8	36.69	36 3/4	0.830	13/16	7/16	12.180	12 1/8	1.360	1 3/8	32 1/8	2 5/16	1 1/4
x194	57.0	36.49	36 1/2	0.765	3/4	3/8	12.115	12 1/8	1.260	1 1/4	32 1/8	2 3/16	1 3/16
x182	53.6	36.33	36 3/8	0.725	3/4	3/8	12.075	12 1/8	1.180	1 3/16	32 1/8	2 1/8	1 3/16
x170	50.0	36.17	36 1/8	0.680	11/16	3/8	12.030	12	1.100	1 1/8	32 1/8	2	1 3/16
x160	47.0	36.01	36	0.650	5/8	5/16	12.000	12	1.020	1	32 1/8	1 15/16	1 1/8
x150	44.2	35.85	35 7/8	0.625	5/8	5/16	11.975	12	0.940	15/16	32 1/8	1 7/8	1 1/8
x135	39.7	35.55	35 1/2	0.600	5/8	5/16	11.950	12	0.790	13/16	32 1/8	1 11/16	1 1/8
W 33x241	70.9	34.18	34 1/8	0.830	13/16	7/16	15.860	15 7/8	1.400	1 3/8	29 3/4	2 3/16	1 3/16
x221	65.0	33.93	33 7/8	0.775	3/4	3/8	15.805	15 3/4	1.275	1 1/4	29 3/4	2 1/16	1 3/16
x201	59.1	33.68	33 5/8	0.715	11/16	3/8	15.745	15 3/4	1.150	1 1/8	29 3/4	1 15/16	1 1/8
W 33x152	44.7	33.49	33 1/2	0.635	5/8	5/16	11.565	11 5/8	1.055	1 1/16	29 3/4	1 7/8	1 1/8
x141	41.6	33.30	33 1/4	0.605	5/8	5/16	11.535	11 1/2	0.960	15/16	29 3/4	1 3/4	1 1/16
x130	38.3	33.09	33 1/8	0.580	9/16	5/16	11.510	11 1/2	0.855	7/8	29 3/4	1 11/16	1 1/16
x118	34.7	32.86	32 7/8	0.550	9/16	5/16	11.480	11 1/2	0.740	3/4	29 3/4	1 9/16	1 1/16
W 30x211	62.0	30.94	31	0.775	3/4	3/8	15.105	15 1/8	1.315	1 5/16	26 3/4	2 1/8	1 1/8
x191	56.1	30.68	30 5/8	0.710	11/16	3/8	15.040	15	1.185	1 3/16	26 3/4	1 15/16	1 1/16
x173	50.8	30.44	30 1/2	0.655	5/8	5/16	14.985	15	1.065	1 1/16	26 3/4	1 7/8	1 1/16
W 30x132	38.9	30.31	30 1/4	0.615	5/8	5/16	10.545	10 1/2	1.000	1	26 3/4	1 3/4	1 1/16
x124	36.5	30.17	30 1/8	0.585	9/16	5/16	10.515	10 1/2	0.930	15/16	26 3/4	1 11/16	1
x116	34.2	30.01	30	0.565	9/16	5/16	10.495	10 1/2	0.850	7/8	26 3/4	1 5/8	1
x108	31.7	29.83	29 7/8	0.545	9/16	5/16	10.475	10 1/2	0.760	3/4	26 3/4	1 9/16	1
x 99	29.1	29.65	29 5/8	0.520	1/2	1/4	10.450	10 1/2	0.670	11/16	26 3/4	1 7/16	1
W 27x178	52.3	27.81	27 3/4	0.725	3/4	3/8	14.085	14 1/8	1.190	1 3/16	24	1 7/8	1 1/16
x161	47.4	27.59	27 5/8	0.660	11/16	3/8	14.020	14	1.080	1 1/16	24	1 13/16	1
x146	42.9	27.38	27 3/8	0.605	5/8	5/16	13.965	14	0.975	1	24	1 11/16	1
W 27x114	33.5	27.29	27 1/4	0.570	9/16	5/16	10.070	10 1/8	0.930	15/16	24	1 5/8	15/16
x102	30.0	27.09	27 1/8	0.515	1/2	1/4	10.015	10	0.830	13/16	24	1 9/16	15/16
x 94	27.7	26.92	26 7/8	0.490	1/2	1/4	9.990	10	0.745	3/4	24	1 7/16	15/16
x 84	24.8	26.71	26 3/4	0.460	7/16	1/4	9.960	10	0.640	5/8	24	1 3/8	15/16

W SHAPES

TABLE A.1 (*Continued*)

Nominal Wt. per Ft.	Compact Section Criteria				r_T	$\frac{d}{A_f}$	Elastic Properties						Torsional constant	Plastic Modulus	
							Axis X-X			Axis Y-Y					
	$\frac{b_f}{2t_f}$	F_y'	$\frac{d}{t_w}$	F_y'''			I	S	r	I	S	r	J	Z_x	Z_y
Lb.		Ksi		Ksi	In.		In.^{4}	In.^{3}	In.	In.^{4}	In.^{3}	In.	In.^{4}	In.^{3}	In.^{3}
300	5.0	—	38.9	43.7	4.39	1.31	20300	1110	15.2	1300	156	3.83	64.2	1260	241
280	5.3	—	41.3	38.8	4.37	1.40	18900	1030	15.1	1200	144	3.81	52.6	1170	223
260	5.7	—	43.2	35.4	4.34	1.52	17300	953	15.0	1090	132	3.78	41.5	1080	204
245	6.1	—	45.1	32.5	4.32	1.62	16100	895	15.0	1010	123	3.75	34.6	1010	190
230	6.5	—	47.2	29.6	4.30	1.73	15000	837	14.9	940	114	3.73	28.6	943	176
210	4.5	—	44.2	33.8	3.09	2.21	13200	719	14.6	411	67.5	2.58	28.0	833	107
194	4.8	—	47.7	29.0	3.07	2.39	12100	664	14.6	375	61.9	2.56	22.2	767	97.7
182	5.1	—	50.1	26.3	3.05	2.55	11300	623	14.5	347	57.6	2.55	18.4	718	90.7
170	5.5	—	53.2	23.3	3.04	2.73	10500	580	14.5	320	53.2	2.53	15.1	668	83.8
160	5.9	—	55.4	21.5	3.02	2.94	9750	542	14.4	295	49.1	2.50	12.4	624	77.3
150	6.4	—	57.4	20.1	2.99	3.18	9040	504	14.3	270	45.1	2.47	10.1	581	70.9
135	7.6	—	59.3	18.8	2.93	3.77	7800	439	14.0	225	37.7	2.38	6.99	509	59.7
241	5.7	—	41.2	38.9	4.17	1.54	14200	829	14.1	932	118	3.63	35.8	939	182
221	6.2	—	43.8	34.5	4.15	1.68	12800	757	14.1	840	106	3.59	27.5	855	164
201	6.8	—	47.1	29.8	4.12	1.86	11500	684	14.0	749	95.2	3.56	20.5	772	147
152	5.5	—	52.7	23.7	2.94	2.74	8160	487	13.5	273	47.2	2.47	12.4	559	73.9
141	6.0	—	55.0	21.8	2.92	3.01	7450	448	13.4	246	42.7	2.43	9.70	514	66.9
130	6.7	—	57.1	20.3	2.88	3.36	6710	406	13.2	218	37.9	2.39	7.37	467	59.5
118	7.8	—	59.7	18.5	2.84	3.87	5900	359	13.0	187	32.6	2.32	5.30	415	51.3
211	5.7	—	39.9	41.4	3.99	1.56	10300	663	12.9	757	100	3.49	27.9	749	154
191	6.3	—	43.2	35.4	3.97	1.72	9170	598	12.8	673	89.5	3.46	20.6	673	138
173	7.0	—	46.5	30.6	3.94	1.91	8200	539	12.7	598	79.8	3.43	15.3	605	123
132	5.3	—	49.3	27.2	2.68	2.87	5770	380	12.2	196	37.2	2.25	9.72	437	58.4
124	5.7	—	51.6	24.8	2.66	3.09	5360	355	12.1	181	34.4	2.23	7.99	408	54.0
116	6.2	—	53.1	23.4	2.64	3.36	4930	329	12.0	164	31.3	2.19	6.43	378	49.2
108	6.9	—	54.7	22.0	2.61	3.75	4470	299	11.9	146	27.9	2.15	4.99	346	43.9
99	7.8	—	57.0	20.3	2.57	4.23	3990	269	11.7	128	24.5	2.10	3.77	312	38.6
178	5.9	—	38.4	44.9	3.72	1.66	6990	502	11.6	555	78.8	3.26	19.5	567	122
161	6.5	—	41.8	37.8	3.70	1.82	6280	455	11.5	497	70.9	3.24	14.7	512	109
146	7.2	—	45.3	32.2	3.68	2.01	5630	411	11.4	443	63.5	3.21	10.9	461	97.5
114	5.4	—	47.9	28.8	2.58	2.91	4090	299	11.0	159	31.5	2.18	7.33	343	49.3
102	6.0	—	52.6	23.9	2.56	3.26	3620	267	11.0	139	27.8	2.15	5.29	305	43.4
94	6.7	—	54.9	21.9	2.53	3.62	3270	243	10.9	124	24.8	2.12	4.03	278	38.8
84	7.8	—	58.1	19.6	2.49	4.19	2850	213	10.7	106	21.2	2.07	2.81	244	33.2

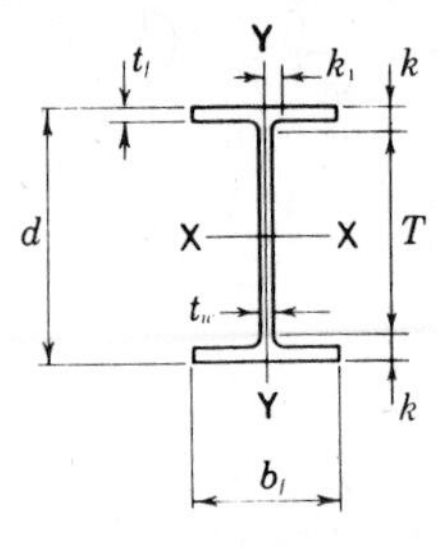

TABLE A.1 Properties of Wide Flange (W) Shapes

Designation	Area A	Depth d		Web			Flange				Distance		
				Thickness t_w		$\frac{t_w}{2}$	Width b_f		Thickness t_f		T	k	k_1
	In.^{2}	In.		In.		In.	In.		In.		In.	In.	In.
W 24x162	47.7	25.00	25	0.705	11/16	3/8	12.955	13	1.220	1 1/4	21	2	1 1/16
x146	43.0	24.74	24 3/4	0.650	5/8	5/16	12.900	12 7/8	1.090	1 1/16	21	1 7/8	1 1/16
x131	38.5	24.48	24 1/2	0.605	5/8	5/16	12.855	12 7/8	0.960	15/16	21	1 3/4	1 1/16
x117	34.4	24.26	24 1/4	0.550	9/16	5/16	12.800	12 3/4	0.850	7/8	21	1 5/8	1
x104	30.6	24.06	24	0.500	1/2	1/4	12.750	12 3/4	0.750	3/4	21	1 1/2	1
W 24x 94	27.7	24.31	24 1/4	0.515	1/2	1/4	9.065	9 1/8	0.875	7/8	21	1 5/8	1
x 84	24.7	24.10	24 1/8	0.470	1/2	1/4	9.020	9	0.770	3/4	21	1 9/16	15/16
x 76	22.4	23.92	23 7/8	0.440	7/16	1/4	8.990	9	0.680	11/16	21	1 7/16	15/16
x 68	20.1	23.73	23 3/4	0.415	7/16	1/4	8.965	9	0.585	9/16	21	1 3/8	15/16
W 24x 62	18.2	23.74	23 3/4	0.430	7/16	1/4	7.040	7	0.590	9/16	21	1 3/8	15/16
x 55	16.2	23.57	23 5/8	0.395	3/8	3/16	7.005	7	0.505	1/2	21	1 5/16	15/16
W 21x147	43.2	22.06	22	0.720	3/4	3/8	12.510	12 1/2	1.150	1 1/8	18 1/4	1 7/8	1 1/16
x132	38.8	21.83	21 7/8	0.650	5/8	5/16	12.440	12 1/2	1.035	1 1/16	18 1/4	1 13/16	1
x122	35.9	21.68	21 5/8	0.600	5/8	5/16	12.390	12 3/8	0.960	15/16	18 1/4	1 11/16	1
x111	32.7	21.51	21 1/2	0.550	9/16	5/16	12.340	12 3/8	0.875	7/8	18 1/4	1 5/8	15/16
x101	29.8	21.36	21 3/8	0.500	1/2	1/4	12.290	12 1/4	0.800	13/16	18 1/4	1 9/16	15/16
W 21x 93	27.3	21.62	21 5/8	0.580	9/16	5/16	8.420	8 3/8	0.930	15/16	18 1/4	1 11/16	1
x 83	24.3	21.43	21 3/8	0.515	1/2	1/4	8.355	8 3/8	0.835	13/16	18 1/4	1 9/16	15/16
x 73	21.5	21.24	21 1/4	0.455	7/16	1/4	8.295	8 1/4	0.740	3/4	18 1/4	1 1/2	15/16
x 68	20.0	21.13	21 1/8	0.430	7/16	1/4	8.270	8 1/4	0.685	11/16	18 1/4	1 7/16	7/8
x 62	18.3	20.99	21	0.400	3/8	3/16	8.240	8 1/4	0.615	5/8	18 1/4	1 3/8	7/8
W 21x 57	16.7	21.06	21	0.405	3/8	3/16	6.555	6 1/2	0.650	5/8	18 1/4	1 3/8	7/8
x 50	14.7	20.83	20 7/8	0.380	3/8	3/16	6.530	6 1/2	0.535	9/16	18 1/4	1 5/16	7/8
x 44	13.0	20.66	20 5/8	0.350	3/8	3/16	6.500	6 1/2	0.450	7/16	18 1/4	1 3/16	7/8
W 18x119	35.1	18.97	19	0.655	5/8	5/16	11.265	11 1/4	1.060	1 1/16	15 1/2	1 3/4	15/16
x106	31.1	18.73	18 3/4	0.590	9/16	5/16	11.200	11 1/4	0.940	15/16	15 1/2	1 5/8	15/16
x 97	28.5	18.59	18 5/8	0.535	9/16	5/16	11.145	11 1/8	0.870	7/8	15 1/2	1 9/16	7/8
x 86	25.3	18.39	18 3/8	0.480	1/2	1/4	11.090	11 1/8	0.770	3/4	15 1/2	1 7/16	7/8
x 76	22.3	18.21	18 1/4	0.425	7/16	1/4	11.035	11	0.680	11/16	15 1/2	1 3/8	13/16
W 18x 71	20.8	18.47	18 1/2	0.495	1/2	1/4	7.635	7 5/8	0.810	13/16	15 1/2	1 1/2	7/8
x 65	19.1	18.35	18 3/8	0.450	7/16	1/4	7.590	7 5/8	0.750	3/4	15 1/2	1 7/16	7/8
x 60	17.6	18.24	18 1/4	0.415	7/16	1/4	7.555	7 1/2	0.695	11/16	15 1/2	1 3/8	13/16
x 55	16.2	18.11	18 1/8	0.390	3/8	3/16	7.530	7 1/2	0.630	5/8	15 1/2	1 5/16	13/16
x 50	14.7	17.99	18	0.355	3/8	3/16	7.495	7 1/2	0.570	9/16	15 1/2	1 1/4	13/16

W SHAPES

TABLE A.1 (*Continued*)

Nominal Wt. per Ft.	Compact Section Criteria				r_T	$\frac{d}{A_f}$	Elastic Properties						Torsional constant	Plastic Modulus	
							Axis X-X			Axis Y-Y					
	$\frac{b_f}{2t_f}$	F_y'	$\frac{d}{t_w}$	F_y'''			I	S	r	I	S	r	J	Z_x	Z_y
Lb.		Ksi		Ksi	In.		In.^{4}	In.^{3}	In.	In.^{4}	In.^{3}	In.	In.^{4}	In.^{3}	In.^{3}
162	5.3	—	35.5	52.5	3.45	1.58	5170	414	10.4	443	68.4	3.05	18.5	468	105.
146	5.9	—	38.1	45.6	3.43	1.76	4580	371	10.3	391	60.5	3.01	13.4	418	93.2
131	6.7	—	40.5	40.3	3.40	1.98	4020	329	10.2	340	53.0	2.97	9.50	370	81.5
117	7.5	—	44.1	33.9	3.37	2.23	3540	291	10.1	297	46.5	2.94	6.72	327	71.4
104	8.5	58.5	48.1	28.5	3.35	2.52	3100	258	10.1	259	40.7	2.91	4.72	289	62.4
94	5.2	—	47.2	29.6	2.33	3.06	2700	222	9.87	109	24.0	1.98	5.26	254	37.5
84	5.9	—	51.3	25.1	2.31	3.47	2370	196	9.79	94.4	20.9	1.95	3.70	224	32.6
76	6.6	—	54.4	22.3	2.29	3.91	2100	176	9.69	82.5	18.4	1.92	2.68	200	28.6
68	7.7	—	57.2	20.2	2.26	4.52	1830	154	9.55	70.4	15.7	1.87	1.87	177	24.5
62	6.0	—	55.2	21.7	1.71	5.72	1550	131	9.23	34.5	9.80	1.38	1.71	153	15.7
55	6.9	—	59.7	18.5	1.68	6.66	1350	114	9.11	29.1	8.30	1.34	1.18	134	13.3
147	5.4	—	30.6	—	3.34	1.53	3630	329	9.17	376	60.1	2.95	15.4	373	92.6
132	6.0	—	33.6	58.6	3.31	1.70	3220	295	9.12	333	53.5	2.93	11.3	333	82.3
122	6.5	—	36.1	50.6	3.30	1.82	2960	273	9.09	305	49.2	2.92	8.98	307	75.6
111	7.1	—	39.1	43.2	3.28	1.99	2670	249	9.05	274	44.5	2.90	6.83	279	68.2
101	7.7	—	42.7	36.2	3.27	2.17	2420	227	9.02	248	40.3	2.89	5.21	253	61.7
93	4.5	—	37.3	47.5	2.17	2.76	2070	192	8.70	92.9	22.1	1.84	6.03	221	34.7
83	5.0	—	41.6	38.1	2.15	3.07	1830	171	8.67	81.4	19.5	1.83	4.34	196	30.5
73	5.6	—	46.7	30.3	2.13	3.46	1600	151	8.64	70.6	17.0	1.81	3.02	172	26.6
68	6.0	—	49.1	27.4	2.12	3.73	1480	140	8.60	64.7	15.7	1.80	2.45	160	24.4
62	6.7	—	52.5	24.0	2.10	4.14	1330	127	8.54	57.5	13.9	1.77	1.83	144	21.7
57	5.0	—	52.0	24.4	1.64	4.94	1170	111	8.36	30.6	9.35	1.35	1.77	129	14.8
50	6.1	—	54.8	22.0	1.60	5.96	984	94.5	8.18	24.9	7.64	1.30	1.14	110	12.2
44	7.2	—	59.0	19.0	1.57	7.06	843	81.6	8.06	20.7	6.36	1.26	0.77	95.4	10.2
119	5.3	—	29.0	—	3.02	1.59	2190	231	7.90	253	44.9	2.69	10.6	261	69.1
106	6.0	—	31.7	—	3.00	1.78	1910	204	7.84	220	39.4	2.66	7.48	230	60.5
97	6.4	—	34.7	54.7	2.99	1.92	1750	188	7.82	201	36.1	2.65	5.86	211	55.3
86	7.2	—	38.3	45.0	2.97	2.15	1530	166	7.77	175	31.6	2.63	4.10	186	48.4
76	8.1	64.2	42.8	36.0	2.95	2.43	1330	146	7.73	152	27.6	2.61	2.83	163	42.2
71	4.7	—	37.3	47.4	1.98	2.99	1170	127	7.50	60.3	15.8	1.70	3.48	145	24.7
65	5.1	—	40.8	39.7	1.97	3.22	1070	117	7.49	54.8	14.4	1.69	2.73	133	22.5
60	5.4	—	44.0	34.2	1.96	3.47	984	108	7.47	50.1	13.3	1.69	2.17	123	20.6
55	6.0	—	46.4	30.6	1.95	3.82	890	98.3	7.41	44.9	11.9	1.67	1.66	112	18.5
50	6.6	—	50.7	25.7	1.94	4.21	800	88.9	7.38	40.1	10.7	1.65	1.24	101	16.6

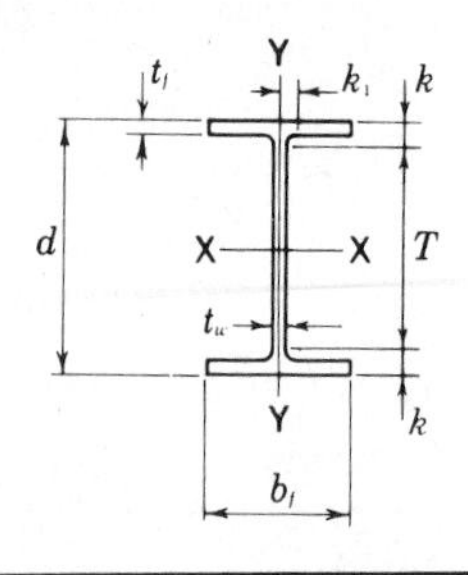

TABLE A.1 Properties of Wide Flange (W) Shapes

Designation	Area A	Depth d		Web Thickness t_w		Web $\frac{t_w}{2}$	Flange Width b_f		Flange Thickness t_f		Distance T	Distance k	Distance k_1
	In.^{2}	In.		In.		In.	In.		In.		In.	In.	In.
W 18x 46	13.5	18.06	18	0.360	3/8	3/16	6.060	6	0.605	5/8	15 1/2	1 1/4	13/16
x 40	11.8	17.90	17 7/8	0.315	5/16	3/16	6.015	6	0.525	1/2	15 1/2	1 3/16	13/16
x 35	10.3	17.70	17 3/4	0.300	5/16	3/16	6.000	6	0.425	7/16	15 1/2	1 1/8	3/4
W 16x100	29.4	16.97	17	0.585	9/16	5/16	10.425	10 3/8	0.985	1	13 5/8	1 11/16	15/16
x 89	26.2	16.75	16 3/4	0.525	1/2	1/4	10.365	10 3/8	0.875	7/8	13 5/8	1 9/16	7/8
x 77	22.6	16.52	16 1/2	0.455	7/16	1/4	10.295	10 1/4	0.760	3/4	13 5/8	1 7/16	7/8
x 67	19.7	16.33	16 3/8	0.395	3/8	3/16	10.235	10 1/4	0.665	11/16	13 5/8	1 3/8	13/16
W 16x 57	16.8	16.43	16 3/8	0.430	7/16	1/4	7.120	7 1/8	0.715	11/16	13 5/8	1 3/8	7/8
x 50	14.7	16.26	16 1/4	0.380	3/8	3/16	7.070	7 1/8	0.630	5/8	13 5/8	1 5/16	13/16
x 45	13.3	16.13	16 1/8	0.345	3/8	3/16	7.035	7	0.565	9/16	13 5/8	1 1/4	13/16
x 40	11.8	16.01	16	0.305	5/16	3/16	6.995	7	0.505	1/2	13 5/8	1 3/16	13/16
x 36	10.6	15.86	15 7/8	0.295	5/16	3/16	6.985	7	0.430	7/16	13 5/8	1 1/8	3/4
W 16x 31	9.12	15.88	15 7/8	0.275	1/4	1/8	5.525	5 1/2	0.440	7/16	13 5/8	1 1/8	3/4
x 26	7.68	15.69	15 3/4	0.250	1/4	1/8	5.500	5 1/2	0.345	3/8	13 5/8	1 1/16	3/4
W 14x730	215.0	22.42	22 3/8	3.070	3 1/16	1 9/16	17.890	17 7/8	4.910	4 15/16	11 1/4	5 9/16	2 3/16
x665	196.0	21.64	21 5/8	2.830	2 13/16	1 7/16	17.650	17 5/8	4.520	4 1/2	11 1/4	5 3/16	2 1/16
x605	178.0	20.92	20 7/8	2.595	2 5/8	1 5/16	17.415	17 3/8	4.160	4 3/16	11 1/4	4 13/16	1 15/16
x550	162.0	20.24	20 1/4	2.380	2 3/8	1 3/16	17.200	17 1/4	3.820	3 13/16	11 1/4	4 1/2	1 13/16
x500	147.0	19.60	19 5/8	2.190	2 3/16	1 1/8	17.010	17	3.500	3 1/2	11 1/4	4 3/16	1 3/4
x455	134.0	19.02	19	2.015	2	1	16.835	16 7/8	3.210	3 3/16	11 1/4	3 7/8	1 5/8
W 14x426	125.0	18.67	18 5/8	1.875	1 7/8	15/16	16.695	16 3/4	3.035	3 1/16	11 1/4	3 11/16	1 9/16
x398	117.0	18.29	18 1/4	1.770	1 3/4	7/8	16.590	16 5/8	2.845	2 7/8	11 1/4	3 1/2	1 1/2
x370	109.0	17.92	17 7/8	1.655	1 5/8	13/16	16.475	16 1/2	2.660	2 11/16	11 1/4	3 5/16	1 7/16
x342	101.0	17.54	17 1/2	1.540	1 9/16	13/16	16.360	16 3/8	2.470	2 1/2	11 1/4	3 1/8	1 3/8
x311	91.4	17.12	17 1/8	1.410	1 7/16	3/4	16.230	16 1/4	2.260	2 1/4	11 1/4	2 15/16	1 5/16
x283	83.3	16.74	16 3/4	1.290	1 5/16	11/16	16.110	16 1/8	2.070	2 1/16	11 1/4	2 3/4	1 1/4
x257	75.6	16.38	16 3/8	1.175	1 3/16	5/8	15.995	16	1.890	1 7/8	11 1/4	2 9/16	1 3/16
x233	68.5	16.04	16	1.070	1 1/16	9/16	15.890	15 7/8	1.720	1 3/4	11 1/4	2 3/8	1 3/16
x211	62.0	15.72	15 3/4	0.980	1	1/2	15.800	15 3/4	1.560	1 9/16	11 1/4	2 1/4	1 1/8
x193	56.8	15.48	15 1/2	0.890	7/8	7/16	15.710	15 3/4	1.440	1 7/16	11 1/4	2 1/8	1 1/16
x176	51.8	15.22	15 1/4	0.830	13/16	7/16	15.650	15 5/8	1.310	1 5/16	11 1/4	2	1 1/16
x159	46.7	14.98	15	0.745	3/4	3/8	15.565	15 5/8	1.190	1 3/16	11 1/4	1 7/8	1
x145	42.7	14.78	14 3/4	0.680	11/16	3/8	15.500	15 1/2	1.090	1 1/16	11 1/4	1 3/4	1

W SHAPES

TABLE A.1 (*Continued*)

Nominal Wt. per Ft.	Compact Section Criteria				r_T	$\frac{d}{A_f}$	Elastic Properties						Torsional constant J	Plastic Modulus	
	$\frac{b_f}{2t_f}$	F_y'	$\frac{d}{t_w}$	F_y'''			Axis X-X			Axis Y-Y				Z_x	Z_y
							I	S	r	I	S	r			
Lb.		Ksi		Ksi	In.		In.^{4}	In.^{3}	In.	In.^{4}	In.^{3}	In.	In.^{4}	In.^{3}	In.^{3}
46	5.0	—	50.2	26.2	1.54	4.93	712	78.8	7.25	22.5	7.43	1.29	1.22	90.7	11.7
40	5.7	—	56.8	20.5	1.52	5.67	612	68.4	7.21	19.1	6.35	1.27	0.81	78.4	9.95
35	7.1	—	59.0	19.0	1.49	6.94	510	57.6	7.04	15.3	5.12	1.22	0.51	66.5	8.06
100	5.3	—	29.0	—	2.81	1.65	1490	175	7.10	186	35.7	2.51	7.73	198	54.9
89	5.9	—	31.9	64.9	2.79	1.85	1300	155	7.05	163	31.4	2.49	5.45	175	48.1
77	6.8	—	36.3	50.1	2.77	2.11	1110	134	7.00	138	26.9	2.47	3.57	150	41.1
67	7.7	—	41.3	38.6	2.75	2.40	954	117	6.96	119	23.2	2.46	2.39	130	35.5
57	5.0	—	38.2	45.2	1.86	3.23	758	92.2	6.72	43.1	12.1	1.60	2.22	105	18.9
50	5.6	—	42.8	36.1	1.84	3.65	659	81.0	6.68	37.2	10.5	1.59	1.52	92.0	16.3
45	6.2	—	46.8	30.2	1.83	4.06	586	72.7	6.65	32.8	9.34	1.57	1.11	82.3	14.5
40	6.9	—	52.5	24.0	1.82	4.53	518	64.7	6.63	28.9	8.25	1.57	0.79	72.9	12.7
36	8.1	64.0	53.8	22.9	1.79	5.28	448	56.5	6.51	24.5	7.00	1.52	0.54	64.0	10.8
31	6.3	—	57.7	19.8	1.39	6.53	375	47.2	6.41	12.4	4.49	1.17	0.46	54.0	7.03
26	8.0	—	62.8	16.8	1.36	8.27	301	38.4	6.26	9.59	3.49	1.12	0.26	44.2	5.48
730	1.8	—	7.3	—	4.99	0.25	14300	1280	8.17	4720	527	4.69	1450	1660	816
665	2.0	—	7.6	—	4.92	0.27	12400	1150	7.98	4170	472	4.62	1120	1480	730
605	2.1	—	8.1	—	4.85	0.29	10800	1040	7.80	3680	423	4.55	870	1320	652
550	2.3	—	8.5	—	4.79	0.31	9430	931	7.63	3250	378	4.49	670	1180	583
500	2.4	—	8.9	—	4.73	0.33	8210	838	7.48	2880	339	4.43	514	1050	522
455	2.6	—	9.4	—	4.68	0.35	7190	756	7.33	2560	304	4.38	395	936	468
426	2.8	—	10.0	—	4.64	0.37	6600	707	7.26	2360	283	4.34	331	869	434
398	2.9	—	10.3	—	4.61	0.39	6000	656	7.16	2170	262	4.31	273	801	402
370	3.1	—	10.8	—	4.57	0.41	5440	607	7.07	1990	241	4.27	222	736	370
342	3.3	—	11.4	—	4.54	0.43	4900	559	6.98	1810	221	4.24	178	672	338
311	3.6	—	12.1	—	4.50	0.47	4330	506	6.88	1610	199	4.20	136	603	304
283	3.9	—	13.0	—	4.46	0.50	3840	459	6.79	1440	179	4.17	104	542	274
257	4.2	—	13.9	—	4.43	0.54	3400	415	6.71	1290	161	4.13	79.1	487	246
233	4.6	—	15.0	—	4.40	0.59	3010	375	6.63	1150	145	4.10	59.5	436	221
211	5.1	—	16.0	—	4.37	0.64	2660	338	6.55	1030	130	4.07	44.6	390	198
193	5.5	—	17.4	—	4.35	0.68	2400	310	6.50	931	119	4.05	34.8	355	180
176	6.0	—	18.3	—	4.32	0.74	2140	281	6.43	838	107	4.02	26.5	320	163
159	6.5	—	20.1	—	4.30	0.81	1900	254	6.38	748	96.2	4.00	19.8	287	146
145	7.1	—	21.7	—	4.28	0.88	1710	232	6.33	677	87.3	3.98	15.2	260	133

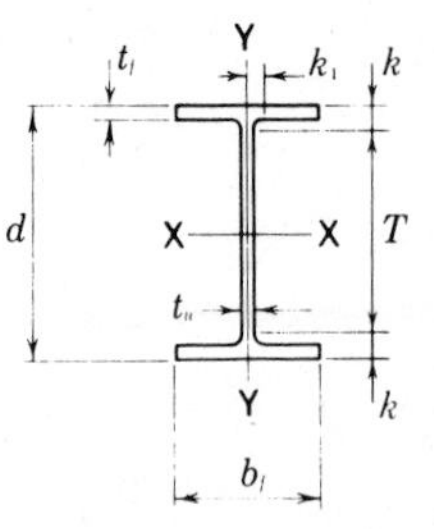

TABLE A.1 Properties of Wide Flange (W) Shapes

Designation	Area A	Depth d		Web			Flange				Distance		
				Thickness t_w		$\frac{t_w}{2}$	Width b_f		Thickness t_f		T	k	k_1
	In.^{2}	In.		In.		In.	In.		In.		In.	In.	In.
W 14x132	38.8	14.66	$14\frac{5}{8}$	0.645	$\frac{5}{8}$	$\frac{5}{16}$	14.725	$14\frac{3}{4}$	1.030	1	$11\frac{1}{4}$	$1\frac{11}{16}$	$\frac{15}{16}$
x120	35.3	14.48	$14\frac{1}{2}$	0.590	$\frac{9}{16}$	$\frac{5}{16}$	14.670	$14\frac{5}{8}$	0.940	$\frac{15}{16}$	$11\frac{1}{4}$	$1\frac{5}{8}$	$\frac{15}{16}$
x109	32.0	14.32	$14\frac{3}{8}$	0.525	$\frac{1}{2}$	$\frac{1}{4}$	14.605	$14\frac{5}{8}$	0.860	$\frac{7}{8}$	$11\frac{1}{4}$	$1\frac{9}{16}$	$\frac{7}{8}$
x 99	29.1	14.16	$14\frac{1}{8}$	0.485	$\frac{1}{2}$	$\frac{1}{4}$	14.565	$14\frac{5}{8}$	0.780	$\frac{3}{4}$	$11\frac{1}{4}$	$1\frac{7}{16}$	$\frac{7}{8}$
x 90	26.5	14.02	14	0.440	$\frac{7}{16}$	$\frac{1}{4}$	14.520	$14\frac{1}{2}$	0.710	$\frac{11}{16}$	$11\frac{1}{4}$	$1\frac{3}{8}$	$\frac{7}{8}$
W 14x 82	24.1	14.31	$14\frac{1}{4}$	0.510	$\frac{1}{2}$	$\frac{1}{4}$	10.130	$10\frac{1}{8}$	0.855	$\frac{7}{8}$	11	$1\frac{5}{8}$	1
x 74	21.8	14.17	$14\frac{1}{8}$	0.450	$\frac{7}{16}$	$\frac{1}{4}$	10.070	$10\frac{1}{8}$	0.785	$\frac{13}{16}$	11	$1\frac{9}{16}$	$\frac{15}{16}$
x 68	20.0	14.04	14	0.415	$\frac{7}{16}$	$\frac{1}{4}$	10.035	10	0.720	$\frac{3}{4}$	11	$1\frac{1}{2}$	$\frac{15}{16}$
x 61	17.9	13.89	$13\frac{7}{8}$	0.375	$\frac{3}{8}$	$\frac{3}{16}$	9.995	10	0.645	$\frac{5}{8}$	11	$1\frac{7}{16}$	$\frac{15}{16}$
W 14x 53	15.6	13.92	$13\frac{7}{8}$	0.370	$\frac{3}{8}$	$\frac{3}{16}$	8.060	8	0.660	$\frac{11}{16}$	11	$1\frac{7}{16}$	$\frac{15}{16}$
x 48	14.1	13.79	$13\frac{3}{4}$	0.340	$\frac{5}{16}$	$\frac{3}{16}$	8.030	8	0.595	$\frac{5}{8}$	11	$1\frac{3}{8}$	$\frac{7}{8}$
x 43	12.6	13.66	$13\frac{5}{8}$	0.305	$\frac{5}{16}$	$\frac{3}{16}$	7.995	8	0.530	$\frac{1}{2}$	11	$1\frac{5}{16}$	$\frac{7}{8}$
W 14x 38	11.2	14.10	$14\frac{1}{8}$	0.310	$\frac{5}{16}$	$\frac{3}{16}$	6.770	$6\frac{3}{4}$	0.515	$\frac{1}{2}$	12	$1\frac{1}{16}$	$\frac{5}{8}$
x 34	10.0	13.98	14	0.285	$\frac{5}{16}$	$\frac{3}{16}$	6.745	$6\frac{3}{4}$	0.455	$\frac{7}{16}$	12	1	$\frac{5}{8}$
x 30	8.85	13.84	$13\frac{7}{8}$	0.270	$\frac{1}{4}$	$\frac{1}{8}$	6.730	$6\frac{3}{4}$	0.385	$\frac{3}{8}$	12	$\frac{15}{16}$	$\frac{5}{8}$
W 14x 26	7.69	13.91	$13\frac{7}{8}$	0.255	$\frac{1}{4}$	$\frac{1}{8}$	5.025	5	0.420	$\frac{7}{16}$	12	$\frac{15}{16}$	$\frac{9}{16}$
x 22	6.49	13.74	$13\frac{3}{4}$	0.230	$\frac{1}{4}$	$\frac{1}{8}$	5.000	5	0.335	$\frac{5}{16}$	12	$\frac{7}{8}$	$\frac{9}{16}$
W 12x336	98.8	16.82	$16\frac{7}{8}$	1.775	$1\frac{3}{4}$	$\frac{7}{8}$	13.385	$13\frac{3}{8}$	2.955	$2\frac{15}{16}$	$9\frac{1}{2}$	$3\frac{11}{16}$	$1\frac{1}{2}$
x305	89.6	16.32	$16\frac{3}{8}$	1.625	$1\frac{5}{8}$	$\frac{13}{16}$	13.235	$13\frac{1}{4}$	2.705	$2\frac{11}{16}$	$9\frac{1}{2}$	$3\frac{7}{16}$	$1\frac{7}{16}$
x279	81.9	15.85	$15\frac{7}{8}$	1.530	$1\frac{1}{2}$	$\frac{3}{4}$	13.140	$13\frac{1}{8}$	2.470	$2\frac{1}{2}$	$9\frac{1}{2}$	$3\frac{3}{16}$	$1\frac{3}{8}$
x252	74.1	15.41	$15\frac{3}{8}$	1.395	$1\frac{3}{8}$	$\frac{11}{16}$	13.005	13	2.250	$2\frac{1}{4}$	$9\frac{1}{2}$	$2\frac{15}{16}$	$1\frac{5}{16}$
x230	67.7	15.05	15	1.285	$1\frac{5}{16}$	$\frac{11}{16}$	12.895	$12\frac{7}{8}$	2.070	$2\frac{1}{16}$	$9\frac{1}{2}$	$2\frac{3}{4}$	$1\frac{1}{4}$
x210	61.8	14.71	$14\frac{3}{4}$	1.180	$1\frac{3}{16}$	$\frac{5}{8}$	12.790	$12\frac{3}{4}$	1.900	$1\frac{7}{8}$	$9\frac{1}{2}$	$2\frac{5}{8}$	$1\frac{1}{4}$
x190	55.8	14.38	$14\frac{3}{8}$	1.060	$1\frac{1}{16}$	$\frac{9}{16}$	12.670	$12\frac{5}{8}$	1.735	$1\frac{3}{4}$	$9\frac{1}{2}$	$2\frac{7}{16}$	$1\frac{3}{16}$
x170	50.0	14.03	14	0.960	$\frac{15}{16}$	$\frac{1}{2}$	12.570	$12\frac{5}{8}$	1.560	$1\frac{9}{16}$	$9\frac{1}{2}$	$2\frac{1}{4}$	$1\frac{1}{8}$
x152	44.7	13.71	$13\frac{3}{4}$	0.870	$\frac{7}{8}$	$\frac{7}{16}$	12.480	$12\frac{1}{2}$	1.400	$1\frac{3}{8}$	$9\frac{1}{2}$	$2\frac{1}{8}$	$1\frac{1}{16}$
x136	39.9	13.41	$13\frac{3}{8}$	0.790	$\frac{13}{16}$	$\frac{7}{16}$	12.400	$12\frac{3}{8}$	1.250	$1\frac{1}{4}$	$9\frac{1}{2}$	$1\frac{15}{16}$	1
x120	35.3	13.12	$13\frac{1}{8}$	0.710	$\frac{11}{16}$	$\frac{3}{8}$	12.320	$12\frac{3}{8}$	1.105	$1\frac{1}{8}$	$9\frac{1}{2}$	$1\frac{13}{16}$	1
x106	31.2	12.89	$12\frac{7}{8}$	0.610	$\frac{5}{8}$	$\frac{5}{16}$	12.220	$12\frac{1}{4}$	0.990	1	$9\frac{1}{2}$	$1\frac{11}{16}$	$\frac{15}{16}$
x 96	28.2	12.71	$12\frac{3}{4}$	0.550	$\frac{9}{16}$	$\frac{5}{16}$	12.160	$12\frac{1}{8}$	0.900	$\frac{7}{8}$	$9\frac{1}{2}$	$1\frac{5}{8}$	$\frac{7}{8}$
x 87	25.6	12.53	$12\frac{1}{2}$	0.515	$\frac{1}{2}$	$\frac{1}{4}$	12.125	$12\frac{1}{8}$	0.810	$\frac{13}{16}$	$9\frac{1}{2}$	$1\frac{1}{2}$	$\frac{7}{8}$
x 79	23.2	12.38	$12\frac{3}{8}$	0.470	$\frac{1}{2}$	$\frac{1}{4}$	12.080	$12\frac{1}{8}$	0.735	$\frac{3}{4}$	$9\frac{1}{2}$	$1\frac{7}{16}$	$\frac{7}{8}$
x 72	21.1	12.25	$12\frac{1}{4}$	0.430	$\frac{7}{16}$	$\frac{1}{4}$	12.040	12	0.670	$\frac{11}{16}$	$9\frac{1}{2}$	$1\frac{3}{8}$	$\frac{7}{8}$
x 65	19.1	12.12	$12\frac{1}{8}$	0.390	$\frac{3}{8}$	$\frac{3}{16}$	12.000	12	0.605	$\frac{5}{8}$	$9\frac{1}{2}$	$1\frac{5}{16}$	$\frac{13}{16}$

W SHAPES

TABLE A.1 (*Continued*)

Nominal Wt. per Ft.	Compact Section Criteria				r_T	$\frac{d}{A_f}$	Elastic Properties						Torsional constant J	Plastic Modulus	
	$\frac{b_f}{2t_f}$	F_y'	$\frac{d}{t_w}$	F_y'''			Axis X-X			Axis Y-Y				Z_x	Z_y
							I	S	r	I	S	r			
Lb.		Ksi		Ksi	In.		In.^{4}	In.^{3}	In.	In.^{4}	In.^{3}	In.	In.^{4}	In.^{3}	In.^{3}
132	7.1	—	22.7	—	4.05	0.97	1530	209	6.28	548	74.5	3.76	12.3	234	113
120	7.8	—	24.5	—	4.04	1.05	1380	190	6.24	495	67.5	3.74	9.37	212	102
109	8.5	58.6	27.3	—	4.02	1.14	1240	173	6.22	447	61.2	3.73	7.12	192	92.7
99	9.3	48.5	29.2	—	4.00	1.25	1110	157	6.17	402	55.2	3.71	5.37	173	83.6
90	10.2	40.4	31.9	—	3.99	1.36	999	143	6.14	362	49.9	3.70	4.06	157	75.6
82	5.9	—	28.1	—	2.74	1.65	882	123	6.05	148	29.3	2.48	5.08	139	44.8
74	6.4	—	31.5	—	2.72	1.79	796	112	6.04	134	26.6	2.48	3.88	126	40.6
68	7.0	—	33.8	57.7	2.71	1.94	723	103	6.01	121	24.2	2.46	3.02	115	36.9
61	7.7	—	37.0	48.1	2.70	2.15	640	92.2	5.98	107	21.5	2.45	2.20	102	32.8
53	6.1	—	37.6	46.7	2.15	2.62	541	77.8	5.89	57.7	14.3	1.92	1.94	87.1	22.0
48	6.7	—	40.6	40.2	2.13	2.89	485	70.3	5.85	51.4	12.8	1.91	1.46	78.4	19.6
43	7.5	—	44.8	32.9	2.12	3.22	428	62.7	5.82	45.2	11.3	1.89	1.05	69.6	17.3
38	6.6	—	45.5	31.9	1.77	4.04	385	54.6	5.87	26.7	7.88	1.55	0.80	61.5	12.1
34	7.4	—	49.1	27.4	1.76	4.56	340	48.6	5.83	23.3	6.91	1.53	0.57	54.6	10.6
30	8.7	55.3	51.3	25.1	1.74	5.34	291	42.0	5.73	19.6	5.82	1.49	0.38	47.3	8.99
26	6.0	—	54.5	22.2	1.28	6.59	245	35.3	5.65	8.91	3.54	1.08	0.36	40.2	5.54
22	7.5	—	59.7	18.5	1.25	8.20	199	29.0	5.54	7.00	2.80	1.04	0.21	33.2	4.39
336	2.3	—	9.5	—	3.71	0.43	4060	483	6.41	1190	177	3.47	243	603	274
305	2.4	—	10.0	—	3.67	0.46	3550	435	6.29	1050	159	3.42	185	537	244
279	2.7	—	10.4	—	3.64	0.49	3110	393	6.16	937	143	3.38	143	481	220
252	2.9	—	11.0	—	3.59	0.53	2720	353	6.06	828	127	3.34	108	428	196
230	3.1	—	11.7	—	3.56	0.56	2420	321	5.97	742	115	3.31	83.8	386	177
210	3.4	—	12.5	—	3.53	0.61	2140	292	5.89	664	104	3.28	64.7	348	159
190	3.7	—	13.6	—	3.50	0.65	1890	263	5.82	589	93.0	3.25	48.8	311	143
170	4.0	—	14.6	—	3.47	0.72	1650	235	5.74	517	82.3	3.22	35.6	275	126
152	4.5	—	15.8	—	3.44	0.79	1430	209	5.66	454	72.8	3.19	25.8	243	111
136	5.0	—	17.0	—	3.41	0.87	1240	186	5.58	398	64.2	3.16	18.5	214	98.0
120	5.6	—	18.5	—	3.38	0.96	1070	163	5.51	345	56.0	3.13	12.9	186	85.4
106	6.2	—	21.1	—	3.36	1.07	933	145	5.47	301	49.3	3.11	9.13	164	75.1
96	6.8	—	23.1	—	3.34	1.16	833	131	5.44	270	44.4	3.09	6.86	147	67.5
87	7.5	—	24.3	—	3.32	1.28	740	118	5.38	241	39.7	3.07	5.10	132	60.4
79	8.2	62.6	26.3	—	3.31	1.39	662	107	5.34	216	35.8	3.05	3.84	119	54.3
72	9.0	52.3	28.5	—	3.29	1.52	597	97.4	5.31	195	32.4	3.04	2.93	108	49.2
65	9.9	43.0	31.1	—	3.28	1.67	533	87.9	5.28	174	29.1	3.02	2.18	96.8	44.1

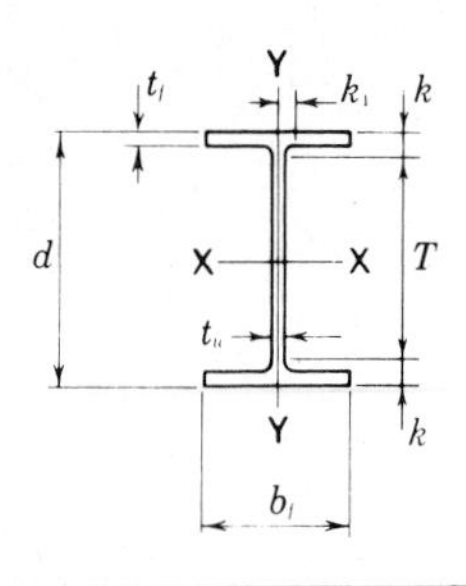

TABLE A.1 Properties of Wide Flange (W) Shapes

Designation	Area A	Depth d		Web Thickness t_w		Web $\frac{t_w}{2}$	Flange Width b_f		Flange Thickness t_f		Distance T	Distance k	Distance k_1
	In.^{2}	In.		In.		In.	In.		In.		In.	In.	In.
W 12x 58	17.0	12.19	12 1/4	0.360	3/8	3/16	10.010	10	0.640	5/8	9 1/2	1 3/8	13/16
x 53	15.6	12.06	12	0.345	3/8	3/16	9.995	10	0.575	9/16	9 1/2	1 1/4	13/16
W 12x 50	14.7	12.19	12 1/4	0.370	3/8	3/16	8.080	8 1/8	0.640	5/8	9 1/2	1 3/8	13/16
x 45	13.2	12.06	12	0.335	5/16	3/16	8.045	8	0.575	9/16	9 1/2	1 1/4	13/16
x 40	11.8	11.94	12	0.295	5/16	3/16	8.005	8	0.515	1/2	9 1/2	1 1/4	3/4
W 12x 35	10.3	12.50	12 1/2	0.300	5/16	3/16	6.560	6 1/2	0.520	1/2	10 1/2	1	9/16
x 30	8.79	12.34	12 3/8	0.260	1/4	1/8	6.520	6 1/2	0.440	7/16	10 1/2	15/16	1/2
x 26	7.65	12.22	12 1/4	0.230	1/4	1/8	6.490	6 1/2	0.380	3/8	10 1/2	7/8	1/2
W 12x 22	6.48	12.31	12 1/4	0.260	1/4	1/8	4.030	4	0.425	7/16	10 1/2	7/8	1/2
x 19	5.57	12.16	12 1/8	0.235	1/4	1/8	4.005	4	0.350	3/8	10 1/2	13/16	1/2
x 16	4.71	11.99	12	0.220	1/4	1/8	3.990	4	0.265	1/4	10 1/2	3/4	1/2
x 14	4.16	11.91	11 7/8	0.200	3/16	1/8	3.970	4	0.225	1/4	10 1/2	11/16	1/2
W 10x112	32.9	11.36	11 3/8	0.755	3/4	3/8	10.415	10 3/8	1.250	1 1/4	7 5/8	1 7/8	15/16
x100	29.4	11.10	11 1/8	0.680	11/16	3/8	10.340	10 3/8	1.120	1 1/8	7 5/8	1 3/4	7/8
x 88	25.9	10.84	10 7/8	0.605	5/8	5/16	10.265	10 1/4	0.990	1	7 5/8	1 5/8	13/16
x 77	22.6	10.60	10 5/8	0.530	1/2	1/4	10.190	10 1/4	0.870	7/8	7 5/8	1 1/2	13/16
x 68	20.0	10.40	10 3/8	0.470	1/2	1/4	10.130	10 1/8	0.770	3/4	7 5/8	1 3/8	3/4
x 60	17.6	10.22	10 1/4	0.420	7/16	1/4	10.080	10 1/8	0.680	11/16	7 5/8	1 5/16	3/4
x 54	15.8	10.09	10 1/8	0.370	3/8	3/16	10.030	10	0.615	5/8	7 5/8	1 1/4	11/16
x 49	14.4	9.98	10	0.340	5/16	3/16	10.000	10	0.560	9/16	7 5/8	1 3/16	11/16
W 10x 45	13.3	10.10	10 1/8	0.350	3/8	3/16	8.020	8	0.620	5/8	7 5/8	1 1/4	11/16
x 39	11.5	9.92	9 7/8	0.315	5/16	3/16	7.985	8	0.530	1/2	7 5/8	1 1/8	11/16
x 33	9.71	9.73	9 3/4	0.290	5/16	3/16	7.960	8	0.435	7/16	7 5/8	1 1/16	11/16
W 10x 30	8.84	10.47	10 1/2	0.300	5/16	3/16	5.810	5 3/4	0.510	1/2	8 5/8	15/16	1/2
x 26	7.61	10.33	10 3/8	0.260	1/4	1/8	5.770	5 3/4	0.440	7/16	8 5/8	7/8	1/2
x 22	6.49	10.17	10 1/8	0.240	1/4	1/8	5.750	5 3/4	0.360	3/8	8 5/8	3/4	1/2
W 10x 19	5.62	10.24	10 1/4	0.250	1/4	1/8	4.020	4	0.395	3/8	8 5/8	13/16	1/2
x 17	4.99	10.11	10 1/8	0.240	1/4	1/8	4.010	4	0.330	5/16	8 5/8	3/4	1/2
x 15	4.41	9.99	10	0.230	1/4	1/8	4.000	4	0.270	1/4	8 5/8	11/16	7/16
x 12	3.54	9.87	9 7/8	0.190	3/16	1/8	3.960	4	0.210	3/16	8 5/8	5/8	7/16

W SHAPES

TABLE A.1 (*Continued*)

Nominal Wt. per Ft.	Compact Section Criteria				r_T	$\frac{d}{A_f}$	Elastic Properties						Torsional constant J	Plastic Modulus	
							Axis X-X			Axis Y-Y					
	$\frac{b_f}{2t_f}$	F_y'	$\frac{d}{t_w}$	F_y'''			I	S	r	I	S	r		Z_x	Z_y
Lb.		Ksi		Ksi	In.		In.^{4}	In.^{3}	In.	In.^{4}	In.^{3}	In.	In.^{4}	In.^{3}	In.^{3}
58	7.8	—	33.9	57.6	2.72	1.90	475	78.0	5.28	107	21.4	2.51	2.10	86.4	32.5
53	8.7	55.9	35.0	54.1	2.71	2.10	425	70.6	5.23	95.8	19.2	2.48	1.58	77.9	29.1
50	6.3	—	32.9	60.9	2.17	2.36	394	64.7	5.18	56.3	13.9	1.96	1.78	72.4	21.4
45	7.0	—	36.0	51.0	2.15	2.61	350	58.1	5.15	50.0	12.4	1.94	1.31	64.7	19.0
40	7.8	—	40.5	40.3	2.14	2.90	310	51.9	5.13	44.1	11.0	1.93	0.95	57.5	16.8
35	6.3	—	41.7	38.0	1.74	3.66	285	45.6	5.25	24.5	7.47	1.54	0.74	51.2	11.5
30	7.4	—	47.5	29.3	1.73	4.30	238	38.6	5.21	20.3	6.24	1.52	0.46	43.1	9.56
26	8.5	57.9	53.1	23.4	1.72	4.95	204	33.4	5.17	17.3	5.34	1.51	0.30	37.2	8.17
22	4.7	—	47.3	29.5	1.02	7.19	156	25.4	4.91	4.66	2.31	0.847	0.29	29.3	3.66
19	5.7	—	51.7	24.7	1.00	8.67	130	21.3	4.82	3.76	1.88	0.822	0.18	24.7	2.98
16	7.5	—	54.5	22.2	0.96	11.3	103	17.1	4.67	2.82	1.41	0.773	0.10	20.1	2.26
14	8.8	54.3	59.6	18.6	0.95	13.3	88.6	14.9	4.62	2.36	1.19	0.753	0.07	17.4	1.90
112	4.2	—	15.0	—	2.88	0.87	716	126	4.66	236	45.3	2.68	15.1	147	69.2
100	4.6	—	16.3	—	2.85	0.96	623	112	4.60	207	40.0	2.65	10.9	130	61.0
88	5.2	—	17.9	—	2.83	1.07	534	98.5	4.54	179	34.8	2.63	7.53	113	53.1
77	5.9	—	20.0	—	2.80	1.20	455	85.9	4.49	154	30.1	2.60	5.11	97.6	45.9
68	6.6	—	22.1	—	2.79	1.33	394	75.7	4.44	134	26.4	2.59	3.56	85.3	40.1
60	7.4	—	24.3	—	2.77	1.49	341	66.7	4.39	116	23.0	2.57	2.48	74.6	35.0
54	8.2	63.5	27.3	—	2.75	1.64	303	60.0	4.37	103	20.6	2.56	1.82	66.6	31.3
49	8.9	53.0	29.4	—	2.74	1.78	272	54.6	4.35	93.4	18.7	2.54	1.39	60.4	28.3
45	6.5	—	28.9	—	2.18	2.03	248	49.1	4.32	53.4	13.3	2.01	1.51	54.9	20.3
39	7.5	—	31.5	—	2.16	2.34	209	42.1	4.27	45.0	11.3	1.98	0.98	46.8	17.2
33	9.1	50.5	33.6	58.7	2.14	2.81	170	35.0	4.19	36.6	9.20	1.94	0.58	38.8	14.0
30	5.7	—	34.9	54.2	1.55	3.53	170	32.4	4.38	16.7	5.75	1.37	0.62	36.6	8.84
26	6.6	—	39.7	41.8	1.54	4.07	144	27.9	4.35	14.1	4.89	1.36	0.40	31.3	7.50
22	8.0	—	42.4	36.8	1.51	4.91	118	23.2	4.27	11.4	3.97	1.33	0.24	26.0	6.10
19	5.1	—	41.0	39.4	1.03	6.45	96.3	18.8	4.14	4.29	2.14	0.874	0.23	21.6	3.35
17	6.1	—	42.1	37.2	1.01	7.64	81.9	16.2	4.05	3.56	1.78	0.844	0.16	18.7	2.80
15	7.4	—	43.4	35.0	0.99	9.25	68.9	13.8	3.95	2.89	1.45	0.810	0.10	16.0	2.30
12	9.4	47.5	51.9	24.5	0.96	11.9	53.8	10.9	3.90	2.18	1.10	0.785	0.06	12.6	1.74

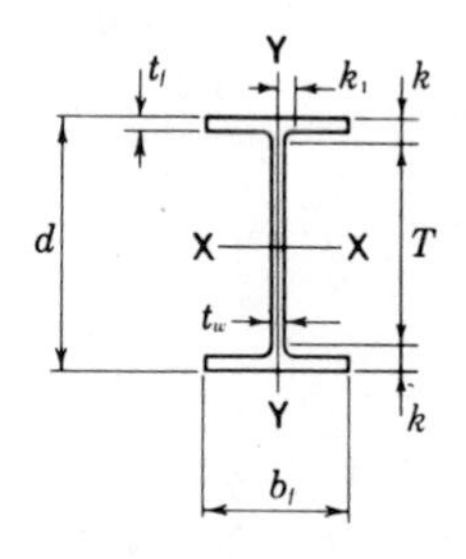

TABLE A.1 Properties of Wide Flange (W) Shapes

Designation	Area A	Depth d		Web: Thickness t_w		Web: $\frac{t_w}{2}$	Flange: Width b_f		Flange: Thickness t_f		Distance: T	Distance: k	Distance: k_1
	In.^{2}	In.		In.		In.	In.		In.		In.	In.	In.
W 8x67	19.7	9.00	9	0.570	9/16	5/16	8.280	8 1/4	0.935	15/16	6 1/8	1 7/16	11/16
x58	17.1	8.75	8 3/4	0.510	1/2	1/4	8.220	8 1/4	0.810	13/16	6 1/8	1 5/16	11/16
x48	14.1	8.50	8 1/2	0.400	3/8	3/16	8.110	8 1/8	0.685	11/16	6 1/8	1 3/16	5/8
x40	11.7	8.25	8 1/4	0.360	3/8	3/16	8.070	8 1/8	0.560	9/16	6 1/8	1 1/16	5/8
x35	10.3	8.12	8 1/8	0.310	5/16	3/16	8.020	8	0.495	1/2	6 1/8	1	9/16
x31	9.13	8.00	8	0.285	5/16	3/16	7.995	8	0.435	7/16	6 1/8	15/16	9/16
W 8x28	8.25	8.06	8	0.285	5/16	3/16	6.535	6 1/2	0.465	7/16	6 1/8	15/16	9/16
x24	7.08	7.93	7 7/8	0.245	1/4	1/8	6.495	6 1/2	0.400	3/8	6 1/8	7/8	9/16
W 8x21	6.16	8.28	8 1/4	0.250	1/4	1/8	5.270	5 1/4	0.400	3/8	6 5/8	13/16	1/2
x18	5.26	8.14	8 1/8	0.230	1/4	1/8	5.250	5 1/4	0.330	5/16	6 5/8	3/4	7/16
W 8x15	4.44	8.11	8 1/8	0.245	1/4	1/8	4.015	4	0.315	5/16	6 5/8	3/4	1/2
x13	3.84	7.99	8	0.230	1/4	1/8	4.000	4	0.255	1/4	6 5/8	11/16	7/16
x10	2.96	7.89	7 7/8	0.170	3/16	1/8	3.940	4	0.205	3/16	6 5/8	5/8	7/16
W 6x25	7.34	6.38	6 3/8	0.320	5/16	3/16	6.080	6 1/8	0.455	7/16	4 3/4	13/16	7/16
x20	5.87	6.20	6 1/4	0.260	1/4	1/8	6.020	6	0.365	3/8	4 3/4	3/4	7/16
x15	4.43	5.99	6	0.230	1/4	1/8	5.990	6	0.260	1/4	4 3/4	5/8	3/8
W 6x16	4.74	6.28	6 1/4	0.260	1/4	1/8	4.030	4	0.405	3/8	4 3/4	3/4	7/16
x12	3.55	6.03	6	0.230	1/4	1/8	4.000	4	0.280	1/4	4 3/4	5/8	3/8
x 9	2.68	5.90	5 7/8	0.170	3/16	1/8	3.940	4	0.215	3/16	4 3/4	9/16	3/8
W 5x19	5.54	5.15	5 1/8	0.270	1/4	1/8	5.030	5	0.430	7/16	3 1/2	13/16	7/16
x16	4.68	5.01	5	0.240	1/4	1/8	5.000	5	0.360	3/8	3 1/2	3/4	7/16
W 4x13	3.83	4.16	4 1/8	0.280	1/4	1/8	4.060	4	0.345	3/8	2 3/4	11/16	7/16

W SHAPES

TABLE A.1 (*Continued*)

Nominal Wt. per Ft.	Compact Section Criteria				r_T	$\frac{d}{A_f}$	Elastic Properties						Torsional constant	Plastic Modulus	
							Axis X-X			Axis Y-Y					
	$\frac{b_f}{2t_f}$	F_y'	$\frac{d}{t_w}$	F_y'''			I	S	r	I	S	r	J	Z_x	Z_y
Lb.		Ksi		Ksi	In.		In.^{4}	In.^{3}	In.	In.^{4}	In.^{3}	In.	In.^{4}	In.^{3}	In.^{3}
67	4.4	—	15.8	—	2.28	1.16	272	60.4	3.72	88.6	21.4	2.12	5.06	70.2	32.7
58	5.1	—	17.2	—	2.26	1.31	228	52.0	3.65	75.1	18.3	2.10	3.34	59.8	27.9
48	5.9	—	21.3	—	2.23	1.53	184	43.3	3.61	60.9	15.0	2.08	1.96	49.0	22.9
40	7.2	—	22.9	—	2.21	1.83	146	35.5	3.53	49.1	12.2	2.04	1.12	39.8	18.5
35	8.1	64.4	26.2	—	2.20	2.05	127	31.2	3.51	42.6	10.6	2.03	0.77	34.7	16.1
31	9.2	50.0	28.1	—	2.18	2.30	110	27.5	3.47	37.1	9.27	2.02	0.54	30.4	14.1
28	7.0	—	28.3	—	1.77	2.65	98.0	24.3	3.45	21.7	6.63	1.62	0.54	27.2	10.1
24	8.1	64.1	32.4	63.0	1.76	3.05	82.8	20.9	3.42	18.3	5.63	1.61	0.35	23.2	8.57
21	6.6	—	33.1	60.2	1.41	3.93	75.3	18.2	3.49	9.77	3.71	1.26	0.28	20.4	5.69
18	8.0	—	35.4	52.7	1.39	4.70	61.9	15.2	3.43	7.97	3.04	1.23	0.17	17.0	4.66
15	6.4	—	33.1	60.3	1.03	6.41	48.0	11.8	3.29	3.41	1.70	0.876	0.14	13.6	2.67
13	7.8	—	34.7	54.7	1.01	7.83	39.6	9.91	3.21	2.73	1.37	0.843	0.09	11.4	2.15
10	9.6	45.8	46.4	30.7	0.99	9.77	30.8	7.81	3.22	2.09	1.06	0.841	0.04	8.87	1.66
25	6.7	—	19.9	—	1.66	2.31	53.4	16.7	2.70	17.1	5.61	1.52	0.46	18.9	8.56
20	8.2	62.1	23.8	—	1.64	2.82	41.4	13.4	2.66	13.3	4.41	1.50	0.24	14.9	6.72
15	11.5	31.8	26.0	—	1.61	3.85	29.1	9.72	2.56	9.32	3.11	1.46	0.10	10.8	4.75
16	5.0	—	24.2	—	1.08	3.85	32.1	10.2	2.60	4.43	2.20	0.966	0.22	11.7	3.39
12	7.1	—	26.2	—	1.05	5.38	22.1	7.31	2.49	2.99	1.50	0.918	0.09	8.30	2.32
9	9.2	50.3	34.7	54.8	1.03	6.96	16.4	5.56	2.47	2.19	1.11	0.905	0.04	6.23	1.72
19	5.8	—	19.1	—	1.38	2.38	26.2	10.2	2.17	9.13	3.63	1.28	0.31	11.6	5.53
16	6.9	—	20.9	—	1.37	2.78	21.3	8.51	2.13	7.51	3.00	1.27	0.19	9.59	4.57
13	5.9	—	14.9	—	1.10	2.97	11.3	5.46	1.72	3.86	1.90	1.00	0.15	6.28	2.92

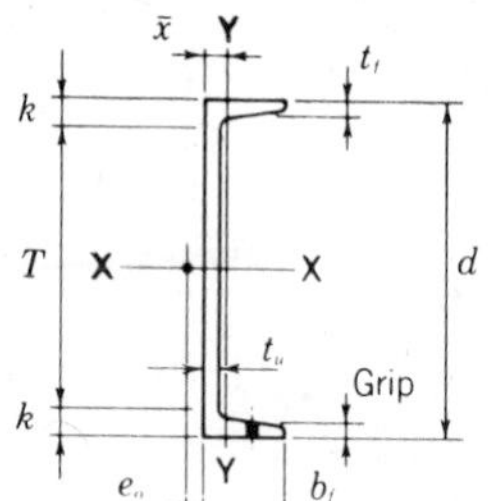

TABLE A.2 Properties of American Standard Channels (C)

Designation	Area A	Depth d	Web: Thickness t_w		Web: $\frac{t_w}{2}$	Flange: Width b_f		Flange: Average thickness t_f		Distance: T	Distance: k	Grip	Max. Flge. Fastener
	In.^{2}	In.	In.		In.	In.		In.		In.	In.	In.	In.
C 15x50	14.7	15.00	0.716	11/16	3/8	3.716	3 3/4	0.650	5/8	12 1/8	1 7/16	5/8	1
x40	11.8	15.00	0.520	1/2	1/4	3.520	3 1/2	0.650	5/8	12 1/8	1 7/16	5/8	1
x33.9	9.96	15.00	0.400	3/8	3/16	3.400	3 3/8	0.650	5/8	12 1/8	1 7/16	5/8	1
C 12x30	8.82	12.00	0.510	1/2	1/4	3.170	3 1/8	0.501	1/2	9 3/4	1 1/8	1/2	7/8
x25	7.35	12.00	0.387	3/8	3/16	3.047	3	0.501	1/2	9 3/4	1 1/8	1/2	7/8
x20.7	6.09	12.00	0.282	5/16	1/8	2.942	3	0.501	1/2	9 3/4	1 1/8	1/2	7/8
C 10x30	8.82	10.00	0.673	11/16	5/16	3.033	3	0.436	7/16	8	1	7/16	3/4
x25	7.35	10.00	0.526	1/2	1/4	2.886	2 7/8	0.436	7/16	8	1	7/16	3/4
x20	5.88	10.00	0.379	3/8	3/16	2.739	2 3/4	0.436	7/16	8	1	7/16	3/4
x15.3	4.49	10.00	0.240	1/4	1/8	2.600	2 5/8	0.436	7/16	8	1	7/16	3/4
C 9x20	5.88	9.00	0.448	7/16	1/4	2.648	2 5/8	0.413	7/16	7 1/8	15/16	7/16	3/4
x15	4.41	9.00	0.285	5/16	1/8	2.485	2 1/2	0.413	7/16	7 1/8	15/16	7/16	3/4
x13.4	3.94	9.00	0.233	1/4	1/8	2.433	2 3/8	0.413	7/16	7 1/8	15/16	7/16	3/4
C 8x18.75	5.51	8.00	0.487	1/2	1/4	2.527	2 1/2	0.390	3/8	6 1/8	15/16	3/8	3/4
x13.75	4.04	8.00	0.303	5/16	1/8	2.343	2 3/8	0.390	3/8	6 1/8	15/16	3/8	3/4
x11.5	3.38	8.00	0.220	1/4	1/8	2.260	2 1/4	0.390	3/8	6 1/8	15/16	3/8	3/4
C 7x14.75	4.33	7.00	0.419	7/16	3/16	2.299	2 1/4	0.366	3/8	5 1/4	7/8	3/8	5/8
x12.25	3.60	7.00	0.314	5/16	3/16	2.194	2 1/4	0.366	3/8	5 1/4	7/8	3/8	5/8
x 9.8	2.87	7.00	0.210	3/16	1/8	2.090	2 1/8	0.366	3/8	5 1/4	7/8	3/8	5/8
C 6x13	3.83	6.00	0.437	7/16	3/16	2.157	2 1/8	0.343	5/16	4 3/8	13/16	5/16	5/8
x10.5	3.09	6.00	0.314	5/16	3/16	2.034	2	0.343	5/16	4 3/8	13/16	3/8	5/8
x 8.2	2.40	6.00	0.200	3/16	1/8	1.920	1 7/8	0.343	5/16	4 3/8	13/16	5/16	5/8
C 5x 9	2.64	5.00	0.325	5/16	3/16	1.885	1 7/8	0.320	5/16	3 1/2	3/4	5/16	5/8
x 6.7	1.97	5.00	0.190	3/16	1/8	1.750	1 3/4	0.320	5/16	3 1/2	3/4	—	—
C 4x 7.25	2.13	4.00	0.321	5/16	3/16	1.721	1 3/4	0.296	5/16	2 5/8	11/16	5/16	5/8
x 5.4	1.59	4.00	0.184	3/16	1/16	1.584	1 5/8	0.296	5/16	2 5/8	11/16	—	—
C 3x 6	1.76	3.00	0.356	3/8	3/16	1.596	1 5/8	0.273	1/4	1 5/8	11/16	—	—
x 5	1.47	3.00	0.258	1/4	1/8	1.498	1 1/2	0.273	1/4	1 5/8	11/16	—	—
x 4.1	1.21	3.00	0.170	3/16	1/16	1.410	1 3/8	0.273	1/4	1 5/8	11/16	—	—

TABLE A.2 (*Continued*)

Nominal Weight per Ft.	$\bar{x}$	Shear Center Location e_o	$\frac{d}{A_f}$	Axis X-X			Axis Y-Y		
				I	S	r	I	S	r
	In.	In.		In.^{4}	In.^{3}	In.	In.^{4}	In.^{3}	In.
50	0.798	0.583	6.21	404	53.8	5.24	11.0	3.78	0.867
40	0.777	0.767	6.56	349	46.5	5.44	9.23	3.37	0.886
33.9	0.787	0.896	6.79	315	42.0	5.62	8.13	3.11	0.904
30	0.674	0.618	7.55	162	27.0	4.29	5.14	2.06	0.763
25	0.674	0.746	7.85	144	24.1	4.43	4.47	1.88	0.780
20.7	0.698	0.870	8.13	129	21.5	4.61	3.88	1.73	0.799
30	0.649	0.369	7.55	103	20.7	3.42	3.94	1.65	0.669
25	0.617	0.494	7.94	91.2	18.2	3.52	3.36	1.48	0.676
20	0.606	0.637	8.36	78.9	15.8	3.66	2.81	1.32	0.692
15.3	0.634	0.796	8.81	67.4	13.5	3.87	2.28	1.16	0.713
20	0.583	0.515	8.22	60.9	13.5	3.22	2.42	1.17	0.642
15	0.586	0.682	8.76	51.0	11.3	3.40	1.93	1.01	0.661
13.4	0.601	0.743	8.95	47.9	10.6	3.48	1.76	0.962	0.669
18.75	0.565	0.431	8.12	44.0	11.0	2.82	1.98	1.01	0.599
13.75	0.553	0.604	8.75	36.1	9.03	2.99	1.53	0.854	0.615
11.5	0.571	0.697	9.08	32.6	8.14	3.11	1.32	0.781	0.625
14.75	0.532	0.441	8.31	27.2	7.78	2.51	1.38	0.779	0.564
12.25	0.525	0.538	8.71	24.2	6.93	2.60	1.17	0.703	0.571
9.8	0.540	0.647	9.14	21.3	6.08	2.72	0.968	0.625	0.581
13	0.514	0.380	8.10	17.4	5.80	2.13	1.05	0.642	0.525
10.5	0.499	0.486	8.59	15.2	5.06	2.22	0.866	0.564	0.529
8.2	0.511	0.599	9.10	13.1	4.38	2.34	0.693	0.492	0.537
9	0.478	0.427	8.29	8.90	3.56	1.83	0.632	0.450	0.489
6.7	0.484	0.552	8.93	7.49	3.00	1.95	0.479	0.378	0.493
7.25	0.459	0.386	7.84	4.59	2.29	1.47	0.433	0.343	0:450
5.4	0.457	0.502	8.52	3.85	1.93	1.56	0.319	0.283	0.449
6	0.455	0.322	6.87	2.07	1.38	1.08	0.305	0.268	0.416
5	0.438	0.392	7.32	1.85	1.24	1.12	0.247	0.233	0.410
4.1	0.436	0.461	7.78	1.66	1.10	1.17	0.197	0.202	0.404

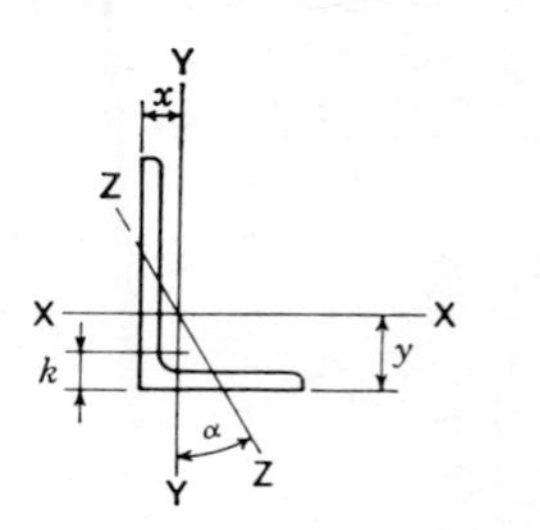

TABLE A.3 Properties of Angles

Size and Thickness	k	Weight per Foot	Area	AXIS X-X				AXIS Y-Y				AXIS Z-Z	
				I	S	r	y	I	S	r	x	r	Tan α
In.	In.	Lb.	In.^{2}	In.^{4}	In.^{3}	In.	In.	In.^{4}	In.^{3}	In.	In.	In.	
L 9 × 4 ×													
L 8 × 8 × 1⅛	1¾	56.9	16.7	98.0	17.5	2.42	2.41	98.0	17.5	2.42	2.41	1.56	1.000
1	1⅝	51.0	15.0	89.0	15.8	2.44	2.37	89.0	15.8	2.44	2.37	1.56	1.000
⅞	1½	45.0	13.2	79.6	14.0	2.45	2.32	79.6	14.0	2.45	2.32	1.57	1.000
¾	1⅜	38.9	11.4	69.7	12.2	2.47	2.28	69.7	12.2	2.47	2.28	1.58	1.000
⅝	1¼	32.7	9.61	59.4	10.3	2.49	2.23	59.4	10.3	2.49	2.23	1.58	1.000
½	1⅛	26.4	7.75	48.6	8.36	2.50	2.19	48.6	8.36	2.50	2.19	1.59	1.000
L 8 × 6 × 1	1½	44.2	13.0	80.8	15.1	2.49	2.65	38.8	8.92	1.73	1.65	1.28	0.543
¾	1¼	33.8	9.94	63.4	11.7	2.53	2.56	30.7	6.92	1.76	1.56	1.29	0.551
½	1	23.0	6.75	44.3	8.02	2.56	2.47	21.7	4.79	1.79	1.47	1.30	0.558
L8 × 4 × 1	1½	37.4	11.0	69.6	14.1	2.52	3.05	11.6	3.94	1.03	1.05	0.846	0.247
¾	1¼	28.7	8.44	54.9	10.9	2.55	2.95	9.36	3.07	1.05	0.953	0.852	0.258
½	1	19.6	5.75	38.5	7.49	2.59	2.86	6.74	2.15	1.08	0.859	0.865	0.267
L 7 × 4 × ¾	1¼	26.2	7.69	37.8	8.42	2.22	2.51	9.05	3.03	1.09	1.01	0.860	0.324
½	1	17.9	5.25	26.7	5.81	2.25	2.42	6.53	2.12	1.11	0.917	0.872	0.335
⅜	⅞	13.6	3.98	20.6	4.44	2.27	2.37	5.10	1.63	1.13	0.870	0.880	0.340

L6 × 6 × 1	1 1/2	37.4	11.0	35.5	8.57	1.80	1.86	35.5	8.57	1.80	1.86	1.17	1.000
7/8	1 3/8	33.1	9.73	31.9	7.63	1.81	1.82	31.9	7.63	1.81	1.82	1.17	1.000
3/4	1 1/4	28.7	8.44	28.2	6.66	1.83	1.78	28.2	6.66	1.83	1.78	1.17	1.000
5/8	1 1/8	24.2	7.11	24.2	5.66	1.84	1.73	24.2	5.66	1.84	1.73	1.18	1.000
1/2	1	19.6	5.75	19.9	4.61	1.86	1.68	19.9	4.61	1.86	1.68	1.18	1.000
3/8	7/8	14.9	4.36	15.4	3.53	1.88	1.64	15.4	3.53	1.88	1.64	1.19	1.000
L 6 × 4 × 3/4	1 1/4	23.6	6.94	24.5	6.25	1.88	2.08	8.68	2.97	1.12	1.08	0.860	0.428
5/8	1 1/8	20.0	5.86	21.1	5.31	1.90	2.03	7.52	2.54	1.13	1.03	0.864	0.435
1/2	1	16.2	4.75	17.4	4.33	1.91	1.99	6.27	2.08	1.15	0.987	0.870	0.440
3/8	7/8	12.3	3.61	13.5	3.32	1.93	1.94	4.90	1.60	1.17	0.941	0.877	0.446
L 6 × 3 1/2 × 3/8	7/8	11.7	3.42	12.9	3.24	1.94	2.04	3.34	1.23	0.988	0.787	0.767	0.350
5/16	13/16	9.8	2.87	10.9	2.73	1.95	2.01	2.85	1.04	0.996	0.763	0.772	0.352
L 5 × 5 × 7/8	1 3/8	27.2	7.98	17.8	5.17	1.49	1.57	17.8	5.17	1.49	1.57	0.973	1.000
3/4	1 1/4	23.6	6.94	15.7	4.53	1.51	1.52	15.7	4.53	1.51	1.52	0.975	1.000
1/2	1	16.2	4.75	11.3	3.16	1.54	1.43	11.3	3.16	1.54	1.43	0.983	1.000
3/8	7/8	12.3	3.61	8.74	2.42	1.56	1.39	8.74	2.42	1.56	1.39	0.990	1.000
5/16	13/16	10.3	3.03	7.42	2.04	1.57	1.37	7.42	2.04	1.57	1.37	0.994	1.000
L 5 × 3 1/2 × 3/4	1 1/4	19.8	5.81	13.9	4.28	1.55	1.75	5.55	2.22	0.977	0.996	0.748	0.464
1/2	1	13.6	4.00	9.99	2.99	1.58	1.66	4.05	1.56	1.01	0.906	0.755	0.479
3/8	7/8	10.4	3.05	7.78	2.29	1.60	1.61	3.18	1.21	1.02	0.861	0.762	0.486
5/16	13/16	8.7	2.56	6.60	1.94	1.61	1.59	2.72	1.02	1.03	0.838	0.766	0.489
L 5 × 3 × 1/2	1	12.8	3.75	9.45	2.91	1.59	1.75	2.58	1.15	0.829	0.750	0.648	0.357
3/8	7/8	9.8	2.86	7.37	2.24	1.61	1.70	2.04	0.888	0.845	0.704	0.654	0.364
5/16	13/16	8.2	2.40	6.26	1.89	1.61	1.68	1.75	0.753	0.853	0.681	0.658	0.368
1/4	3/4	6.6	1.94	5.11	1.53	1.62	1.66	1.44	0.614	0.861	0.657	0.663	0.371
L 4 × 4 × 3/4	1 1/8	18.5	5.44	7.67	2.81	1.19	1.27	7.67	2.81	1.19	1.27	0.778	1.000
5/8	1	15.7	4.61	6.66	2.40	1.20	1.23	6.66	2.40	1.20	1.23	0.779	1.000
1/2	7/8	12.8	3.75	5.56	1.97	1.22	1.18	5.56	1.97	1.22	1.18	0.782	1.000
3/8	3/4	9.8	2.86	4.36	1.52	1.23	1.14	4.36	1.52	1.23	1.14	0.788	1.000
5/16	11/16	8.2	2.40	3.71	1.29	1.24	1.12	3.71	1.29	1.24	1.12	0.791	1.000
1/4	5/8	6.6	1.94	3.04	1.05	1.25	1.09	3.04	1.05	1.25	1.09	0.795	1.000

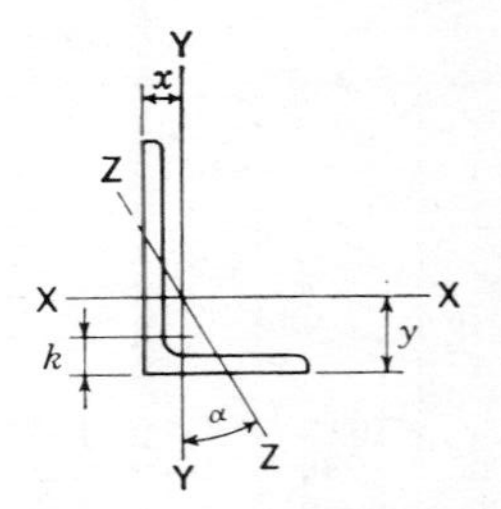

Size and Thickness	k	Weight per Foot	Area	AXIS X-X				AXIS Y-Y				AXIS Z-Z	
				I	S	r	y	I	S	r	x	r	Tan α
In.	In.	Lb.	In.^{2}	In.^{4}	In.^{3}	In.	In.	In.^{4}	In.^{3}	In.	In.	In.	
L 4 × $3\frac{1}{2}$ × $\frac{1}{2}$	$\frac{15}{16}$	11.9	3.50	5.32	1.94	1.23	1.25	3.79	1.52	1.04	1.00	0.722	0.750
$\frac{3}{8}$	$\frac{13}{16}$	9.1	2.67	4.18	1.49	1.25	1.21	2.95	1.17	1.06	0.955	0.727	0.755
$\frac{5}{16}$	$\frac{3}{4}$	7.7	2.25	3.56	1.26	1.26	1.18	2.55	0.994	1.07	0.932	0.730	0.757
$\frac{1}{4}$	$\frac{11}{16}$	6.2	1.81	2.91	1.03	1.27	1.16	2.09	0.808	1.07	0.909	0.734	0.759
L 4 × 3 × $\frac{1}{2}$	$\frac{15}{16}$	11.1	3.25	5.05	1.89	1.25	1.33	2.42	1.12	0.864	0.827	0.639	0.543
$\frac{3}{8}$	$\frac{13}{16}$	8.5	2.48	3.96	1.46	1.26	1.28	1.92	0.866	0.879	0.782	0.644	0.551
$\frac{5}{16}$	$\frac{3}{4}$	7.2	2.09	3.38	1.23	1.27	1.26	1.65	0.734	0.887	0.759	0.647	0.554
$\frac{1}{4}$	$\frac{11}{16}$	5.8	1.69	2.77	1.00	1.28	1.24	1.36	0.599	0.896	0.736	0.651	0.558
L $3\frac{1}{2}$ × $3\frac{1}{2}$ × $\frac{3}{8}$	$\frac{3}{4}$	8.5	2.48	2.87	1.15	1.07	1.01	2.87	1.15	1.07	1.01	0.687	1.000
$\frac{5}{16}$	$\frac{11}{16}$	7.2	2.09	2.45	0.976	1.08	0.990	2.45	0.976	1.08	0.990	0.690	1.000
$\frac{1}{4}$	$\frac{5}{8}$	5.8	1.69	2.01	0.794	1.09	0.968	2.01	0.794	1.09	0.968	0.694	1.000
L $3\frac{1}{2}$ × 3 × $\frac{3}{8}$	$\frac{13}{16}$	7.9	2.30	2.72	1.13	1.09	1.08	1.85	0.851	0.897	0.830	0.625	0.721
$\frac{5}{16}$	$\frac{3}{4}$	6.6	1.93	2.33	0.954	1.10	1.06	1.58	0.722	0.905	0.808	0.627	0.724
$\frac{1}{4}$	$\frac{11}{16}$	5.4	1.56	1.91	0.776	1.11	1.04	1.30	0.589	0.914	0.785	0.631	0.727

L 3½ × 2½ × 3/8	13/16	7.2	2.11	2.56	1.09	1.10	1.16	1.09	0.592	0.719	0.660	0.537	0.496
5/16	3/4	6.1	1.78	2.19	0.927	1.11	1.14	0.939	0.504	0.727	0.637	0.540	0.501
1/4	11/16	4.9	1.44	1.80	0.755	1.12	1.11	0.777	0.412	0.735	0.614	0.544	0.506
L 3 × 3 × 1/2	13/16	9.4	2.75	2.22	1.07	0.898	0.932	2.22	1.07	0.898	0.932	0.584	1.000
3/8	11/16	7.2	2.11	1.76	0.833	0.913	0.888	1.76	0.833	0.913	0.888	0.587	1.000
5/16	5/8	6.1	1.78	1.51	0.707	0.922	0.865	1.51	0.707	0.922	0.865	0.589	1.000
1/4	9/16	4.9	1.44	1.24	0.577	0.930	0.842	1.24	0.577	0.930	0.842	0.592	1.000
3/16	1/2	3.71	1.09	0.962	0.441	0.939	0.820	0.962	0.441	0.939	0.820	0.596	1.000
L 3 × 2½ × 3/8	3/4	6.6	1.92	1.66	0.810	0.928	0.956	1.04	0.581	0.736	0.706	0.522	0.676
1/4	5/8	4.5	1.31	1.17	0.561	0.945	0.911	0.743	0.404	0.753	0.661	0.528	0.684
3/16	9/16	3.89	0.996	0.907	0.430	0.954	0.888	0.577	0.310	0.761	0.638	0.533	0.688
L 3 × 2 × 3/8	11/16	5.9	1.73	1.53	0.781	0.940	1.04	0.543	0.371	0.559	0.539	0.430	0.428
5/16	5/8	5.0	1.46	1.32	0.664	0.948	1.02	0.470	0.317	0.567	0.516	0.432	0.435
1/4	9/16	4.1	1.19	1.09	0.542	0.957	0.993	0.392	0.260	0.574	0.493	0.435	0.440
3/16	1/2	3.07	0.902	0.842	0.415	0.966	0.970	0.307	0.200	0.583	0.470	0.439	0.446
L 2'× 2½ × 3/8	11/16	5.9	1.73	0.984	0.566	0.753	0.762	0.984	0.566	0.753	0.762	0.487	1.000
5/16	5/8	5.0	1.46	0.849	0.482	0.761	0.740	0.849	0.482	0.761	0.740	0.489	1.000
1/4	9/16	4.1	1.19	0.703	0.394	0.769	0.717	0.703	0.394	0.769	0.717	0.491	1.000
3/16	1/2	3.07	0.902	0.547	0.303	0.778	0.694	0.547	0.303	0.778	0.694	0.495	1.000
L 2½ × 2 × 3/8	11/16	5.3	1.55	0.912	0.547	0.768	0.831	0.514	0.363	0.577	0.581	0.420	0.614
5/16	5/8	4.5	1.31	0.788	0.466	0.776	0.809	0.46	0.310	0.584	0.559	0.422	0.620
1/4	9/16	3.62	1.06	0.654	0.381	0.784	0.787	0.372	0.254	0.592	0.537	0.424	0.626
3/16	1/2	2.75	0.809	0.509	0.293	0.793	0.764	0.291	0.196	0.600	0.514	0.427	0.631
L 2 × 2 × 3/8	11/16	4.7	1.36	0.479	0.351	0.594	0.636	0.479	0.351	0.594	0.636	0.389	1.00
5/16	5/8	3.92	1.15	0.416	0.300	0.601	0.614	0.416	0.300	0.601	0.614	0.390	1.000
1/4	9/16	3.19	0.938	0.348	0.247	0.609	0.592	0.348	0.247	0.609	0.592	0.391	1.000
3/16	1/2	2.44	0.715	0.272	0.190	0.617	0.569	0.272	0.190	0.617	0.569	0.394	1.000
1/8	7/16	1.65	0.484	0.190	0.131	0.626	0.546	0.190	0.131	0.626	0.546	0.398	1.000

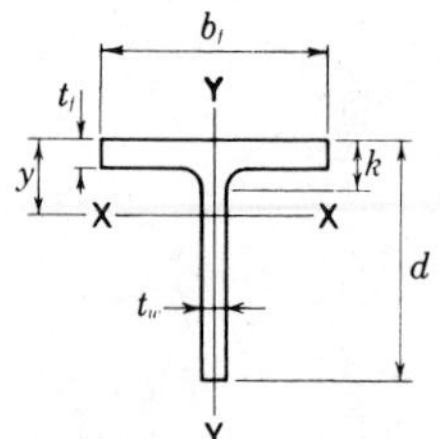

TABLE A.4 Properties of Structural Tees Cut from W Shapes (WT)

Designation	Area	Depth of Tee d		Stem Thickness t_w		$\frac{t_w}{2}$	Area of Stem	Flange Width b_f		Flange Thickness t_f		Distance k
	In.^{2}	In.		In.		In.	In.^{2}	In.		In.		In.
WT 18 x150	44.1	18.370	18 3/8	0.945	15/16	1/2	17.4	16.655	16 5/8	1.680	1 11/16	2 13/16
x140	41.2	18.260	18 1/4	0.885	7/8	7/16	16.2	16.595	16 5/8	1.570	1 9/16	2 11/16
x130	38.2	18.130	18 1/8	0.840	13/16	7/16	15.2	16.550	16 1/2	1.440	1 7/16	2 9/16
x122.5	36.0	18.040	18	0.800	13/16	7/16	14.4	16.510	16 1/2	1.350	1 3/8	2 1/2
x115	33.8	17.950	18	0.760	3/4	3/8	13.6	16.470	16 1/2	1.260	1 1/4	2 3/8
WT 18 x105	30.9	18.345	18 3/8	0.830	13/16	7/16	15.2	12.180	12 1/8	1.360	1 3/8	2 5/16
x 97	28.5	18.245	18 1/4	0.765	3/4	3/8	14.0	12.115	12 1/8	1.260	1 1/4	2 3/16
x 91	26.8	18.165	18 1/8	0.725	3/4	3/8	13.2	12.075	12 1/8	1.180	1 3/16	2 1/8
x 85	25.0	18.085	18 1/8	0.680	11/16	3/8	12.3	12.030	12	1.100	1 1/8	2
x 80	23.5	18.005	18	0.650	5/8	5/16	11.7	12.000	12	1.020	1	1 15/16
x 75	22.1	17.925	17 7/8	0.625	5/8	5/16	11.2	11.975	12	0.940	15/16	1 7/8
x 67.5	19.9	17.775	17 3/4	0.600	5/8	5/16	10.7	11.950	12	0.790	13/16	1 11/16
WT 16.5x120.5	35.4	17.090	17 1/8	0.830	13/16	7/16	14.2	15.860	15 7/8	1.400	1 3/8	2 3/16
x110.5	32.5	16.965	17	0.775	3/4	3/8	13.1	15.805	15 3/4	1.275	1 1/4	2 1/16
x100.5	29.5	16.840	16 7/8	0.715	11/16	3/8	12.0	15.745	15 3/4	1.150	1 1/8	1 15/16
WT 16.5x 76	22.4	16.745	16 3/4	0.635	5/8	5/16	10.6	11.565	11 5/8	1.055	1 1/16	1 7/8
x 70.5	20.8	16.650	16 5/8	0.605	5/8	5/16	10.1	11.535	11 1/2	0.960	15/16	1 3/4
x 65	19.2	16.545	16 1/2	0.580	9/16	5/16	9.60	11.510	11 1/2	0.855	7/8	1 11/16
x 59	17.3	16.430	16 3/8	0.550	9/16	5/16	9.04	11.480	11 1/2	0.740	3/4	1 9/16
WT 15 x105.5	31.0	15.470	15 1/2	0.775	3/4	3/8	12.0	15.105	15 1/8	1.315	1 5/16	2 1/8
x 95.5	28.1	15.340	15 3/8	0.710	11/16	3/8	10.9	15.040	15	1.185	1 3/16	1 15/16
x 86.5	25.4	15.220	15 1/4	0.655	5/8	5/16	9.97	14.985	15	1.065	1 1/16	1 7/8
WT 15 x 66	19.4	15.155	15 1/8	0.615	5/8	5/16	9.32	10.545	10 1/2	1.000	1	1 3/4
x 62	18.2	15.085	15 1/8	0.585	9/16	5/16	8.82	10.515	10 1/2	0.930	15/16	1 11/16
x 58	17.1	15.005	15	0.565	9/16	5/16	8.48	10.495	10 1/2	0.850	7/8	1 5/8
x 54	15.9	14.915	14 7/8	0.545	9/16	5/16	8.13	10.475	10 1/2	0.760	3/4	1 9/16
x 49.5	14.5	14.825	14 7/8	0.520	1/2	1/4	7.71	10.450	10 1/2	0.670	11/16	1 7/16

TABLE A.4 (*Continued*)

Nominal Weight per Ft.	$\frac{d}{t_w}$	AXIS X-X				AXIS Y-Y			$C_c' = \sqrt{\frac{2\pi^2 E}{Q_s Q_a F_y}}$, $Q_a = 1.0$			
		I	S	r	y	I	S	r	F_y = 36 ksi		**F_y = 50 ksi**	
Lb.		In.[4]	In.[3]	In.	In.	In.[4]	In.[3]	In.	Q_s	C_c'	**Q_s**	**C_c'**
150	19.4	1230	86.1	5.27	4.13	648	77.8	3.83	—	—	**0.927**	**111**
140	20.6	1140	80.0	5.25	4.07	599	72.2	3.81	—	—	**0.867**	**115**
130	21.6	1060	75.1	5.26	4.05	545	65.9	3.78	0.981	127	**0.816**	**118**
122.5	22.5	995	71.0	5.26	4.03	507	61.4	3.75	0.943	130	**0.770**	**122**
115.	23.6	934	67.0	5.25	4.01	470	57.1	3.73	0.896	133	**0.715**	**127**
105	22.1	985	73.1	5.65	4.87	206	33.8	2.58	0.960	129	**0.791**	**120**
97	23.8	901	67.0	5.62	4.80	187	30.9	2.56	0.887	134	**0.705**	**127**
91	25.1	845	63.1	5.62	4.77	174	28.8	2.55	0.831	138	**0.635**	**134**
85	26.6	786	58.9	5.61	4.73	160	26.6	2.53	0.767	144	**0.565**	**142**
80	27.7	740	55.8	5.61	4.74	147	24.6	2.50	0.720	149	**0.521**	**148**
75	28.7	698	53.1	5.62	4.78	135	22.5	2.47	0.677	153	**0.486**	**154**
67.5	29.6	636	49.7	5.66	4.96	113	18.9	2.38	0.634	158	**0.457**	**158**
120.5	20.6	871	65.8	4.96	3.85	466	58.8	3.63	—	—	**0.867**	**115**
110.5	21.9	799	60.8	4.96	3.81	420	53.2	3.59	0.968	128	**0.801**	**120**
100.5	23.6	725	55.5	4.95	3.78	375	47.6	3.56	0.896	133	**0.715**	**127**
76	26.4	592	47.4	5.14	4.26	136	23.6	2.47	0.775	143	**0.574**	**141**
70.5	27.5	552	44.7	5.15	4.29	123	21.3	2.43	0.728	148	**0.529**	**147**
65	28.5	513	42.1	5.18	4.36	109	18.9	2.39	0.685	152	**0.492**	**152**
59	29.9	469	39.2	5.20	4.47	93.6	16.3	2.32	0.621	160	**0.447**	**160**
105.5	20.0	610	50.5	4.43	3.40	378	50.1	3.49	—	—	**0.897**	**113**
95.5	21.6	549	45.7	4.42	3.35	336	44.7	3.46	0.981	127	**0.816**	**118**
86.5	23.2	497	41.7	4.42	3.31	299	39.9	3.43	0.913	132	**0.735**	**125**
66	24.6	421	37.4	4.66	3.90	98.0	18.6	2.25	0.853	137	**0.664**	**131**
62	25.8	396	35.3	4.66	3.90	90.4	17.2	2.23	0.801	141	**0.601**	**138**
58	26.6	373	33.7	4.67	3.94	82.1	15.7	2.19	0.767	144	**0.565**	**142**
54	27.4	349	32.0	4.69	4.01	73.0	13.9	2.15	0.733	147	**0.533**	**147**
49.5	28.5	322	30.0	4.71	4.09	63.9	12.2	2.10	0.685	152	**0.492**	**152**

Where no value of C_c' or Q_s is shown, the Tee complies with Specification Sect. 1.9.1.2.

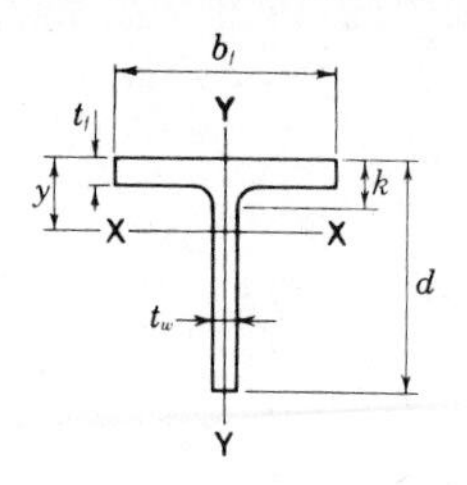

TABLE A.4 Properties of Structural Tees Cut from W Shapes (WT)

Designation	Area	Depth of Tee d		Stem			Area of Stem	Flange				Distance k
				Thickness t_w		$\frac{t_w}{2}$		Width b_f		Thickness t_f		
	In.^{2}	In.		In.		In.	In.^{2}	In.		In.		In.
WT 13.5x89	26.1	13.905	13⅞	0.725	¾	⅜	10.1	14.085	14⅛	1.190	1³⁄₁₆	1⅞
x80.5	23.7	13.795	13¾	0.660	¹¹⁄₁₆	⅜	9.10	14.020	14	1.080	1¹⁄₁₆	1¹³⁄₁₆
x73	21.5	13.690	13¾	0.605	⅝	⁵⁄₁₆	8.28	13.965	14	0.975	1	1¹¹⁄₁₆
WT 13.5x57	16.8	13.645	13⅝	0.570	⁹⁄₁₆	⁵⁄₁₆	7.78	10.070	10⅛	0.930	¹⁵⁄₁₆	1⅝
x51	15.0	13.545	13½	0.515	½	¼	6.98	10.015	10	0.830	¹³⁄₁₆	1⁹⁄₁₆
x47	13.8	13.460	13½	0.490	½	¼	6.60	9.990	10	0.745	¾	1⁷⁄₁₆
x42	12.4	13.355	13⅜	0.460	⁷⁄₁₆	¼	6.14	9.960	10	0.640	⅝	1⅜
WT 12 x81	23.9	12.500	12½	0.705	¹¹⁄₁₆	⅜	8.81	12.955	13	1.220	1¼	2
x73	21.5	12.370	12⅜	0.650	⅝	⁵⁄₁₆	8.04	12.900	12⅞	1.090	1¹⁄₁₆	1⅞
x65.5	19.3	12.240	12¼	0.605	⅝	⁵⁄₁₆	7.41	12.855	12⅞	0.960	¹⁵⁄₁₆	1¾
x58.5	17.2	12.130	12⅛	0.550	⁹⁄₁₆	⁵⁄₁₆	6.67	12.800	12¾	0.850	⅞	1⅝
x52	15.3	12.030	12	0.500	½	¼	6.01	12.750	12¾	0.750	¾	1½
WT 12 x47	13.8	12.155	12⅛	0.515	½	¼	6.26	9.065	9⅛	0.875	⅞	1⅝
x42	12.4	12.050	12	0.470	½	¼	5.66	9.020	9	0.770	¾	1⁹⁄₁₆
x38	11.2	11.960	12	0.440	⁷⁄₁₆	¼	5.26	8.990	9	0.680	¹¹⁄₁₆	1⁷⁄₁₆
x34	10.0	11.865	11⅞	0.415	⁷⁄₁₆	¼	4.92	8.965	9	0.585	⁹⁄₁₆	1⅜
WT 12 x31	9.11	11.870	11⅞	0.430	⁷⁄₁₆	¼	5.10	7.040	7	0.590	⁹⁄₁₆	1⅜
x27.5	8.10	11.785	11¾	0.395	⅜	³⁄₁₆	4.66	7.005	7	0.505	½	1⁵⁄₁₆
WT 10.5x73.5	21.6	11.030	11	0.720	¾	⅜	7.94	12.510	12½	1.150	1⅛	1⅞
x66	19.4	10.915	10⅞	0.650	⅝	⁵⁄₁₆	7.09	12.440	12½	1.035	1¹⁄₁₆	1¹³⁄₁₆
x61	17.9	10.840	10⅞	0.600	⅝	⁵⁄₁₆	6.50	12.390	12⅜	0.960	¹⁵⁄₁₆	1¹¹⁄₁₆
x55.5	16.3	10.755	10¾	0.550	⁹⁄₁₆	⁵⁄₁₆	5.92	12.340	12⅜	0.875	⅞	1⅝
x50.5	14.9	10.680	10⅝	0.500	½	¼	5.34	12.290	12¼	0.800	¹³⁄₁₆	1⁹⁄₁₆
WT 10.5x46.5	13.7	10.810	10¾	0.580	⁹⁄₁₆	⁵⁄₁₆	6.27	8.420	8⅜	0.930	¹⁵⁄₁₆	1¹¹⁄₁₆
x41.5	12.2	10.715	10¾	0.515	½	¼	5.52	8.355	8⅜	0.835	¹³⁄₁₆	1⁹⁄₁₆
x36.5	10.7	10.620	10⅝	0.455	⁷⁄₁₆	¼	4.83	8.295	8¼	0.740	¾	1½
x34	10.0	10.565	10⅝	0.430	⁷⁄₁₆	¼	4.54	8.270	8¼	0.685	¹¹⁄₁₆	1⁷⁄₁₆
x31	9.13	10.495	10½	0.400	⅜	³⁄₁₆	4.20	8.240	8¼	0.615	⅝	1⅜
WT 10.5x28.5	8.37	10.530	10½	0.405	⅜	³⁄₁₆	4.26	6.555	6½	0.650	⅝	1⅜
x25	7.36	10.415	10⅜	0.380	⅜	³⁄₁₆	3.96	6.530	6½	0.535	⁹⁄₁₆	1⁵⁄₁₆
x22	6.49	10.330	10⅜	0.350	⅜	³⁄₁₆	3.62	6.500	6½	0.450	⁷⁄₁₆	1³⁄₁₆
WT 9x59.5	17.5	9.485	9½	0.655	⅝	⁵⁄₁₆	6.21	11.265	11¼	1.060	1¹⁄₁₆	1¾
x53	15.6	9.365	9⅜	0.590	⁹⁄₁₆	⁵⁄₁₆	5.53	11.200	11¼	0.940	¹⁵⁄₁₆	1⅝
x48.5	14.3	9.295	9¼	0.535	⁹⁄₁₆	⁵⁄₁₆	4.97	11.145	11⅛	0.870	⅞	1⁹⁄₁₆
x43	12.7	9.195	9¼	0.480	½	¼	4.41	11.090	11⅛	0.770	¾	1⁷⁄₁₆
x38	11.2	9.105	9⅛	0.425	⁷⁄₁₆	¼	3.87	11.035	11	0.680	¹¹⁄₁₆	1⅜

TABLE A.4 (*Continued*)

Nominal Weight per Ft.	$\frac{d}{t_w}$	AXIS X-X				AXIS Y-Y			$C_c' = \sqrt{\frac{2\pi^2 E}{Q_s Q_a F_y}}$, $Q_a = 1.0$			
		I	S	r	y	I	S	r	$F_y = 36$ ksi		$F_y = 50$ ksi	
Lb.		In.[4]	In.[3]	In.	In.	In.[4]	In.[3]	In.	Q_s	C_c'	Q_s	C_c'
89	19.2	414	38.2	3.98	3.05	278	39.4	3.26	—	—	0.937	111
80.5	20.9	372	34.4	3.96	2.99	248	35.4	3.24	—	—	0.851	116
73	22.6	336	31.2	3.95	2.95	222	31.7	3.21	0.938	130	0.765	122
57	23.9	289	28.3	4.15	3.42	79.4	15.8	2.18	0.883	134	0.700	128
51	26.3	258	25.3	4.14	3.37	69.6	13.9	2.15	0.780	143	0.578	141
47	27.5	239	23.8	4.16	3.41	62.0	12.4	2.12	0.728	148	0.529	147
42	29.0	216	21.9	4.18	3.48	52.8	10.6	2.07	0.664	155	0.476	155
81	17.7	293	29.9	3.50	2.70	221	34.2	3.05	—	—	—	—
73	19.0	264	27.2	3.50	2.66	195	30.3	3.01	—	—	0.947	110
65.5	20.2	238	24.8	3.52	2.65	170	26.5	2.97	—	—	0.887	114
58.5	22.1	212	22.3	3.51	2.62	149	23.2	2.94	0.960	129	0.791	120
52	24.1	189	20.0	3.51	2.59	130	20.3	2.91	0.874	135	0.690	129
47	23.6	186	20.3	3.67	2.99	54.5	12.0	1.98	0.896	133	0.715	127
42	25.6	166	18.3	3.67	2.97	47.2	10.5	1.95	0.810	140	0.610	137
38	27.2	151	16.9	3.68	3.00	41.3	9.18	1.92	0.741	146	0.541	146
34	28.6	137	15.6	3.70	3.06	35.2	7.85	1.87	0.681	153	0.489	153
31	27.6	131	15.6	3.79	3.46	17.2	4.90	1.38	0.724	148	0.525	148
27.5	29.8	117	14.1	3.80	3.50	14.5	4.15	1.34	0.626	159	0.450	159
73.5	15.3	204	23.7	3.08	2.39	188	30.0	2.95	—	—	—	—
66	16.8	181	21.1	3.06	2.33	166	26.7	2.93	—	—	—	—
61	18.1	166	19.3	3.04	2.28	152	24.6	2.92	—	—	0.993	107
55.5	19.6	150	17.5	3.03	2.23	137	22.2	2.90	—	—	0.917	112
50.5	21.4	135	15.8	3.01	2.18	124	20.2	2.89	0.990	127	0.826	118
46.5	18.6	144	17.9	3.25	2.74	46.4	11.0	1.84	—	—	0.968	109
41.5	20.8	127	15.7	3.22	2.66	40.7	9.75	1.83	—	—	0.856	116
36.5	23.3	110	13.8	3.21	2.60	35.3	8.51	1.81	0.908	132	0.730	125
34	24.6	103	12.9	3.20	2.59	32.4	7.83	1.80	0.853	137	0.664	131
31	26.2	93.8	11.9	3.21	2.58	28.7	6.97	1.77	0.784	142	0.583	140
28.5	26.0	90.4	11.8	3.29	2.85	15.3	4.67	1.35	0.793	142	0.592	139
25	27.4	80.3	10.7	3.30	2.93	12.5	3.82	1.30	0.733	147	0.533	147
22	29.5	71.1	9.68	3.31	2.98	10.3	3.18	1.26	0.638	158	0.460	158
59.5	14.5	119	15.9	2.60	2.03	126	22.5	2.69	—	—	—	—
53	15.9	104	14.1	2.59	1.97	110	19.7	2.66	—	—	—	—
48.5	17.4	93.8	12.7	2.56	1.91	100	18.0	2.65	—	—	—	—
43	19.2	82.4	11.2	2.55	1.86	87.6	15.8	2.63	—	—	0.937	111
38	21.4	71.8	9.83	2.54	1.80	76.2	13.8	2.61	0.990	127	0.826	118

Where no value of C_c' or Q_s is shown, the Tee complies with Specification Sect. 1.9.1.2.

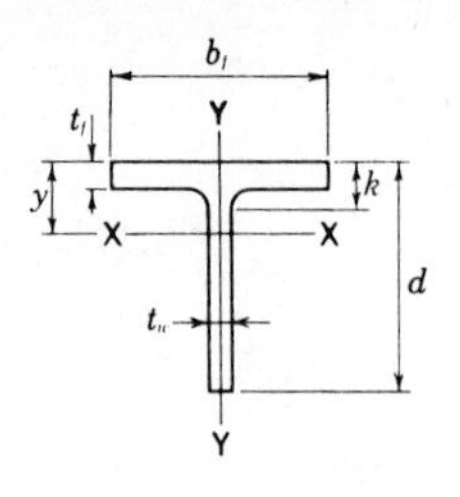

TABLE A.4 Properties of Structural Tees Cut from W Shapes (WT)

Designation	Area	Depth of Tee d		Stem Thickness t_w		Stem $\frac{t_w}{2}$	Area of Stem	Flange Width b_f		Flange Thickness t_f		Distance k
	In.^{2}	In.		In.		In.	In.^{2}	In.		In.		In.
WT 9x35.5	10.4	9.235	9 1/4	0.495	1/2	1/4	4.57	7.635	7 5/8	0.810	13/16	1 1/2
x32.5	9.55	9.175	9 1/8	0.450	7/16	1/4	4.13	7.590	7 5/8	0.750	3/4	1 7/16
x30	8.82	9.120	9 1/8	0.415	7/16	1/4	3.78	7.555	7 1/2	0.695	11/16	1 3/8
x27.5	8.10	9.055	9	0.390	3/8	3/16	3.53	7.530	7 1/2	0.630	5/8	1 5/16
x25	7.33	8.995	9	0.355	3/8	3/16	3.19	7.495	7 1/2	0.570	9/16	1 1/4
WT 9x23	6.77	9.030	9	0.360	3/8	3/16	3.25	6.060	6	0.605	5/8	1 1/4
x20	5.88	8.950	9	0.315	5/16	3/16	2.82	6.015	6	0.525	1/2	1 3/16
x17.5	5.15	8.850	8 7/8	0.300	5/16	3/16	2.65	6.000	6	0.425	7/16	1 1/8
WT 8x50	14.7	8.485	8 1/2	0.585	9/16	5/16	4.96	10.425	10 3/8	0.985	1	1 11/16
x44.5	13.1	8.375	8 3/8	0.525	1/2	1/4	4.40	10.365	10 3/8	0.875	7/8	1 9/16
x38.5	11.3	8.260	8 1/4	0.455	7/16	1/4	3.76	10.295	10 1/4	0.760	3/4	1 7/16
x33.5	9.84	8.165	8 1/8	0.395	3/8	3/16	3.23	10.235	10 1/4	0.665	11/16	1 3/8
WT 8x28.5	8.38	8.215	8 1/4	0.430	7/16	1/4	3.53	7.120	7 1/8	0.715	11/16	1 3/8
x25	7.37	8.130	8 1/8	0.380	3/8	3/16	3.09	7.070	7 1/8	0.630	5/8	1 5/16
x22.5	6.63	8.065	8 1/8	0.345	3/8	3/16	2.78	7.035	7	0.565	9/16	1 1/4
x20	5.89	8.005	8	0.305	5/16	3/16	2.44	6.995	7	0.505	1/2	1 3/16
x18	5.28	7.930	7 7/8	0.295	5/16	3/16	2.34	6.985	7	0.430	7/16	1 1/8
WT 8x15.5	4.56	7.940	8	0.275	1/4	1/8	2.18	5.525	5 1/2	0.440	7/16	1 1/8
x13	3.84	7.845	7 7/8	0.250	1/4	1/8	1.96	5.500	5 1/2	0.345	3/8	1 1/16
WT 7x365	107	11.210	11 1/4	3.070	3 1/16	1 9/16	34.4	17.890	17 7/8	4.910	4 15/16	5 9/16
x332.5	97.8	10.820	10 7/8	2.830	2 13/16	1 7/16	30.6	17.650	17 5/8	4.520	4 1/2	5 3/16
x302.5	88.9	10.460	10 1/2	2.595	2 5/8	1 5/16	27.1	17.415	17 3/8	4.160	4 3/16	4 13/16
x275	80.9	10.120	10 1/8	2.380	2 3/8	1 3/16	24.1	17.200	17 1/4	3.820	3 13/16	4 1/2
x250	73.5	9.800	9 3/4	2.190	2 3/16	1 1/8	21.5	17.010	17	3.500	3 1/2	4 3/16
x227.5	66.9	9.510	9 1/2	2.015	2	1	19.2	16.835	16 7/8	3.210	3 3/16	3 7/8
x213	62.6	9.335	9 3/8	1.875	1 7/8	15/16	17.5	16.695	16 3/4	3.035	3 1/16	3 11/16
x199	58.5	9.145	9 1/8	1.770	1 3/4	7/8	16.2	16.590	16 5/8	2.845	2 7/8	3 1/2
x185	54.4	8.960	9	1.655	1 5/8	13/16	14.8	16.475	16 1/2	2.660	2 11/16	3 5/16
x171	50.3	8.770	8 3/4	1.540	1 9/16	13/16	13.5	16.360	16 3/8	2.470	2 1/2	3 1/8
x155.5	45.7	8.560	8 1/2	1.410	1 7/16	3/4	12.1	16.230	16 1/4	2.260	2 1/4	2 15/16
x141.5	41.6	8.370	8 3/8	1.290	1 5/16	11/16	10.8	16.110	16 1/8	2.070	2 1/16	2 3/4
x128.5	37.8	8.190	8 1/4	1.175	1 3/16	5/8	9.62	15.995	16	1.890	1 7/8	2 9/16
x116.5	34.2	8.020	8	1.070	1 1/16	9/16	8.58	15.890	15 7/8	1.720	1 3/4	2 3/8
x105.5	31.0	7.860	7 7/8	0.980	1	1/2	7.70	15.800	15 3/4	1.560	1 9/16	2 1/4
x 96.5	28.4	7.740	7 3/4	0.890	7/8	7/16	6.89	15.710	15 3/4	1.440	1 7/16	2 1/8
x 88	25.9	7.610	7 5/8	0.830	13/16	7/16	6.32	15.650	15 5/8	1.310	1 5/16	2
x 79.5	23.4	7.490	7 1/2	0.745	3/4	3/8	5.58	15.565	15 5/8	1.190	1 3/16	1 7/8
x 72.5	21.3	7.390	7 3/8	0.680	11/16	3/8	5.03	15.500	15 1/2	1.090	1 1/16	1 3/4

TABLE A.4 (*Continued*)

Nominal Weight per Ft.	$\frac{d}{t_w}$	AXIS X-X				AXIS Y-Y			$C_c' = \sqrt{\frac{2\pi^2 E}{Q_s Q_a F_y}}$, $Q_a = 1.0$			
		I	S	r	y	I	S	r	F_y = 36 ksi		**F_y = 50 ksi**	
Lb.		In.^{4}	In.^{3}	In.	In.	In.^{4}	In.^{3}	In.	Q_s	C_c'	**Q_s**	**C_c'**
35.5	18.7	78.2	11.2	2.74	2.26	30.1	7.89	1.70	—	—	**0.963**	**109**
32.5	20.4	70.7	10.1	2.72	2.20	27.4	7.22	1.69	—	—	**0.877**	**114**
30	22.0	64.7	9.29	2.71	2.16	25.0	6.63	1.69	0.964	128	**0.796**	**120**
27.5	23.2	59.5	8.63	2.71	2.16	22.5	5.97	1.67	0.913	132	**0.735**	**125**
25	25.3	53.5	7.79	2.70	2.12	20.0	5.35	1.65	0.823	139	**0.625**	**135**
23	25.1	52.1	7.77	2.77	2.33	11.3	3.72	1.29	0.831	138	**0.635**	**134**
20	28.4	44.8	6.73	2.76	2.29	9.55	3.17	1.27	0.690	152	**0.496**	**152**
17.5	29.5	40.1	6.21	2.79	2.39	7.67	2.56	1.22	0.638	158	**0.460**	**158**
50	14.5	76.8	11.4	2.28	1.76	93.1	17.9	2.51	—	—	**—**	**—**
44.5	16.0	67.2	10.1	2.27	1.70	81.3	15.7	2.49	—	—	**—**	**—**
38.5	18.2	56.9	8.59	2.24	1.63	69.2	13.4	2.47	—	—	**0.988**	**108**
33.5	20.7	48.6	7.36	2.22	1.56	59.5	11.6	2.46	—	—	**0.861**	**115**
28.5	19.1	48.7	7.77	2.41	1.94	21.6	6.06	1.60	—	—	**0.942**	**110**
25	21.4	42.3	6.78	2.40	1.89	18.6	5.26	1.59	0.990	127	**0.826**	**118**
22.5	23.4	37.8	6.10	2.39	1.86	16.4	4.67	1.57	0.904	133	**0.725**	**126**
20	26.2	33.1	5.35	2.37	1.81	14.4	4.12	1.57	0.784	142	**0.583**	**140**
18	26.9	30.6	5.05	2.41	1.88	12.2	3.50	1.52	0.754	145	**0.553**	**144**
15.5	28.9	27.4	4.64	2.45	2.02	6.20	2.24	1.17	0.668	154	**0.479**	**155**
13	31.4	23.5	4.09	2.47	2.09	4.80	1.74	1.12	0.563	168	**0.406**	**168**
365	3.7	739	95.4	2.62	3.47	2360	264	4.69	—	—	—	—
332.5	3.8	622	82.1	2.52	3.25	2080	236	4.62	—	—	—	—
302.5	4.0	524	70.6	2.43	3.05	1840	211	4.55	—	—	—	—
275	4.3	442	60.9	2.34	2.85	1630	189	4.49	—	—	—	—
250	4.5	375	52.7	2.26	2.67	1440	169	4.43	—	—	—	—
227.5	4.7	321	45.9	2.19	2.51	1280	152	4.38	—	—	—	—
213	5.0	287	41.4	2.14	2.40	1180	141	4.34	—	—	—	—
199	5.2	257	37.6	2.10	2.30	1090	131	4.31	—	—	—	—
185	5.4	229	33.9	2.05	2.19	994	121	4.27	—	—	—	—
171	5.7	203	30.4	2.01	2.09	903	110	4.24	—	—	—	—
155.5	6.1	176	26.7	1.96	1.97	807	99.4	4.20	—	—	—	—
141.5	6.5	153	23.5	1.92	1.86	722	89.7	4.17	—	—	—	—
128.5	7.0	133	20.7	1.88	1.75	645	80.7	4.13	—	—	—	—
116.5	7.5	116	18.2	1.84	1.65	576	72.5	4.10	—	—	—	—
105.5	8.0	102	16.2	1.81	1.57	513	65.0	4.07	—	—	—	—
96.5	8.7	89.8	14.4	1.78	1.49	466	59.3	4.05	—	—	—	—
88	9.2	80.5	13.0	1.76	1.43	419	53.5	4.02	—	—	—	—
79.5	10.1	70.2	11.4	1.73	1.35	374	48.1	4.00	—	—	—	—
72.5	10.9	62.5	10.2	1.71	1.29	338	43.7	3.98	—	—	—	—

Where no value of C_c' or Q_s is shown, the Tee complies with Specification Sect. 1.9.1.2.

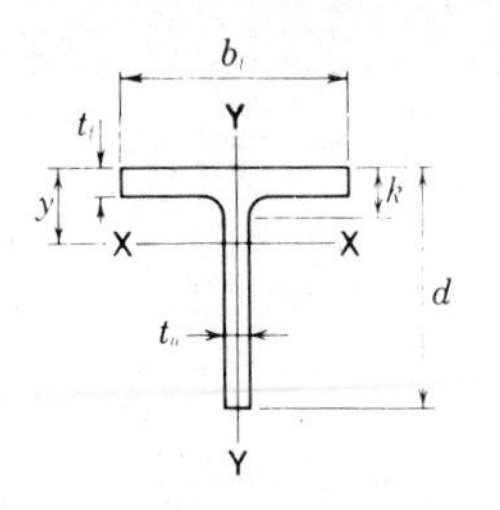

TABLE A.4 Properties of Structural Tees Cut from W Shapes (WT)

Designation	Area	Depth of Tee d		Stem Thickness t_w		$\frac{t_w}{2}$	Area of Stem	Flange Width b_f		Flange Thickness t_f		Distance k
	In.^{2}	In.		In.		In.	In.^{2}	In.		In.		In.
WT 7x66	19.4	7.330	7 3/8	0.645	5/8	5/16	4.73	14.725	14 3/4	1.030	1	1 11/16
x60	17.7	7.240	7 1/4	0.590	9/16	5/16	4.27	14.670	14 5/8	0.940	15/16	1 5/8
x54.5	16.0	7.160	7 1/8	0.525	1/2	1/4	3.76	14.605	14 5/8	0.860	7/8	1 9/16
x49.5	14.6	7.080	7 1/8	0.485	1/2	1/4	3.43	14.565	14 5/8	0.780	3/4	1 7/16
x45	13.2	7.010	7	0.440	7/16	1/4	3.08	14.520	14 1/2	0.710	11/16	1 3/8
WT 7x41	12.0	7.155	7 1/8	0.510	1/2	1/4	3.65	10.130	10 1/8	0.855	7/8	1 5/8
x37	10.9	7.085	7 1/8	0.450	7/16	1/4	3.19	10.070	10 1/8	0.785	13/16	1 9/16
x34	9.99	7.020	7	0.415	7/16	1/4	2.91	10.035	10	0.720	3/4	1 1/2
x30.5	8.96	6.945	7	0.375	3/8	3/16	2.60	9.995	10	0.645	5/8	1 7/16
WT 7x26.5	7.81	6.960	7	0.370	3/8	3/16	2.58	8.060	8	0.660	11/16	1 7/16
x24	7.07	6.895	6 7/8	0.340	5/16	3/16	2.34	8.030	8	0.595	5/8	1 3/8
x21.5	6.31	6.830	6 7/8	0.305	5/16	3/16	2.08	7.995	8	0.530	1/2	1 5/16
WT 7x19	5.58	7.050	7	0.310	5/16	3/16	2.19	6.770	6 3/4	0.515	1/2	1 1/16
x17	5.00	6.990	7	0.285	5/16	3/16	1.99	6.745	6 3/4	0.455	7/16	1
x15	4.42	6.920	6 7/8	0.270	1/4	1/8	1.87	6.730	6 3/4	0.385	3/8	15/16
WT 7x13	3.85	6.955	7	0.255	1/4	1/8	1.77	5.025	5	0.420	7/16	15/16
x11	3.25	6.870	6 7/8	0.230	1/4	1/8	1.58	5.000	5	0.335	5/16	7/8
WT 6x168	49.4	8.410	8 3/8	1.775	1 3/4	7/8	14.9	13.385	13 3/8	2.955	2 15/16	3 11/16
x152.5	44.8	8.160	8 1/8	1.625	1 5/8	13/16	13.3	13.235	13 1/4	2.705	2 11/16	3 7/16
x139.5	41.0	7.925	7 7/8	1.530	1 1/2	3/4	12.1	13.140	13 1/8	2.470	2 1/2	3 3/16
x126	37.0	7.705	7 3/4	1.395	1 3/8	11/16	10.7	13.005	13	2.250	2 1/4	2 15/16
x115	33.9	7.525	7 1/2	1.285	1 5/16	11/16	9.67	12.895	12 7/8	2.070	2 1/16	2 3/4
x105	30.9	7.355	7 3/8	1.180	1 3/16	5/8	8.68	12.790	12 3/4	1.900	1 7/8	2 5/8
x 95	27.9	7.190	7 1/4	1.060	1 1/16	9/16	7.62	12.670	12 5/8	1.735	1 3/4	2 7/16
x 85	25.0	7.015	7	0.960	15/16	1/2	6.73	12.570	12 5/8	1.560	1 9/16	2 1/4
x 76	22.4	6.855	6 7/8	0.870	7/8	7/16	5.96	12.480	12 1/2	1.400	1 3/8	2 1/8
x 68	20.0	6.705	6 3/4	0.790	13/16	7/16	5.30	12.400	12 3/8	1.250	1 1/4	1 15/16
x 60	17.6	6.560	6 1/2	0.710	11/16	3/8	4.66	12.320	12 3/8	1.105	1 1/8	1 13/16
x 53	15.6	6.445	6 1/2	0.610	5/8	5/16	3.93	12.220	12 1/4	0.990	1	1 11/16
x 48	14.1	6.355	6 3/8	0.550	9/16	5/16	3.50	12.160	12 1/8	0.900	7/8	1 5/8
x 43.5	12.8	6.265	6 1/4	0.515	1/2	1/4	3.23	12.125	12 1/8	0.810	13/16	1 1/2
x 39.5	11.6	6.190	6 1/4	0.470	1/2	1/4	2.91	12.080	12 1/8	0.735	3/4	1 7/16
x 36	10.6	6.125	6 1/8	0.430	7/16	1/4	2.63	12.040	12	0.670	11/16	1 3/8
x 32.5	9.54	6.060	6	0.390	3/8	3/16	2.36	12.000	12	0.605	5/8	1 5/16

TABLE A.4 (*Continued*)

Nominal Weight per Ft.	$\frac{d}{t_w}$	AXIS X-X				AXIS Y-Y			$C_c' = \sqrt{\frac{2\pi^2 E}{Q_s Q_a F_y}}$, $Q_a = 1.0$			
		I	S	r	y	I	S	r	F_y = 36 ksi		F_y = 50 ksi	
Lb.		In.^{4}	In.^{3}	In.	In.	In.^{4}	In.^{3}	In.	Q_s	C_c'	Q_s	C_c'
66	11.4	57.8	9.57	1.73	1.29	274	37.2	3.76	—	—	—	—
60	12.3	51.7	8.61	1.71	1.24	247	33.7	3.74	—	—	—	—
54.5	13.6	45.3	7.56	1.68	1.17	223	30.6	3.73	—	—	—	—
49.5	14.6	40.9	6.88	1.67	1.14	201	27.6	3.71	—	—	—	—
45	15.9	36.4	6.16	1.66	1.09	181	25.0	3.70	—	—	—	—
41	14.0	41.2	7.14	1.85	1.39	74.2	14.6	2.48	—	—	—	—
37	15.7	36.0	6.25	1.82	1.32	66.9	13.3	2.48	—	—	—	—
34	16.9	32.6	5.69	1.81	1.29	60.7	12.1	2.46	—	—	—	—
30.5	18.5	28.9	5.07	1.80	1.25	53.7	10.7	2.45	—	—	0.973	108
26.5	18.8	27.6	4.94	1.88	1.38	28.8	7.16	1.92	—	—	0.958	109
24	20.3	24.9	4.48	1.87	1.35	25.7	6.40	1.91	—	—	0.882	114
21.5	22.4	21.9	3.98	1.86	1.31	22.6	5.65	1.89	0.947	130	0.775	122
19	22.7	23.3	4.22	2.04	1.54	13.3	3.94	1.55	0.934	130	0.760	123
17	24.5	20.9	3.83	2.04	1.53	11.7	3.45	1.53	0.857	136	0.669	131
15	25.6	19.0	3.55	2.07	1.58	9.79	2.91	1.49	0.810	140	0.610	137
13	27.3	17.3	3.31	2.12	1.72	4.45	1.77	1.08	0.737	147	0.537	146
11	29.9	14.8	2.91	2.14	1.76	3.50	1.40	1.04	0.621	160	0.447	160
168	4.7	190	31.2	1.96	2.31	593	88.6	3.47	—	—	—	—
152.5	5.0	162	27.0	1.90	2.16	525	79.3	3.42	—	—	—	—
139.5	5.2	141	24.1	1.86	2.05	469	71.3	3.38	—	—	—	—
126	5.5	121	20.9	1.81	1.92	414	63.6	3.34	—	—	—	—
115	5.9	106	18.5	1.77	1.82	371	57.5	3.31	—	—	—	—
105	6.2	92.1	16.4	1.73	1.72	332	51.9	3.28	—	—	—	—
95	6.8	79.0	14.2	1.68	1.62	295	46.5	3.25	—	—	—	—
85	7.3	67.8	12.3	1.65	1.52	259	41.2	3.22	—	—	—	—
76	7.9	58.5	10.8	1.62	1.43	227	36.4	3.19	—	—	—	—
68	8.5	50.6	9.46	1.59	1.35	199	32.1	3.16	—	—	—	—
60	9.2	43.4	8.22	1.57	1.28	172	28.0	3.13	—	—	—	—
53	10.6	36.3	6.91	1.53	1.19	151	24.7	3.11	—	—	—	—
48	11.6	32.0	6.12	1.51	1.13	135	22.2	3.09	—	—	—	—
43.5	12.2	28.9	5.60	1.50	1.10	120	19.9	3.07	—	—	—	—
39.5	13.2	25.8	5.03	1.49	1.06	108	17.9	3.05	—	—	—	—
36	14.2	23.2	4.54	1.48	1.02	97.5	16.2	3.04	—	—	—	—
32.5	15.5	20.6	4.06	1.47	0.985	87.2	14.5	3.02	—	—	—	—

Where no value of C_c' or Q_s is shown, the Tee complies with Specification Sect. 1.9.1.2.

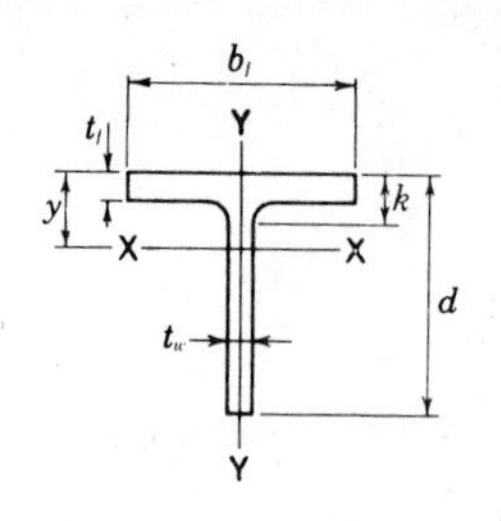

TABLE A.4 Properties of Structural Tees Cut from W Shapes (WT)

Designation	Area	Depth of Tee d		Stem Thickness t_w		Stem $\frac{t_w}{2}$	Area of Stem	Flange Width b_f		Flange Thickness t_f		Distance k
	In.^{2}	In.		In.		In.	In.^{2}	In.		In.		In.
WT 6x 29	8.52	6.095	6 1/8	0.360	3/8	3/16	2.19	10.010	10	0.640	5/8	1 3/8
x 26.5	7.78	6.030	6	0.345	3/8	3/16	2.08	9.995	10	0.575	9/16	1 1/4
WT 6x 25	7.34	6.095	6 1/8	0.370	3/8	3/16	2.26	8.080	8 1/8	0.640	5/8	1 3/8
x 22.5	6.61	6.030	6	0.335	5/16	3/16	2.02	8.045	8	0.575	9/16	1 1/4
x 20	5.89	5.970	6	0.295	5/16	3/16	1.76	8.005	8	0.515	1/2	1 1/4
WT 6x 17.5	5.17	6.250	6 1/4	0.300	5/16	3/16	1.88	6.560	6 1/2	0.520	1/2	1
x 15	4.40	6.170	6 1/8	0.260	1/4	1/8	1.60	6.520	6 1/2	0.440	7/16	15/16
x 13	3.82	6.110	6 1/8	0.230	1/4	1/8	1.41	6.490	6 1/2	0.380	3/8	7/8
WT 6x 11	3.24	6.155	6 1/8	0.260	1/4	1/8	1.60	4.030	4	0.425	7/16	7/8
x 9.5	2.79	6.080	6 1/8	0.235	1/4	1/8	1.43	4.005	4	0.350	3/8	13/16
x 8	2.36	5.995	6	0.220	1/4	1/8	1.32	3.990	4	0.265	1/4	3/4
x 7	2.08	5.955	6	0.200	3/16	1/8	1.19	3.970	4	0.225	1/4	11/16
WT 5x56	16.5	5.680	5 5/8	0.755	3/4	3/8	4.29	10.415	10 3/8	1.250	1 1/4	1 7/8
x50	14.7	5.550	5 1/2	0.680	11/16	3/8	3.77	10.340	10 3/8	1.120	1 1/8	1 3/4
x44	12.9	5.420	5 3/8	0.605	5/8	5/16	3.28	10.265	10 1/4	0.990	1	1 5/8
x38.5	11.3	5.300	5 1/4	0.530	1/2	1/4	2.81	10.190	10 1/4	0.870	7/8	1 1/2
x34	9.99	5.200	5 1/4	0.470	1/2	1/4	2.44	10.130	10 1/8	0.770	3/4	1 3/8
x30	8.82	5.110	5 1/8	0.420	7/16	1/4	2.15	10.080	10 1/8	0.680	11/16	1 5/16
x27	7.91	5.045	5	0.370	3/8	3/16	1.87	10.030	10	0.615	5/8	1 1/4
x24.5	7.21	4.990	5	0.340	5/16	3/16	1.70	10.000	10	0.560	9/16	1 3/16
WT 5x22.5	6.63	5.050	5	0.350	3/8	3/16	1.77	8.020	8	0.620	5/8	1 1/4
x19.5	5.73	4.960	5	0.315	5/16	3/16	1.56	7.985	8	0.530	1/2	1 1/8
x16.5	4.85	4.865	4 7/8	0.290	5/16	3/16	1.41	7.960	8	0.435	7/16	1 1/16
WT 5x15	4.42	5.235	5 1/4	0.300	5/16	3/16	1.57	5.810	5 3/4	0.510	1/2	15/16
x13	3.81	5.165	5 1/8	0.260	1/4	1/8	1.34	5.770	5 3/4	0.440	7/16	7/8
x11	3.24	5.085	5 1/8	0.240	1/4	1/8	1.22	5.750	5 3/4	0.360	3/8	3/4
WT 5x 9.5	2.81	5.120	5 1/8	0.250	1/4	1/8	1.28	4.020	4	0.395	3/8	13/16
x 8.5	2.50	5.055	5	0.240	1/4	1/8	1.21	4.010	4	0.330	5/16	3/4
x 7.5	2.21	4.995	5	0.230	1/4	1/8	1.15	4.000	4	0.270	1/4	11/16
x 6	1.77	4.935	4 7/8	0.190	3/16	1/8	0.938	3.960	4	0.210	3/16	5/8

TABLE A.4 (Continued)

Nominal Weight per Ft.	$\frac{d}{t_w}$	AXIS X-X				AXIS Y-Y			$C_c' = \sqrt{\frac{2\pi^2 E}{Q_s Q_a F_y}}$, $Q_a = 1.0$			
		I	S	r	y	I	S	r	F_y = 36 ksi		F_y = 50 ksi	
Lb.		In.[4]	In.[3]	In.	In.	In.[4]	In.[3]	In.	Q_s	C_c'	Q_s	C_c'
29	16.9	19.1	3.76	1.50	1.03	53.5	10.7	2.51	—	—	—	—
26.5	17.5	17.7	3.54	1.51	1.02	47.9	9.58	2.48	—	—	—	—
25	16.5	18.7	3.79	1.60	1.17	28.2	6.97	1.96	—	—	—	—
22.5	18.0	16.6	3.39	1.58	1.13	25.0	6.21	1.94	—	—	0.998	107
20	20.2	14.4	2.95	1.57	1.08	22.0	5.51	1.93	—	—	0.887	114
17.5	20.8	16.0	3.23	1.76	1.30	12.2	3.73	1.54	—	—	0.856	116
15	23.7	13.5	2.75	1.75	1.27	10.2	3.12	1.52	0.891	134	0.710	127
13	26.6	11.7	2.40	1.75	1.25	8.66	2.67	1.51	0.767	144	0.565	142
11	23.7	11.7	2.59	1.90	1.63	2.33	1.16	0.847	0.891	134	0.710	127
9.5	25.9	10.1	2.28	1.90	1.65	1.88	0.939	0.822	0.797	141	0.596	139
8	27.2	8.70	2.04	1.92	1.74	1.41	0.706	0.773	0.741	146	0.541	146
7	29.8	7.67	1.83	1.92	1.76	1.18	0.594	0.753	0.626	159	0.450	159
56	7.5	28.6	6.40	1.32	1.21	118	22.6	2.68	—	—	—	—
50	8.2	24.5	5.56	1.29	1.13	103	20.0	2.65	—	—	—	—
44	9.0	20.8	4.77	1.27	1.06	89.3	17.4	2.63	—	—	—	—
38.5	10.0	17.4	4.04	1.24	0.990	76.8	15.1	2.60	—	—	—	—
34	11.1	14.9	3.49	1.22	0.932	66.8	13.2	2.59	—	—	—	—
30	12.2	12.9	3.04	1.21	0.884	58.1	11.5	2.57	—	—	—	—
27	13.6	11.1	2.64	1.19	0.836	51.7	10.3	2.56	—	—	—	—
24.5	14.7	10.0	2.39	1.18	0.807	46.7	9.34	2.54	—	—	—	—
22.5	14.4	10.2	2.47	1.24	0.907	26.7	6.65	2.01	—	—	—	—
19.5	15.7	8.84	2.16	1.24	0.876	22.5	5.64	1.98	—	—	—	—
16.5	16.8	7.71	1.93	1.26	0.869	18.3	4.60	1.94	—	—	—	—
15	17.4	9.28	2.24	1.45	1.10	8.35	2.87	1.37	—	—	—	—
13	19.9	7.86	1.91	1.44	1.06	7.05	2.44	1.36	—	—	0.902	113
11	21.2	6.88	1.72	1.46	1.07	5.71	1.99	1.33	0.999	126	0.836	117
9.5	20.5	6.68	1.74	1.54	1.28	2.15	1.07	0.874	—	—	0.872	115
8.5	21.1	6.06	1.62	1.56	1.32	1.78	0.888	0.844	—	—	0.841	117
7.5	21.7	5.45	1.50	1.57	1.37	1.45	0.723	0.810	0.977	128	0.811	119
6	26.0	4.35	1.22	1.57	1.36	1.09	0.551	0.785	0.793	142	0.592	139

Where no value of C_c' or Q_s is shown, the Tee complies with Specification Sect. 1.9.1.2.

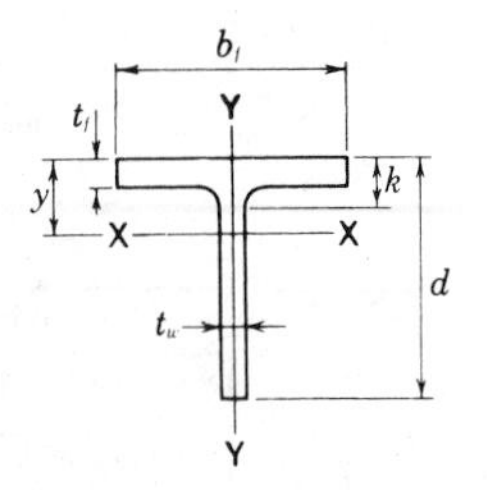

TABLE A.4 Properties of Structural Tees Cut from W Shapes (WT)

Designation	Area	Depth of Tee d		Stem Thickness t_w		Stem $\frac{t_w}{2}$	Area of Stem	Flange Width b_f		Flange Thickness t_f		Distance k
	In.^{2}	In.		In.		In.	In.^{2}	In.		In.		In.
WT 4 x33.5	9.84	4.500	$4\frac{1}{2}$	0.570	$\frac{9}{16}$	$\frac{5}{16}$	2.56	8.280	$8\frac{1}{4}$	0.935	$\frac{15}{16}$	$1\frac{7}{16}$
x29	8.55	4.375	$4\frac{3}{8}$	0.510	$\frac{1}{2}$	$\frac{1}{4}$	2.23	8.220	$8\frac{1}{4}$	0.810	$\frac{13}{16}$	$1\frac{5}{16}$
x24	7.05	4.250	$4\frac{1}{4}$	0.400	$\frac{3}{8}$	$\frac{3}{16}$	1.70	8.110	$8\frac{1}{8}$	0.685	$\frac{11}{16}$	$1\frac{3}{16}$
x20	5.87	4.125	$4\frac{1}{8}$	0.360	$\frac{3}{8}$	$\frac{3}{16}$	1.48	8.070	$8\frac{1}{8}$	0.560	$\frac{9}{16}$	$1\frac{1}{16}$
x17.5	5.14	4.060	4	0.310	$\frac{5}{16}$	$\frac{3}{16}$	1.26	8.020	8	0.495	$\frac{1}{2}$	1
x15.5	4.56	4.000	4	0.285	$\frac{5}{16}$	$\frac{3}{16}$	1.14	7.995	8	0.435	$\frac{7}{16}$	$\frac{15}{16}$
WT 4 x14	4.12	4.030	4	0.285	$\frac{5}{16}$	$\frac{3}{16}$	1.15	6.535	$6\frac{1}{2}$	0.465	$\frac{7}{16}$	$\frac{15}{16}$
x12	3.54	3.965	4	0.245	$\frac{1}{4}$	$\frac{1}{8}$	0.971	6.495	$6\frac{1}{2}$	0.400	$\frac{3}{8}$	$\frac{7}{8}$
WT 4 x10.5	3.08	4.140	$4\frac{1}{8}$	0.250	$\frac{1}{4}$	$\frac{1}{8}$	1.03	5.270	$5\frac{1}{4}$	0.400	$\frac{3}{8}$	$\frac{13}{16}$
x 9	2.63	4.070	$4\frac{1}{8}$	0.230	$\frac{1}{4}$	$\frac{1}{8}$	0.936	5.250	$5\frac{1}{4}$	0.330	$\frac{5}{16}$	$\frac{3}{4}$
WT 4 x 7.5	2.22	4.055	4	0.245	$\frac{1}{4}$	$\frac{1}{8}$	0.993	4.015	4	0.315	$\frac{5}{16}$	$\frac{3}{4}$
x 6.5	1.92	3.995	4	0.230	$\frac{1}{4}$	$\frac{1}{8}$	0.919	4.000	4	0.255	$\frac{1}{4}$	$\frac{11}{16}$
x 5	1.48	3.945	4	0.170	$\frac{3}{16}$	$\frac{1}{8}$	0.671	3.940	4	0.205	$\frac{3}{16}$	$\frac{5}{8}$
WT 3 x12.5	3.67	3.190	$3\frac{1}{4}$	0.320	$\frac{5}{16}$	$\frac{3}{16}$	1.02	6.080	$6\frac{1}{8}$	0.455	$\frac{7}{16}$	$\frac{13}{16}$
x10	2.94	3.100	$3\frac{1}{8}$	0.260	$\frac{1}{4}$	$\frac{1}{8}$	0.806	6.020	6	0.365	$\frac{3}{8}$	$\frac{3}{4}$
x 7.5	2.21	2.995	3	0.230	$\frac{1}{4}$	$\frac{1}{8}$	0.689	5.990	6	0.260	$\frac{1}{4}$	$\frac{5}{8}$
WT 3 x 8	2.37	3.140	$3\frac{1}{8}$	0.260	$\frac{1}{4}$	$\frac{1}{8}$	0.816	4.030	4	0.405	$\frac{3}{8}$	$\frac{3}{4}$
x 6	1.78	3.015	3	0.230	$\frac{1}{4}$	$\frac{1}{8}$	0.693	4.000	4	0.280	$\frac{1}{4}$	$\frac{5}{8}$
x 4.5	1.34	2.950	3	0.170	$\frac{3}{16}$	$\frac{1}{8}$	0.502	3.940	4	0.215	$\frac{3}{16}$	$\frac{9}{16}$
WT 2.5x 9.5	2.77	2.575	$2\frac{5}{8}$	0.270	$\frac{1}{4}$	$\frac{1}{8}$	0.695	5.030	5	0.430	$\frac{7}{16}$	$\frac{13}{16}$
x 8	2.34	2.505	$2\frac{1}{2}$	0.240	$\frac{1}{4}$	$\frac{1}{8}$	0.601	5.000	5	0.360	$\frac{3}{8}$	$\frac{3}{4}$
WT 2 x 6.5	1.91	2.080	$2\frac{1}{8}$	0.280	$\frac{1}{4}$	$\frac{1}{8}$	0.582	4.060	4	0.345	$\frac{3}{8}$	$\frac{11}{16}$

TABLE A.4 (*Continued*)

Nominal Weight per Ft.	$\frac{d}{t_w}$	AXIS X-X				AXIS Y-Y			$C_c' = \sqrt{\frac{2\pi^2 E}{Q_s Q_a F_y}}$, $Q_a = 1.0$			
		I	S	r	y	I	S	r	F_y = 36 ksi		F_y = 50 ksi	
Lb.		In.^{4}	In.^{3}	In.	In.	In.^{4}	In.^{3}	In.	Q_s	C_c'	Q_s	C_c'
33.5	7.9	10.9	3.05	1.05	0.936	44.3	10.7	2.12	—	—	—	—
29	8.6	9.12	2.61	1.03	0.874	37.5	9.13	2.10	—	—	—	—
24	10.6	6.85	1.97	0.986	0.777	30.5	7.52	2.08	—	—	—	—
20	11.5	5.73	1.69	0.988	0.735	24.5	6.08	2.04	—	—	—	—
17.5	13.1	4.81	1.43	0.967	0.688	21.3	5.31	2.03	—	—	—	—
15.5	14.0	4.28	1.28	0.968	0.667	18.5	4.64	2.02	—	—	—	—
14	14.1	4.22	1.28	1.01	0.734	10.8	3.31	1.62	—	—	—	—
12	16.2	3.53	1.08	0.999	0.695	9.14	2.81	1.61	—	—	—	—
10.5	16.6	3.90	1.18	1.12	0.831	4.89	1.85	1.26	—	—	—	—
9	17.7	3.41	1.05	1.14	0.834	3.98	1.52	1.23	—	—	—	—
7.5	16.6	3.28	1.07	1.22	0.998	1.70	0.849	0.876	—	—	—	—
6.5	17.4	2.89	0.974	1.23	1.03	1.37	0.683	0.843	—	—	—	—
5	23.2	2.15	0.717	1.20	0.953	1.05	0.532	0.841	0.913	132	0.735	125
12.5	10.0	2.28	0.886	0.789	0.610	8.53	2.81	1.52	—	—	—	—
10	11.9	1.76	0.693	0.774	0.560	6.64	2.21	1.50	—	—	—	—
7.5	13.0	1.41	0.577	0.797	0.558	4.66	1.56	1.45	—	—	—	—
8	12.1	1.69	0.685	0.844	0.676	2.21	1.10	0.966	—	—	—	—
6	13.1	1.32	0.564	0.861	0.677	1.50	0.748	0.918	—	—	—	—
4.5	17.4	0.950	0.408	0.842	0.623	1.10	0.557	0.905	—	—	—	—
9.5	9.5	1.01	0.485	0.605	0.487	4.56	1.82	1.28	—	—	—	—
8	10.4	0.845	0.413	0.601	0.458	3.75	1.50	1.27	—	—	—	—
6.5	7.4	0.526	0.321	0.524	0.440	1.93	0.950	1.00	—	—	—	—

Where no value of C_c' or Q_s is shown, the Tee complies with Specification Sect. 1.9.1.2.

TABLE A.5 Properties of Double Angles

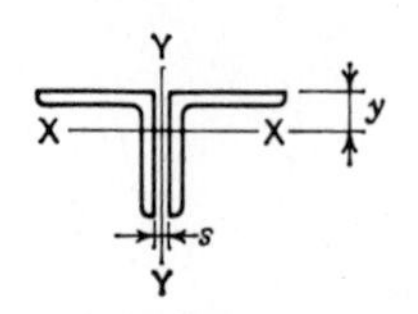

Two equal leg angles

Designation	Wt. per Ft. 2 Angles	Area of 2 Angles	AXIS X – X				AXIS Y – Y Radii of Gyration Back to Back of Angles, Inches			Q_s* Angles in Contact		Angles Separated	
			I	S	r	y				F_y = 36 ksi	F_y = 50 ksi	F_y = 36 ksi	F_y = 50 ksi
	Lb.	In.^{2}	In.^{4}	In.^{3}	In.	In.	0	3/8	3/4				
L 8 x8 x1 1/8	113.8	33.5	195.0	35.1	2.42	2.41	3.42	3.55	3.69	—	—	—	—
1	102.0	30.0	177.0	31.6	2.44	2.37	3.40	3.53	3.67	—	—	—	—
7/8	90.0	26.5	159.0	28.0	2.45	2.32	3.38	3.51	3.64	—	—	—	—
3/4	77.8	22.9	139.0	24.4	2.47	2.28	3.36	3.49	3.62	—	—	—	—
5/8	65.4	19.2	118.0	20.6	2.49	2.23	3.34	3.47	3.60	—	—	.997	.935
1/2	52.8	15.5	97.3	16.7	2.50	2.19	3.32	3.45	3.58	.995	.921	.911	.834
L 6 x6 x1	74.8	22.0	70.9	17.1	1.80	1.86	2.59	2.73	2.87	—	—	—	—
7/8	66.2	19.5	63.8	15.3	1.81	1.82	2.57	2.70	2.85	—	—	—	—
3/4	57.4	16.9	56.3	13.3	1.83	1.78	2.55	2.68	2.82	—	—	—	—
5/8	48.4	14.2	48.3	11.3	1.84	1.73	2.53	2.66	2.80	—	—	—	—
1/2	39.2	11.5	39.8	9.23	1.86	1.68	2.51	2.64	2.78	—	—	—	.961
3/8	29.8	8.72	30.8	7.06	1.88	1.64	2.49	2.62	2.75	.995	.921	.911	.834
L 5 x5 x 7/8	54.4	16.0	35.5	10.3	1.49	1.57	2.16	2.30	2.45	—	—	—	—
3/4	47.2	13.9	31.5	9.06	1.51	1.52	2.14	2.28	2.42	—	—	—	—
1/2	32.4	9.50	22.5	6.31	1.54	1.43	2.10	2.24	2.38	—	—	—	—
3/8	24.6	7.22	17.5	4.84	1.56	1.39	2.09	2.22	2.35	—	—	.982	.919
5/16	20.6	6.05	14.8	4.08	1.57	1.37	2.08	2.21	2.34	.995	.921	.911	.834
L 4 x4 x 3/4	37.0	10.9	15.3	5.62	1.19	1.27	1.74	1.88	2.83	—	—	—	—
5/8	31.4	9.22	13.3	4.80	1.20	1.23	1.72	1.86	2.00	—	—	—	—
1/2	25.6	7.50	11.1	3.95	1.22	1.18	1.70	1.83	1.98	—	—	—	—
3/8	19.6	5.72	8.72	3.05	1.23	1.14	1.68	1.81	1.95	—	—	—	—
5/16	16.4	4.80	7.43	2.58	1.24	1.12	1.67	1.80	1.94	—	—	.997	.935
1/4	13.2	3.88	6.08	2.09	1.25	1.09	1.66	1.79	1.93	.995	.921	.911	.834
L 3 1/2x3 1/2x 3/8	17.0	4.97	5.73	2.30	1.07	1.01	1.48	1.61	1.75	—	—	—	—
5/16	14.4	4.18	4.90	1.95	1.08	.990	1.47	1.60	1.74	—	—	—	.986
1/4	11.6	3.38	4.02	1.59	1.09	.968	1.46	1.59	1.73	—	.982	.965	.897
L 3 x3 x 1/2	18.8	5.50	4.43	2.14	.898	.932	1.29	1.43	1.59	—	—	—	—
3/8	14.4	4.22	3.52	1.67	.913	.888	1.27	1.41	1.56	—	—	—	—
5/16	12.2	3.55	3.02	1.41	.922	.865	1.26	1.40	1.55	—	—	—	—
1/4	9.8	2.88	2.49	1.15	.930	.842	1.26	1.39	1.53	—	—	—	.961
3/16	7.42	2.18	1.92	.882	.939	.820	1.25	1.38	1.52	.995	.921	.911	.834
L 2 1/2x2 1/2x 3/8	11.8	3.47	1.97	1.13	.753	.762	1.07	1.21	1.36	—	—	—	—
5/16	10.0	2.93	1.70	.964	.761	.740	1.06	1.20	1.35	—	—	—	—
1/4	8.2	2.38	1.41	.789	.769	.717	1.05	1.19	1.34	—	—	—	—
3/16	6.14	1.80	1.09	.685	.778	.694	1.04	1.18	1.32	—	—	.982	.919
L 2 x2 x 3/8	9.4	2.72	.958	.702	.594	.636	.870	1.01	1.17	—	—	—	—
5/16	7.84	2.30	.832	.681	.601	.614	.859	1.00	1.16	—	—	—	—
1/4	6.38	1.88	.695	.494	.609	.592	.849	.989	1.14	—	—	—	—
3/16	4.88	1.43	.545	.381	.617	.569	.840	.977	1.13	—	—	—	—
1/8	3.30	.960	.380	.261	.626	.546	.831	.965	1.11	.995	.921	.911	.834

* Where no value of Q_s is shown, the angles comply with Specification Sect. 1.9.1.2 and may be considered fully effective.

For F_y = 36 ksi: $C'_c = 126.1/\sqrt{Q_s}$

For F_y = 50 ksi: $C'_c = 107.0/\sqrt{Q_s}$

TABLE A.5 ***(Continued)***

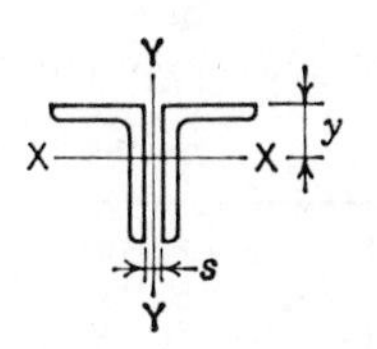

Two unequal leg angles

Long legs back to back

Designation	Wt. per Ft. 2 Angles	Area of 2 Angles	AXIS X − X				AXIS Y − Y			Q_s*			
							Radii of Gyration Back to Back of Angles, Inches			Angles in Contact		Angles Separated	
			I	S	r	y				F_y = 36 ksi	F_y = 50 ksi	F_y = 36 ksi	F_y = 50 ksi
	Lb.	In.^{2}	In.^{4}	In.^{3}	In.	In.	0	3/8	3/4				
L 8 x6 x1	88.4	26.0	161.0	30.2	2.49	2.65	2.39	2.52	2.66	—	—	—	—
3/4	67.6	19.9	126.0	23.3	2.53	2.56	2.35	2.48	2.62	—	—	—	—
1/2	46.0	13.5	88.6	16.0	2.56	2.47	2.32	2.44	2.57	—	—	.911	.834
L 8 x4 x1	74.8	22.0	139.0	28.1	2.52	3.05	1.47	1.61	1.75	—	—	—	—
3/4	57.4	16.9	109.0	21.8	2.55	2.95	1.42	1.55	1.69	—	—	—	—
1/2	39.2	11.5	77.0	15.0	2.59	2.86	1.38	1.51	1.64	—	—	.911	.834
L 7 x4 x 3/4	52.4	15.4	75.6	16.8	2.22	2.51	1.48	1.62	1.76	—	—	—	—
1/2	35.8	10.5	53.3	11.6	2.25	2.42	1.44	1.57	1.71	—	—	.965	.897
3/8	27.2	7.97	41.1	8.88	2.27	2.37	1.43	1.55	1.68	—	—	.839	.750
L 6 x4 x 3/4	47.2	13.9	49.0	12.5	1.88	2.08	1.55	1.69	1.83	—	—	—	—
5/8	40.0	11.7	42.1	10.6	1.90	2.03	1.53	1.67	1.81	—	—	—	—
1/2	32.4	9.50	34.8	8.67	1.91	1.99	1.51	1.64	1.78	—	—	—	.961
3/8	24.6	7.22	26.9	6.64	1.93	1.94	1.50	1.62	1.76	—	—	.911	.834
L 6 x3½x 3/8	23.4	6.84	25.7	6.49	1.94	2.04	1.26	1.39	1.53	—	—	.911	.834
5/16	19.6	5.74	21.8	5.47	1.95	2.01	1.26	1.38	1.51	—	—	.825	.733
L 5 x3½x 3/4	39.6	11.6	27.8	8.55	1.55	1.75	1.40	1.53	1.68	—	—	—	—
1/2	27.2	8.00	20.0	5.97	1.58	1.66	1.35	1.49	1.63	—	—	—	—
3/8	20.8	6.09	15.6	4.59	1.60	1.61	1.34	1.46	1.60	—	—	.982	.919
5/16	17.4	5.12	13.2	3.87	1.61	1.59	1.33	1.45	1.59	—	—	.911	.834
L 5 x3 x 1/2	25.6	7.50	18.9	5.82	1.59	1.75	1.12	1.25	1.40	—	—	—	—
3/8	19.6	5.72	14.7	4.47	1.61	1.70	1.10	1.23	1.37	—	—	.982	.919
5/16	16.4	4.80	12.5	3.77	1.61	1.68	1.09	1.22	1.36	—	—	.911	.834
1/4	13.2	3.88	10.2	3.06	1.62	1.66	1.08	1.21	1.34	—	—	.804	.708
L 4 x3½x 1/2	23.8	7.00	10.6	3.87	1.23	1.25	1.44	1.58	1.72	—	—	—	—
3/8	18.2	5.34	8.35	2.99	1.25	1.21	1.42	1.56	1.70	—	—	—	—
5/16	15.4	4.49	7.12	2.53	1.26	1.18	1.42	1.55	1.69	—	—	.997	.935
1/4	12.4	3.63	5.83	2.05	1.27	1.16	1.41	1.54	1.67	—	.982	.911	.834
L 4 x3 x 1/2	22.2	6.50	10.1	3.78	1.25	1.33	1.20	1.33	1.48	—	—	—	—
3/8	17.0	4.97	7.93	2.92	1.26	1.28	1.18	1.31	1.45	—	—	—	—
5/16	14.4	4.18	6.76	2.47	1.27	1.26	1.17	1.30	1.44	—	—	.997	.935
1/4	11.6	3.38	5.54	2.00	1.28	1.24	1.16	1.29	1.43	—	—	.911	.834

* Where no value of Q_s is shown, the angles comply with Specification Sect. 1.9.1.2 and may be considered fully effective.

For F_y = 36 ksi: $C'_c = 126.1/\sqrt{Q_s}$

For F_y = 50 ksi: $C'_c = 107.0/\sqrt{Q_s}$

Two unequal leg angles

TABLE A.5 (*Continued*)

Long legs back to back

Designation												
L 3½x3 x ⅜	15.8	4.59	5.45	2.25	1.09	1.08	1.22	1.36	1.50	—	—	—	—
5/16	13.2	3.87	4.66	1.91	1.10	1.06	1.21	1.35	1.49	—	—	—	.986
¼	10.8	3.13	3.83	1.55	1.11	1.04	1.20	1.33	1.48	—	—	.965	.897
L 3½x2½x ⅜	14.4	4.22	5.12	2.19	1.10	1.16	.976	1.11	1.26	—	—	—	—
5/16	12.2	3.55	4.38	1.85	1.11	1.14	.966	1.10	1.25	—	—	—	.986
¼	9.8	2.88	3.60	1.51	1.12	1.11	.958	1.09	1.23	—	—	.965	.897
L 3 x2½x ⅜	13.2	3.84	3.31	1.62	.928	.956	1.02	1.16	1.31	—	—	—	—
¼	9.0	2.63	2.35	1.12	.945	.911	1.00	1.13	1.28	—	—	—	.961
3/16	6.77	1.99	1.81	.859	.954	.888	.993	1.12	1.27	—	—	.911	.834
L 3 x2 x ⅜	11.8	3.47	3.06	1.56	.940	1.04	.777	.917	1.07	—	—	—	—
5/16	10.0	2.93	2.63	1.33	.948	1.02	.767	.903	1.06	—	—	—	—
¼	8.2	2.38	2.17	1.08	.957	.993	.757	.891	1.04	—	—	—	.961
3/16	6.1	1.80	1.68	.830	.966	.970	.749	.879	1.03	—	—	.911	.834
L 2½x2 x ⅜	10.6	3.09	1.82	1.09	.768	.831	.819	.961	1.12	—	—	—	—
5/16	9.0	2.62	1.58	.932	.776	.809	.809	.948	1.10	—	—	—	—
¼	7.2	2.13	1.31	.763	.784	.787	.799	.935	1.09	—	—	—	—
3/16	5.5	1.62	1.02	.586	.793	.764	.790	.923	1.07	—	—	.982	.919

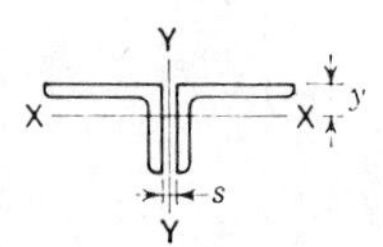

Two unequal leg angles

Short legs back to back

Designation	Wt. per Ft. 2 Angles	Area of 2 Angles	AXIS X – X				AXIS Y – Y Radii of Gyration Back to Back of Angles, Inches			Q_s*			
										Angles in Contact		Angles Separated	
			I	S	r	y				F_y = 36 ksi	F_y = 50 ksi	F_y = 36 ksi	F_y = 50 ksi
	Lb.	In.²	In.⁴	In.³	In.	In.	0	⅜	¾				
L 8 x6 x1	88.4	26.0	77.6	17.8	1.73	1.65	3.64	3.78	3.92	—	—	—	—
¾	67.6	19.9	61.4	13.8	1.76	1.56	3.60	3.74	3.88	—	—	—	—
½	46.0	13.5	43.4	9.58	1.79	1.47	3.56	3.69	3.83	.995	.921	.911	.834
L 8 x4 x1	74.8	22.0	23.3	7.88	1.03	1.05	3.95	4.10	4.25	—	—	—	—
¾	57.4	16.9	18.7	6.14	1.05	.953	3.90	4.05	4.19	—	—	—	—
½	39.2	11.5	13.5	4.29	1.08	.859	3.86	4.00	4.14	.995	.921	.911	.834
L 7 x4 x ¾	52.4	15.4	16.1	6.05	1.09	1.01	3.35	3.49	3.64	—	—	—	—
½	35.8	10.5	13.1	4.23	1.11	.917	3.30	3.44	3.59	—	.982	.965	.897
⅜	27.2	7.97	10.2	3.26	1.13	.870	3.28	3.42	3.56	.926	.838	.839	.750
L 6 x4 x ¾	47.2	13.9	17.4	5.94	1.12	1.08	2.80	2.94	3.09	—	—	—	—
⅝	40.0	11.7	15.0	5.07	1.13	1.03	2.78	2.92	3.06	—	—	—	—
½	32.4	9.50	12.5	4.16	1.15	.987	2.76	2.90	3.04	—	—	—	.961
⅜	24.6	7.22	9.81	3.21	1.17	.941	2.74	2.87	3.02	.995	.921	.911	.834
L 6 x3½x ⅜	23.4	6.84	6.68	2.46	.988	.787	2.81	2.95	3.09	.995	.921	.911	.834
5/16	19.6	5.74	5.70	2.08	.996	.763	2.80	2.94	3.08	.912	.822	.825	.733

* Where no value of Q_s is shown, the angles comply with Specification Sect. 1.9.1.2 and may be considered fully effective.

For F_y = 36 ksi: $C'_c = 126.1/\sqrt{Q_s}$

For F_y = 50 ksi: $C'_c = 107.0/\sqrt{Q_s}$

TABLE A.5 (*Continued*)

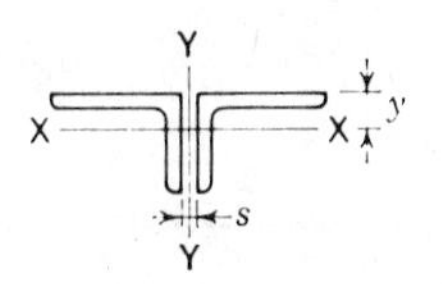

Two unequal leg angles

Short legs back to back

Designation	Wt. per Ft. 2 Angles	Area of 2 Angles	AXIS X – X				AXIS Y – Y Radii of Gyration Back to Back of Angles, Inches			Q_s* Angles in Contact		Q_s* Angles Separated	
			I	S	r	y				F_y =	F_y =	F_y =	F_y =
	Lb.	In.^{2}	In.^{4}	In.^{3}	In.	In.	0	3/8	3/4	36 ksi	50 ksi	36 ksi	50 ksi
L 5 x3½x ¾	39.6	11.6	11.1	4.43	.977	.996	2.33	2.48	2.63	—	—	—	—
½	27.2	8.00	8.10	3.12	1.01	.906	2.29	2.43	2.57	—	—	—	—
3/8	20.8	6.09	6.37	2.41	1.02	.861	2.27	2.41	2.55	—	—	.982	.919
5/16	17.4	5.12	5.44	2.04	1.03	.838	2.26	2.39	2.54	.995	.921	.911	.834
L 5 x3 x ½	25.6	7.50	5.16	2.29	.829	.750	2.36	2.50	2.65	—	—	—	—
3/8	19.6	5.72	4.08	1.78	.845	.704	2.34	2.48	2.63	—	—	.982	.919
5/16	16.4	4.80	3.49	1.51	.853	.681	2.33	2.47	2.61	.995	.921	.911	.834
¼	13.2	3.88	2.88	1.23	.861	.657	2.32	2.46	2.60	.891	.797	.804	.708
L 4 x3½x ½	23.8	7.00	7.58	3.03	1.04	1.00	1.76	1.89	2.04	—	—	—	—
3/8	18.2	5.34	5.97	2.35	1.06	.955	1.74	1.87	2.01	—	—	—	—
5/16	15.4	4.49	5.10	1.99	1.07	.932	1.73	1.86	2.00	—	—	.997	.935
¼	12.4	3.63	4.19	1.62	1.07	.909	1.72	1.85	1.99	.995	.921	.911	.834
L 4 x3 x ½	22.2	6.50	4.85	2.23	.864	.827	1.82	1.96	2.11	—	—	—	—
3/8	17.0	4.97	3.84	1.73	.879	.782	1.80	1.94	2.08	—	—	—	—
5/16	14.4	4.18	3.29	1.47	.887	.759	1.79	1.93	2.07	—	—	.997	.935
¼	11.6	3.38	2.71	1.20	.896	.736	1.78	1.92	2.06	.995	.921	.911	.834
L 3½x3 x 3/8	15.8	4.59	3.69	1.70	.897	.830	1.53	1.67	1.82	—	—	—	—
5/16	13.2	3.87	3.17	1.44	.905	.808	1.52	1.66	1.80	—	—	—	.986
¼	10.8	3.13	2.61	1.18	.914	.785	1.52	1.65	1.79	—	.982	.965	.897
L 3½x2½x 3/8	14.4	4.22	2.18	1.18	.719	.660	1.60	1.74	1.89	—	—	—	—
5/16	12.2	3.55	1.88	1.01	.727	.637	1.59	1.73	1.88	—	—	—	.986
¼	9.8	2.88	1.55	.824	.735	.614	1.58	1.72	1.86	—	.982	.965	.897
L 3 x2½x 3/8	13.2	3.84	2.08	1.16	.736	.706	1.33	1.47	1.62	—	—	—	—
¼	9.0	2.63	1.49	.808	.753	.661	1.31	1.45	1.60	—	—	—	.961
3/16	6.77	1.99	1.15	.620	.761	.638	1.30	1.44	1.58	.995	.921	.911	.834
L 3 x2 x 3/8	11.8	3.47	1.09	.743	.559	.539	1.40	1.55	1.70	—	—	—	—
5/16	10.0	2.93	.941	.634	.567	.516	1.39	1.53	1.68	—	—	—	—
¼	8.2	2.38	.784	.520	.574	.493	1.38	1.52	1.67	—	—	—	.961
3/16	6.1	1.80	.613	.401	.583	.470	1.37	1.51	1.66	.995	.921	.911	.834
L 2½x2 x 3/8	10.6	3.09	1.03	.725	.577	.581	1.13	1.28	1.43	—	—	—	—
5/16	9.0	2.62	.893	.620	.584	.559	1.12	1.26	1.42	—	—	—	—
¼	7.2	2.13	.745	.509	.592	.537	1.11	1.25	1.40	—	—	—	—
3/16	5.5	1.62	.583	.392	.600	.514	1.10	1.24	1.39	—	—	.982	.919

* Where no value of Q_s is shown, the angles comply with Specification Sect. 1.9.1.2 and may be considered fully effective.

For F_y = 36 ksi: $C'_c = 126.1/\sqrt{Q_s}$

For F_y = 50 ksi: $C'_c = 107.0/\sqrt{Q_s}$

TABLE A.6 Properties of Round Steel Pipe

Dimensions				Weight per Foot Lbs. Plain Ends	Properties			
Nominal Diameter In.	Outside Diameter In.	Inside Diameter In.	Wall Thickness In.		A In.^{2}	I In.^{4}	S In.^{3}	r In.
Standard Weight								
½	.840	.622	.109	.85	.250	.017	.041	.261
¾	1.050	.824	.113	1.13	.333	.037	.071	.334
1	1.315	1.049	.133	1.68	.494	.087	.133	.421
1¼	1.660	1.380	.140	2.27	.669	.195	.235	.540
1½	1.900	1.610	.145	2.72	.799	.310	.326	.623
2	2.375	2.067	.154	3.65	1.07	.666	.561	.787
2½	2.875	2.469	.203	5.79	1.70	1.53	1.06	.947
3	3.500	3.068	.216	7.58	2.23	3.02	1.72	1.16
3½	4.000	3.548	.226	9.11	2.68	4.79	2.39	1.34
4	4.500	4.026	.237	10.79	3.17	7.23	3.21	1.51
5	5.563	5.047	.258	14.62	4.30	15.2	5.45	1.88
6	6.625	6.065	.280	18.97	5.58	28.1	8.50	2.25
8	8.625	7.981	.322	28.55	8.40	72.5	16.8	2.94
10	10.750	10.020	.365	40.48	11.9	161	29.9	3.67
12	12.750	12.000	.375	49.56	14.6	279	43.8	4.38
Extra Strong								
½	.840	.546	.147	1.09	.320	.020	.048	.250
¾	1.050	.742	.154	1.47	.433	.045	.085	.321
1	1.315	.957	.179	2.17	.639	.106	.161	.407
1¼	1.660	1.278	.191	3.00	.881	.242	.291	.524
1½	1.900	1.500	.200	3.63	1.07	.391	.412	.605
2	2.375	1.939	.218	5.02	1.48	.868	.731	.766
2½	2.875	2.323	.276	7.66	2.25	1.92	1.34	.924
3	3.500	2.900	.300	10.25	3.02	3.89	2.23	1.14
3½	4.000	3.364	.318	12.50	3.68	6.28	3.14	1.31
4	4.500	3.826	.337	14.98	4.41	9.61	4.27	1.48
5	5.563	4.813	.375	20.78	6.11	20.7	7.43	1.84
6	6.625	5.761	.432	28.57	8.40	40.5	12.2	2.19
8	8.625	7.625	.500	43.39	12.8	106	24.5	2.88
10	10.750	9.750	.500	54.74	16.1	212	39.4	3.63
12	12.750	11.750	.500	65.42	19.2	362	56.7	4.33
Double-Extra Strong								
2	2.375	1.503	.436	9.03	2.66	1.31	1.10	.703
2½	2.875	1.771	.552	13.69	4.03	2.87	2.00	.844
3	3.500	2.300	.600	18.58	5.47	5.99	3.42	1.05
4	4.500	3.152	.674	27.54	8.10	15.3	6.79	1.37
5	5.563	4.063	.750	38.55	11.3	33.6	12.1	1.72
6	6.625	4.897	.864	53.16	15.6	66.3	20.0	2.06
8	8.625	6.875	.875	72.42	21.3	162	37.6	2.76

The listed sections are available in conformance with ASTM Specification A53 Grade B or A501. Other sections are made to these specifications. Consult with pipe manufacturers or distributors for availability.

TABLE A.7 Properties of Rectangular Steel Tubing

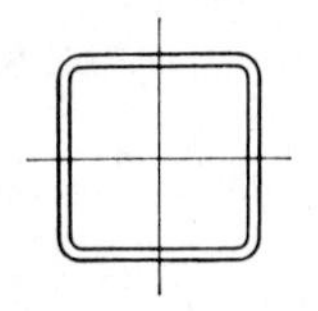

Square

DIMENSIONS				PROPERTIES**			
Nominal* Size	Wall Thickness		Weight per Foot	Area	I	S	r
In.	In.		Lb.	In.^{2}	In.^{4}	In.^{3}	In.
16 x 16	.5000	½	103.30	30.4	1200	150	6.29
	.3750	⅜	78.52	23.1	931	116	6.35
	.3125	5/16	65.87	19.4	789	98.6	6.38
14 x 14	.5000	½	89.68	26.4	791	113	5.48
	.3750	⅜	68.31	20.1	615	87.9	5.54
	.3125	5/16	57.36	16.9	522	74.6	5.57
12 x 12	.5000	½	76.07	22.4	485	80.9	4.66
	.3750	⅜	58.10	17.1	380	63.4	4.72
	.3125	5/16	48.86	14.4	324	54.0	4.75
	.2500	¼	39.43	11.6	265	44.1	4.78
10 x 10	.6250	⅝	76.33	22.4	321	64.2	3.78
	.5000	½	62.46	18.4	271	54.2	3.84
	.3750	⅜	47.90	14.1	214	42.9	3.90
	.3125	5/16	40.35	11.9	183	36.7	3.93
	.2500	¼	32.63	9.59	151	30.1	3.96
8 x 8	.6250	⅝	59.32	17.4	153	38.3	2.96
	.5000	½	48.85	14.4	131	32.9	3.03
	.3750	⅜	37.69	11.1	106	26.4	3.09
	.3125	5/16	31.84	9.36	90.9	22.7	3.12
	.2500	¼	25.82	7.59	75.1	18.8	3.15
	.1875	3/16	19.63	5.77	58.2	14.6	3.18
7 x 7	.5000	½	42.05	12.4	84.6	24.2	2.62
	.3750	⅜	32.58	9.58	68.7	19.6	2.68
	.3125	5/16	27.59	8.11	59.5	17.0	2.71
	.2500	¼	22.42	6.59	49.4	14.1	2.74
	.1875	3/16	17.08	5.02	38.5	11.0	2.77
6 x 6	.5000	½	35.24	10.4	50.5	16.8	2.21
	.3750	⅜	27.48	8.08	41.6	13.9	2.27
	.3125	5/16	23.34	6.86	36.3	12.1	2.30
	.2500	¼	19.02	5.59	30.3	10.1	2.33
	.1875	3/16	14.53	4.27	23.8	7.93	2.36
5 x 5	.5000	½	28.43	8.36	27.0	10.8	1.80
	.3750	⅜	22.37	6.58	22.8	9.11	1.86
	.3125	5/16	19.08	5.61	20.1	8.02	1.89
	.2500	¼	15.62	4.59	16.9	6.78	1.92
	.1875	3/16	11.97	3.52	13.4	5.36	1.95

* Outside dimensions across flat sides.

** Properties are based upon a nominal outside corner radius equal to two times the wall thickness.

TABLE A.7 (*Continued*)

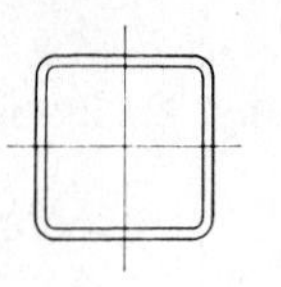

Square

DIMENSIONS				PROPERTIES**			
Nominal* Size	Wall Thickness		Weight per Foot	Area	I	S	r
In.	In.		Lb.	In.^{2}	In.^{4}	In.^{3}	In.
4 x 4	.5000	½	21.63	6.36	12.3	6.13	1.39
	.3750	⅜	17.27	5.08	10.7	5.35	1.45
	.3125	5/16	14.83	4.36	9.58	4.79	1.48
	.2500	¼	12.21	3.59	8.22	4.11	1.51
	.1875	3/16	9.42	2.77	6.59	3.30	1.54
3.5 x 3.5	.3125	5/16	12.70	3.73	6.09	3.48	1.28
	.2500	¼	10.51	3.09	5.29	3.02	1.31
	.1875	3/16	8.15	2.39	4.29	2.45	1.34
3 x 3	.3125	5/16	10.58	3.11	3.58	2.39	1.07
	.2500	¼	8.81	2.59	3.16	2.10	1.10
	.1875	3/16	6.87	2.02	2.60	1.73	1.13
2.5 x 2.5	.2500	¼	7.11	2.09	1.69	1.35	.899
	.1875	3/16	5.59	1.64	1.42	1.14	.930
2 x 2	.2500	¼	5.41	1.59	.766	.766	.694
	.1875	3/16	4.32	1.27	.668	.668	.726

* Outside dimensions across flat sides.
** Properties are based upon a nominal outside corner radius equal to two times the wall thickness.

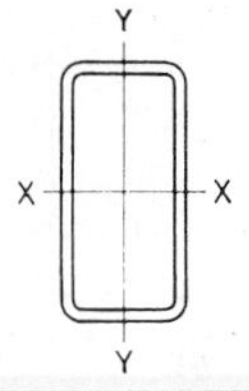

Rectangular

DIMENSIONS				PROPERTIES**						
					X-X AXIS			Y-Y AXIS		
Nominal* Size	Wall Thickness		Weight per Foot	Area	I_x	S_x	r_x	I_y	S_y	r_y
In.	In.		Lb.	In.^{2}	In.^{4}	In.^{3}	In.	In.^{4}	In.^{3}	In.
20 x 12	.5000	½	103.30	30.4	1650	165	7.37	750	125.0	4.97
	.3750	⅜	78.52	23.1	1280	128	7.44	583	97.2	5.03
	.3125	5/16	65.87	19.4	1080	108	7.47	495	82.5	5.06
20 x 8	.5000	½	89.68	26.4	1270	127	6.94	300	75.1	3.38
	.3750	⅜	68.31	20.1	988	98.8	7.02	236	59.1	3.43
	.3125	5/16	57.36	16.9	838	83.8	7.05	202	50.4	3.46
20 x 4	.5000	½	76.07	22.4	889	88.9	6.31	61.6	30.8	1.66
	.3750	⅜	58.10	17.1	699	69.9	6.40	50.3	25.1	1.72
	.3125	5/16	48.86	14.4	596	59.6	6.44	43.7	21.8	1.74

* Outside dimensions across flat sides.
** Properties are based upon a nominal outside corner radius equal to two times the wall thickness.

TABLE A.7 (*Continued*)

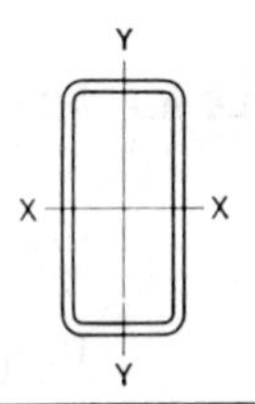

Rectangular

DIMENSIONS				PROPERTIES**						
					X-X AXIS			Y-Y AXIS		
Nominal* Size	Wall Thickness		Weight per Foot	Area	I_x	S_x	r_x	I_y	S_y	r_y
In.	In.		In.	In.^{2}	In.^{4}	In.^{3}	In.	In.^{4}	In.^{3}	In.
18 x 6	.5000	½	76.07	22.4	818	90.9	6.05	141	47.2	2.52
	.3750	⅜	58.10	17.1	641	71.3	6.13	113	37.6	2.57
	.3125	5/16	48.86	14.4	546	60.7	6.17	97.0	32.3	2.60
16 x 12	.5000	½	89.68	26.4	962	120	6.04	618	103.0	4.84
	.3750	⅜	68.31	20.1	748	93.5	6.11	482	80.3	4.90
	.3125	5/16	57.36	16.9	635	79.4	6.14	409	68.2	4.93
16 x 8	.5000	½	76.07	22.4	722	90.2	5.68	244	61.0	3.30
	.3750	⅜	58.10	17.1	565	70.6	5.75	193	48.2	3.36
	.3125	5/16	48.86	14.4	481	60.1	5.79	165	41.2	3.39
16 x 4	.5000	½	62.46	18.4	481	60.2	5.12	49.3	24.6	1.64
	.3750	⅜	47.90	14.1	382	47.8	5.21	40.4	20.2	1.69
	.3125	5/16	40.35	11.9	327	40.9	5.25	35.1	17.6	1.72
14 x 10	.5000	½	76.07	22.4	608	86.9	5.22	361	72.3	4.02
	.3750	⅜	58.10	17.1	476	68.0	5.28	284	56.8	4.08
	.3125	5/16	48.86	14.4	405	57.9	5.31	242	48.4	4.11
14 x 6	.5000	½	62.46	18.4	426	60.8	4.82	111	37.1	2.46
	.3750	⅜	47.90	14.1	337	48.1	4.89	89.1	29.7	2.52
	.3125	5/16	40.35	11.9	288	41.2	4.93	76.7	25.6	2.54
	.2500	¼	32.63	9.59	237	33.8	4.97	63.4	21.1	2.57
14 x 4	.5000	½	55.66	16.4	335	47.8	4.52	43.1	21.5	1.62
	.3750	⅜	42.79	12.6	267	38.2	4.61	35.4	17.7	1.68
	.3125	5/16	36.10	10.6	230	32.8	4.65	30.9	15.4	1.71
	.2500	¼	29.23	8.59	189	27.0	4.69	25.8	12.9	1.73
12 x 8	.6250	⅝	76.33	22.4	418	69.7	4.32	221	55.3	3.14
	.5000	½	62.46	18.4	353	58.9	4.39	188	46.9	3.20
	.3750	⅜	47.90	14.1	279	46.5	4.45	149	37.3	3.26
	.3125	5/16	40.35	11.9	239	39.8	4.49	128	32.0	3.28
	.2500	¼	32.63	9.59	196	32.6	4.52	105	26.3	3.31
12 x 6	.5000	½	55.66	16.4	287	47.8	4.19	96.0	32.0	2.42
	.3750	⅜	42.79	12.6	228	38.1	4.26	77.2	25.7	2.48
	.3125	5/16	36.10	10.6	196	32.6	4.30	66.6	22.2	2.51
	.2500	¼	29.23	8.59	161	26.9	4.33	55.2	18.4	2.53
	.1875	3/16	22.18	6.52	124	20.7	4.37	42.8	14.3	2.56
12 x 4	.5000	½	48.85	14.4	221	36.8	3.92	36.9	18.5	1.60
	.3750	⅜	37.69	11.1	178	29.6	4.01	30.5	15.2	1.66
	.3125	5/16	31.84	9.36	153	25.5	4.05	26.6	13.3	1.69
	.2500	¼	25.82	7.59	127	21.1	4.09	22.3	11.1	1.71
	.1875	3/16	19.63	5.77	98.2	16.4	4.13	17.5	8.75	1.74
12 x 2	.2500	¼	22.42	6.59	92.2	15.4	3.74	4.62	4.62	.837
	.1875	3/16	17.08	5.02	72.0	12.0	3.79	3.76	3.76	.865

* Outside dimensions across flat sides.

** Properties are based upon a nominal outside corner radius equal to two times the wall thickness.

TABLE A.7 (*Continued*)

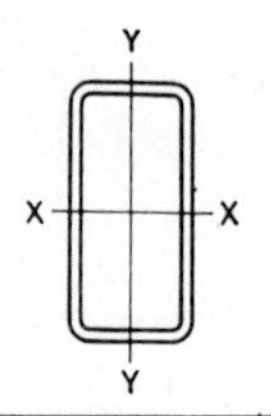

Rectangular

DIMENSIONS				PROPERTIES**						
					X-X AXIS			Y-Y AXIS		
Nominal* Size	Wall Thickness		Weight per Foot	Area	I_x	S_x	r_x	I_y	S_y	r_y
In.	In.		Lb.	In.^{2}	In.^{4}	In.^{3}	In.	In.^{4}	In.^{3}	In.
10 x 6	.5000	½	48.85	14.4	181	36.2	3.55	80.8	26.9	2.37
	.3750	⅜	37.69	11.1	145	29.0	3.62	65.4	21.8	2.43
	.3125	5/16	31.84	9.36	125	25.0	3.65	56.5	18.8	2.46
	.2500	¼	25.82	7.59	103	20.6	3.69	46.9	15.6	2.49
	.1875	3/16	19.63	5.77	79.8	16.0	3.72	36.5	12.2	2.51
10 x 4	.5000	½	42.05	12.4	136	27.1	3.31	30.8	15.4	1.58
	.3750	⅜	32.58	9.58	110	22.0	3.39	25.5	12.8	1.63
	.3125	5/16	27.59	8.11	95.0	19.1	3.43	22.4	11.2	1.66
	.2500	¼	22.42	6.59	79.3	15.9	3.47	18.8	9.39	1.69
	.1875	3/16	17.08	5.02	61.7	12.3	3.51	14.8	7.39	1.72
10 x 2	.3750	⅜	27.48	8.08	75.4	15.1	3.06	4.85	4.85	.775
	.3125	5/16	23.34	6.86	66.1	13.2	3.10	4.42	4.42	.802
	.2500	¼	19.02	5.59	55.5	11.1	3.15	3.85	3.85	.830
	.1875	3/16	14.53	4.27	43.7	8.74	3.20	3.14	3.14	.858
8 x 6	.5000	½	42.05	12.4	103	25.8	2.89	65.7	21.9	2.31
	.3750	⅜	32.58	9.58	83.7	20.9	2.96	53.5	17.8	2.36
	.3125	5/16	27.59	8.11	72.4	18.1	2.99	46.4	15.5	2.39
	.2500	¼	22.42	6.59	60.1	15.0	3.02	38.6	12.9	2.42
	.1875	3/16	17.08	5.02	46.8	11.7	3.05	30.1	10.0	2.45
8 x 4	.5000	½	35.24	10.4	75.1	18.8	2.69	24.6	12.3	1.54
	.3750	⅜	27.48	8.08	61.9	15.5	2.77	20.6	10.3	1.60
	.3125	5/16	23.34	6.86	53.9	13.5	2.80	18.1	9.05	1.62
	.2500	¼	19.02	5.59	45.1	11.3	2.84	15.3	7.63	1.65
	.1875	3/16	14.53	4.27	35.3	8.83	2.88	12.0	6.02	1.68
8 x 3	.3750	⅜	24.93	7.33	51.0	12.7	2.64	10.4	6.92	1.19
	.3125	5/16	21.21	6.23	44.7	11.2	2.68	9.25	6.16	1.22
	.2500	¼	17.32	5.09	37.6	9.40	2.72	7.90	5.26	1.25
	.1875	3/16	13.25	3.89	29.6	7.40	2.76	6.31	4.21	1.27
8 x 2	.3750	⅜	22.37	6.58	40.1	10.0	2.47	3.85	3.85	.765
	.3125	5/16	19.08	5.61	35.5	8.87	2.51	3.52	3.52	.792
	.2500	¼	15.62	4.59	30.1	7.52	2.56	3.08	3.08	.819
	.1875	3/16	11.97	3.52	23.9	5.97	2.60	2.52	2.52	.847
7 x 5	.5000	½	35.24	10.4	63.5	18.1	2.48	37.2	14.9	1.90
	.3750	⅜	27.48	8.08	52.2	14.9	2.54	30.8	12.3	1.95
	.3125	5/16	23.34	6.86	45.5	13.0	2.58	26.9	10.8	1.98
	.2500	¼	19.02	5.59	38.0	10.9	2.61	22.6	9.04	2.01
	.1875	3/16	14.53	4.27	29.8	8.50	2.64	17.7	7.10	2.04
7 x 4	.3750	⅜	24.93	7.33	44.0	12.6	2.45	18.1	9.06	1.57
	.3125	5/16	21.21	6.23	38.5	11.0	2.49	16.0	7.98	1.60
	.2500	¼	17.32	5.09	32.3	9.23	2.52	13.5	6.75	1.63
	.1875	3/16	13.25	3.89	25.4	7.26	2.55	10.7	5.34	1.66

* Outside dimensions across flat sides.

** Properties are based upon a nominal outside corner radius equal to two times the wall thickness.

TABLE A.7 (*Continued*)

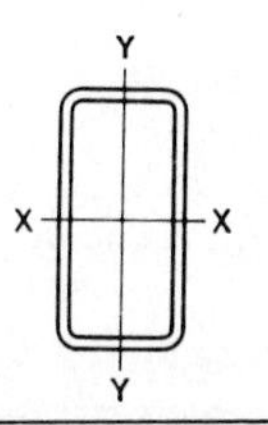

Rectangular

DIMENSIONS				PROPERTIES**						
					X-X AXIS			Y-Y AXIS		
Nominal* Size	Wall Thickness		Weight per Foot	Area	I_x	S_x	r_x	I_y	S_y	r_y
In.	In.		In.	In.^{2}	In.^{4}	In.^{3}	In.	In.^{4}	In.^{3}	In.
7 x 3	.3750	3/8	22.37	6.58	35.7	10.2	2.33	9.08	6.05	1.18
	.3125	5/16	19.08	5.61	31.5	9.00	2.37	8.11	5.41	1.20
	.2500	1/4	15.62	4.59	26.6	7.61	2.41	6.95	4.63	1.23
	.1875	3/16	11.97	3.52	21.1	6.02	2.45	5.57	3.71	1.26
6 x 4	.5000	1/2	28.43	8.36	35.3	11.8	2.06	18.4	9.21	1.48
	.3750	3/8	22.37	6.58	29.7	9.90	2.13	15.6	7.82	1.54
	.3125	5/16	19.08	5.61	26.2	8.72	2.16	13.8	6.92	1.57
	.2500	1/4	15.62	4.59	22.1	7.36	2.19	11.7	5.87	1.60
	.1875	3/16	11.97	3.52	17.4	5.81	2.23	9.32	4.66	1.63
6 x 3	.3750	3/8	19.82	5.83	23.8	7.92	2.02	7.78	5.19	1.16
	.3125	5/16	16.96	4.98	21.1	7.03	2.06	6.98	4.65	1.18
	.2500	1/4	13.91	4.09	17.9	5.98	2.09	6.00	4.00	1.21
	.1875	3/16	10.70	3.14	14.3	4.76	2.13	4.83	3.22	1.24
6 x 2	.3750	3/8	17.27	5.08	17.8	5.94	1.87	2.84	2.84	.748
	.3125	5/16	14.83	4.36	16.0	5.34	1.92	2.62	2.62	.775
	.2500	1/4	12.21	3.59	13.8	4.60	1.96	2.31	2.31	.802
	.1875	3/16	9.42	2.77	11.1	3.70	2.00	1.90	1.90	.829
5 x 4	.3750	3/8	19.82	5.83	18.7	7.50	1.79	13.2	6.58	1.50
	.3125	5/16	16.96	4.98	16.6	6.65	1.83	11.7	5.85	1.53
	.2500	1/4	13.91	4.09	14.1	5.65	1.86	9.98	4.99	1.56
	.1875	3/16	10.70	3.14	11.2	4.49	1.89	7.96	3.98	1.59
5 x 3	.5000	1/2	21.63	6.36	16.9	6.75	1.63	7.33	4.88	1.07
	.3750	3/8	17.27	5.08	14.7	5.89	1.70	6.48	4.32	1.13
	.3125	5/16	14.83	4.36	13.2	5.27	1.74	5.85	3.90	1.16
	.2500	1/4	12.21	3.59	11.3	4.52	1.77	5.05	3.37	1.19
	.1875	3/16	9.42	2.77	9.06	3.62	1.81	4.08	2.72	1.21
5 x 2	.3125	5/16	12.70	3.73	9.74	3.90	1.62	2.16	2.16	.76
	.2500	1/4	10.51	3.09	8.48	3.39	1.66	1.92	1.92	.78
	.1875	3/16	8.15	2.39	6.89	2.75	1.70	1.60	1.60	.81
4 x 3	.3125	5/16	12.70	3.73	7.45	3.72	1.41	4.71	3.14	1.12
	.2500	1/4	10.51	3.09	6.45	3.23	1.45	4.10	2.74	1.15
	.1875	3/16	8.15	2.39	5.23	2.62	1.48	3.34	2.23	1.18
4 x 2	.3125	5/16	10.58	3.11	5.32	2.66	1.31	1.71	1.71	.743
	.2500	1/4	8.81	2.59	4.69	2.35	1.35	1.54	1.54	.770
	.1875	3/16	6.87	2.02	3.87	1.93	1.38	1.29	1.29	.798
3 x 2	.2500	1/4	7.11	2.09	2.21	1.47	1.03	1.15	1.15	.742
	.1875	3/16	5.59	1.64	1.86	1.24	1.06	.977	.977	.771

* Outside dimensions across flat sides.

** Properties are based upon a nominal outside corner radius equal to two times the wall thickness.

APPENDIX B

SECTION MODULUS AND MOMENT RESISTANCE FOR SELECTED ROLLED STRUCTURAL SHAPES

The following tables list the rolled structural shapes that are commonly used as beams. Shapes are listed in sequence in accordance with the value of the section modulus about their principal axes (S_x). Moment values are the total resisting moment of the section with a steel yield strength of 36 ksi [250 MPa] and are based on a maximum bending stress of 24 ksi [165 MPa] for compact sections, 22 ksi [150 MPa] for noncompact sections.

Beams whose designations appear in boldface type are *least weight members*. These members have a section modulus value that is higher then other members with greater weight. When bending resistance alone is the critical design factor, these members constitute highly efficient shapes.

To assist in the identification of concern for buckling the limiting lateral unsupported length limits of L_c and L_u are given for each shape. If the actual unsupported lengths exceed these values, the graphs in Appendix C should be used rather than the tables in this section.

These data have been reproduced from the *Manual of Steel Construction,* 8th ed. (Ref. 1) with the permission of the publishers, the American Institute of Steel Construction. It should be noted that these tables do not contain data for the deeper W shapes (40 and 44 in.) or the "jumbo shapes" which are contained in the tables in the 9th edition of the manual. See the discussion in Sec. 1.2.

TABLE B.1 Section Modulus and Moment of Resistance for Selected Rolled Structural Shapes

S_x	Shape	F_y = 36 ksi		
		L_c	L_u	M_R
In.^{3}		Ft.	Ft.	Kip-ft.
1110	**W 36x300**	**17.6**	**35.3**	**2220**
1030	**W 36x280**	**17.5**	**33.1**	**2060**
953	**W 36x260**	**17.5**	**30.5**	**1910**
895	**W 36x245**	**17.4**	**28.6**	**1790**
837	**W 36x230**	**17.4**	**26.8**	**1670**
829	W 33x241	16.7	30.1	1660
757	**W 33x221**	**16.7**	**27.6**	**1510**
719	**W 36x210**	**12.9**	**20.9**	**1440**
684	**W 33x201**	**16.6**	**24.9**	**1370**
664	**W 36x194**	**12.8**	**19.4**	**1330**
663	W 30x211	15.9	29.7	1330
623	**W 36x182**	**12.7**	**18.2**	**1250**
598	W 30x191	15.9	26.9	1200
580	**W 36x170**	**12.7**	**17.0**	**1160**
542	**W 36x160**	**12.7**	**15.7**	**1080**
539	W 30x173	15.8	24.2	1080
504	**W 36x150**	**12.6**	**14.6**	**1010**
502	W 27x178	14.9	27.9	1000
487	W 33x152	12.2	16.9	974
455	W 27x161	14.8	25.4	910
448	**W 33x141**	**12.2**	**15.4**	**896**
439	**W 36x135**	**12.3**	**13.0**	**878**
414	W 24x162	13.7	29.3	828
411	W 27x146	14.7	23.0	822
406	**W 33x130**	**12.1**	**13.8**	**812**
380	W 30x132	11.1	16.1	760
371	W 24x146	13.6	26.3	742
359	**W 33x118**	**12.0**	**12.6**	**718**
355	W 30x124	11.1	15.0	710
329	**W 30x116**	**11.1**	**13.8**	**658**
329	W 24x131	13.6	23.4	658
329	W 21x147	13.2	30.3	658
299	**W 30x108**	**11.1**	**12.3**	**598**
299	W 27x114	10.6	15.9	598
295	W 21x132	13.1	27.2	590
291	W 24x117	13.5	20.8	582
273	W 21x122	13.1	25.4	546
269	**W 30x 99**	**10.9**	**11.4**	**538**
267	W 27x102	10.6	14.2	534
258	W 24x104	13.5	18.4	516
249	W 21x111	13.0	23.3	498
243	**W 27x 94**	**10.5**	**12.8**	**486**
231	W 18x119	11.9	29.1	462
227	W 21x101	13.0	21.3	454
222	**W 24x 94**	**9.6**	**15.1**	**444**
213	**W 27x 84**	**10.5**	**11.0**	**426**
204	W 18x106	11.8	26.0	408
196	**W 24x 84**	**9.5**	**13.3**	**392**
192	W 21x 93	8.9	16.8	384
190	W 14x120	15.5	44.1	380
188	W 18x 97	11.8	24.1	376
176	**W 24x 76**	**9.5**	**11.8**	**352**
175	W 16x100	11.0	28.1	350
173	W 14x109	15.4	40.6	346
171	W 21x 83	8.8	15.1	342
166	W 18x 86	11.7	21.5	332
157	W14x 99	15.4	37.0	314
155	W 16x 89	10.9	25.0	310
154	**W 24x 68**	**9.5**	**10.2**	**308**
151	W 21x 73	8.8	13.4	302
146	W 18x 76	11.6	19.1	292
143	W 14x 90	15.3	34.0	286
140	**W 21x 68**	**8.7**	**12.4**	**280**
134	W 16x 77	10.9	21.9	268
131	**W 24x 62**	**7.4**	**8.1**	**262**
127	**W 21x 62**	**8.7**	**11.2**	**254**
127	W 18x 71	8.1	15.5	254
123	W 14x 82	10.7	28.1	246
118	W 12x 87	12.8	36.2	236
117	W 18x 65	8.0	14.4	234
117	W 16x 67	10.8	19.3	234
114	**W 24x 55**	**7.0**	**7.5**	**228**
112	W 14x 74	10.6	25.9	224
111	W 21x 57	6.9	9.4	222
108	W 18x 60	8.0	13.3	216
107	W 12x 79	12.8	33.3	214
103	W 14x 68	10.6	23.9	206
98.3	**W 18x 55**	**7.9**	**12.1**	**197**
97.4	W 12x 72	12.7	30.5	195

TABLE B.1 ***(Continued)***

S_x	Shape	F_y = 36 ksi		
		L_c	L_u	M_R
In.^{3}		Ft.	Ft.	Kip-ft.
94.5	**W 21x50**	**6.9**	**7.8**	**189**
92.2	W 16x57	7.5	14.3	184
92.2	W 14x61	10.6	21.5	184
88.9	**W 18x50**	**7.9**	**11.0**	**178**
87.9	W 12x65	12.7	27.7	176
81.6	**W 21x44**	**6.6**	**7.0**	**163**
81.0	W 16x50	7.5	12.7	162
78.8	W 18x46	6.4	9.4	158
78.0	W 12x58	10.6	24.4	156
77.8	W 14x53	8.5	17.7	156
72.7	W 16x45	7.4	11.4	145
70.6	W 12x53	10.6	22.0	141
70.3	W 14x48	8.5	16.0	141
68.4	**W 18x40**	**6.3**	**8.2**	**137**
66.7	W 10x60	10.6	31.1	133
64.7	**W 16x40**	**7.4**	**10.2**	**129**
64.7	W 12x50	8.5	19.6	129
62.7	W 14x43	8.4	14.4	125
60.0	W 10x54	10.6	28.2	120
58.1	W 12x45	8.5	17.7	116
57.6	**W 18x35**	**6.3**	**6.7**	**115**
56.5	W 16x36	7.4	8.8	113
54.6	W 14x38	7.1	11.5	109
54.6	W 10x49	10.6	26.0	109
51.9	W 12x40	8.4	16.0	104
49.1	W 10x45	8.5	22.8	98
48.6	**W 14x34**	**7.1**	**10.2**	**97**
47.2	**W 16x31**	**5.8**	**7.1**	**94**
45.6	W 12x35	6.9	12.6	91
42.1	W 10x39	8.4	19.8	84
42.0	**W 14x30**	**7.1**	**8.7**	**84**
38.6	**W 12x30**	**6.9**	**10.8**	**77**
38.4	**W 16x26**	**5.6**	**6.0**	**77**
35.3	**W 14x26**	**5.3**	**7.0**	**71**
35.0	W 10x33	8.4	16.5	70
33.4	**W 12x26**	**6.9**	**9.4**	**67**
32.4	W 10x30	6.1	13.1	65
31.2	W 8x35	8.5	22.6	62

S_x	Shape	F_y = 36 ksi		
		L_c	L_u	M_R
In.^{3}		Ft.	Ft.	Kip-ft.
29.0	**W 14x22**	**5.3**	**5.6**	**58**
27.9	W 10x26	6.1	11.4	56
27.5	W 8x31	8.4	20.1	55
25.4	**W 12x22**	**4.3**	**6.4**	**51**
24.3	W 8x28	6.9	17.5	49
23.2	**W 10x22**	**6.1**	**9.4**	**46**
21.3	**W 12x19**	**4.2**	**5.3**	**43**
21.1	**M 14x18**	**3.6**	**4.0**	**42**
20.9	W 8x24	6.9	15.2	42
18.8	W 10x19	4.2	7.2	38
18.2	W 8x21	5.6	11.8	36
17.1	**W 12x16**	**4.1**	**4.3**	**34**
16.7	W 6x25	6.4	20.0	33
16.2	W 10x17	4.2	6.1	32
15.2	W 8x18	5.5	9.9	30
14.9	**W 12x14**	**3.5**	**4.2**	**30**
13.8	W 10x15	4.2	5.0	28
13.4	W 6x20	6.4	16.4	27
13.0	M 6x20	6.3	17.4	26
12.0	**M 12x11.8**	**2.7**	**3.0**	**24**
11.8	W 8x15	4.2	7.2	24
10.9	W 10x12	3.9	4.3	22
10.2	W 6x16	4.3	12.0	20
10.2	W 5x19	5.3	19.5	20
9.91	W 8x13	4.2	5.9	20
9.72	W 6x15	6.3	12.0	19
9.63	M 5x18.9	5.3	19.3	19
8.51	W 5x16	5.3	16.7	17
7.81	**W 8x10**	**4.2**	**4.7**	**16**
7.76	**M 10x 9**	**2.6**	**2.7**	**16**
7.31	W 6x12	4.2	8.6	15
5.56	**W 6x 9**	**4.2**	**6.7**	**11**
5.46	W 4x13	4.3	15.6	11
5.24	M 4x13	4.2	16.9	10
4.62	**M 8x 6.5**	**2.4**	**2.5**	**9**
2.40	**M 6x 4.4**	**1.9**	**2.4**	**5**

APPENDIX C

ALLOWABLE BENDING MOMENT FOR BEAMS WITH VARIOUS UNBRACED LENGTHS

These charts may be used to determine the allowable bending moment for shapes used as beams. Graphs apply to beams of steel with a yield stress of 36 ksi [250 MPa]. Dashed lines on the graphs indicate that the shape is heavier than one that has a higher bending resistance. To find the most efficient shapes proceed on the chart upward and toward the right until the first solid line graph is encountered.

These charts have been reproduced and adapted from the *Manual of Steel Construction,* 8th ed. (Ref. 1), with permission of the publishers, American Institute of Steel Construction. It should be noted that the charts presented here do not contain data for the deeper W shapes (40 and 44 in.) or the "jumbo shapes" which are contained in the tables in the 9th edition of the manual. See the discussion in Sec. 1.2.

Example. Find the lightest shape that may be used to resist a moment of 325 kip-ft [441 kN-m] if the unbraced length is 10 ft [3.05 m].

Solution. Find the graph page that includes the value for the given moment. Proceed across the horizontal line for the moment of 325 kip-ft until you come to the vertical line for an unbraced length of 10 ft. The first shape whose graph is above and to the right of this point is a W 16 × 96. However, the line for this shape is dashed to indicate that there is a lighter shape with greater moment of resistance. Proceed upward and to the right and you will encounter the solid line for the shape W 24 × 76, which is the lightest beam for this situation.

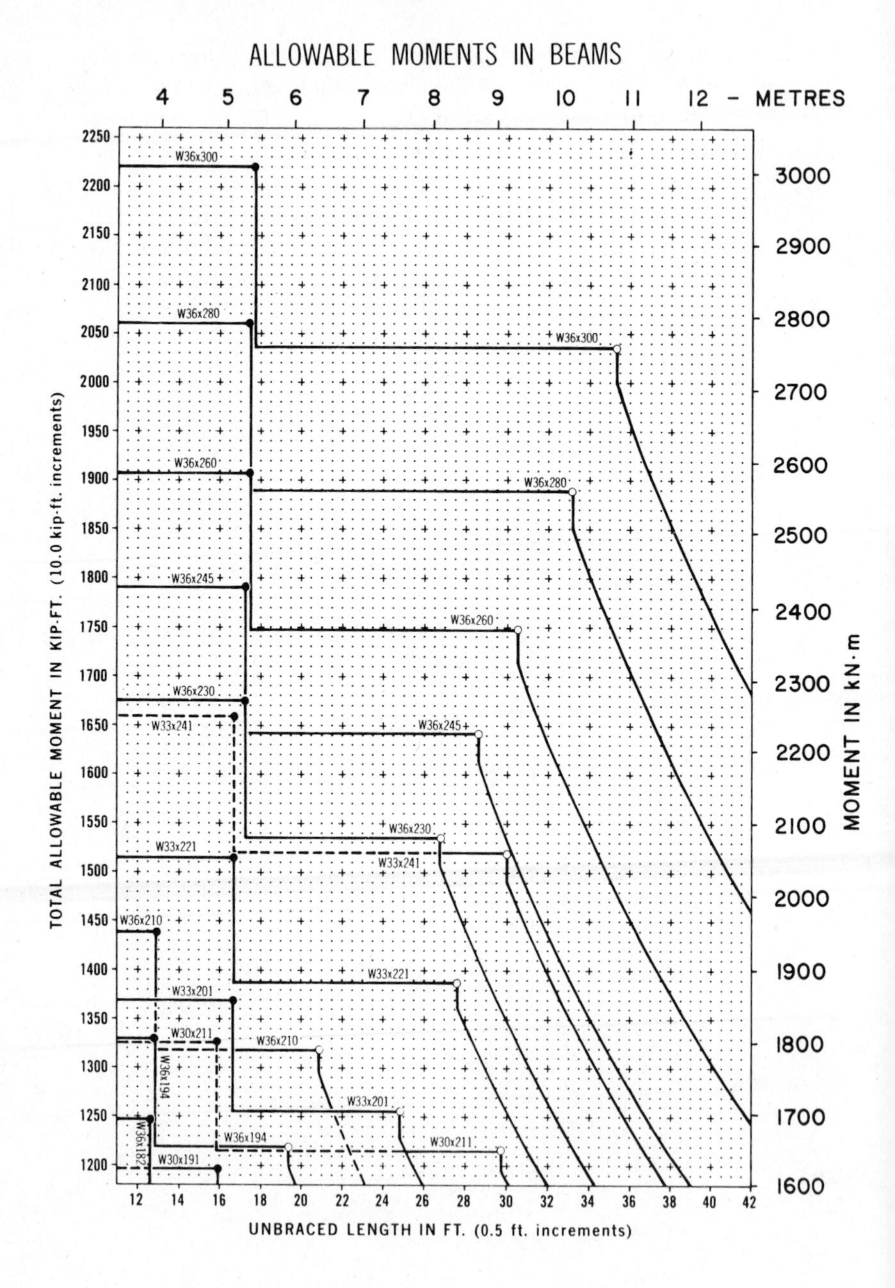
ALLOWABLE MOMENTS IN BEAMS
4 5 6 7 8 9 10 11 12 - METRES
TOTAL ALLOWABLE MOMENT IN KIP-FT. (10.0 kip-ft. increments)
MOMENT IN kN·m
UNBRACED LENGTH IN FT. (0.5 ft. increments)
2250
2200
2150
2100
2050
2000
1950
1900
1850
1800
1750
1700
1650
1600
1550
1500
1450
1400
1350
1300
1250
1200
3000
2900
2800
2700
2600
2500
2400
2300
2200
2100
2000
1900
1800
1700
1600
12 14 16 18 20 22 24 26 28 30 32 34 36 38 40 42
W36x300
W36x280
W36x260
W36x245
W36x230
W33x241
W33x221
W36x210
W33x201
W30x211
W36x194
W36x182
W30x191

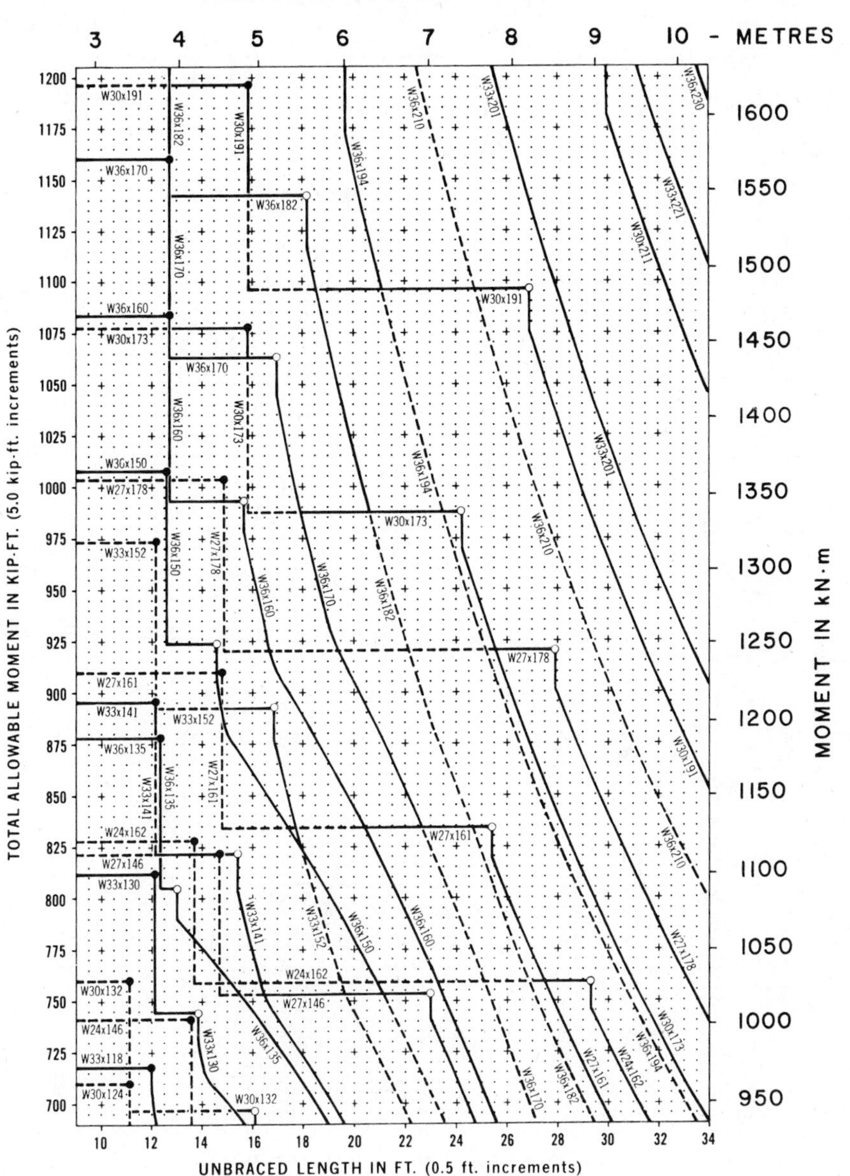
ALLOWABLE MOMENTS IN BEAMS
METRES
TOTAL ALLOWABLE MOMENT IN KIP-FT. (5.0 kip-ft. increments)
MOMENT IN kN·m
UNBRACED LENGTH IN FT. (0.5 ft. increments)

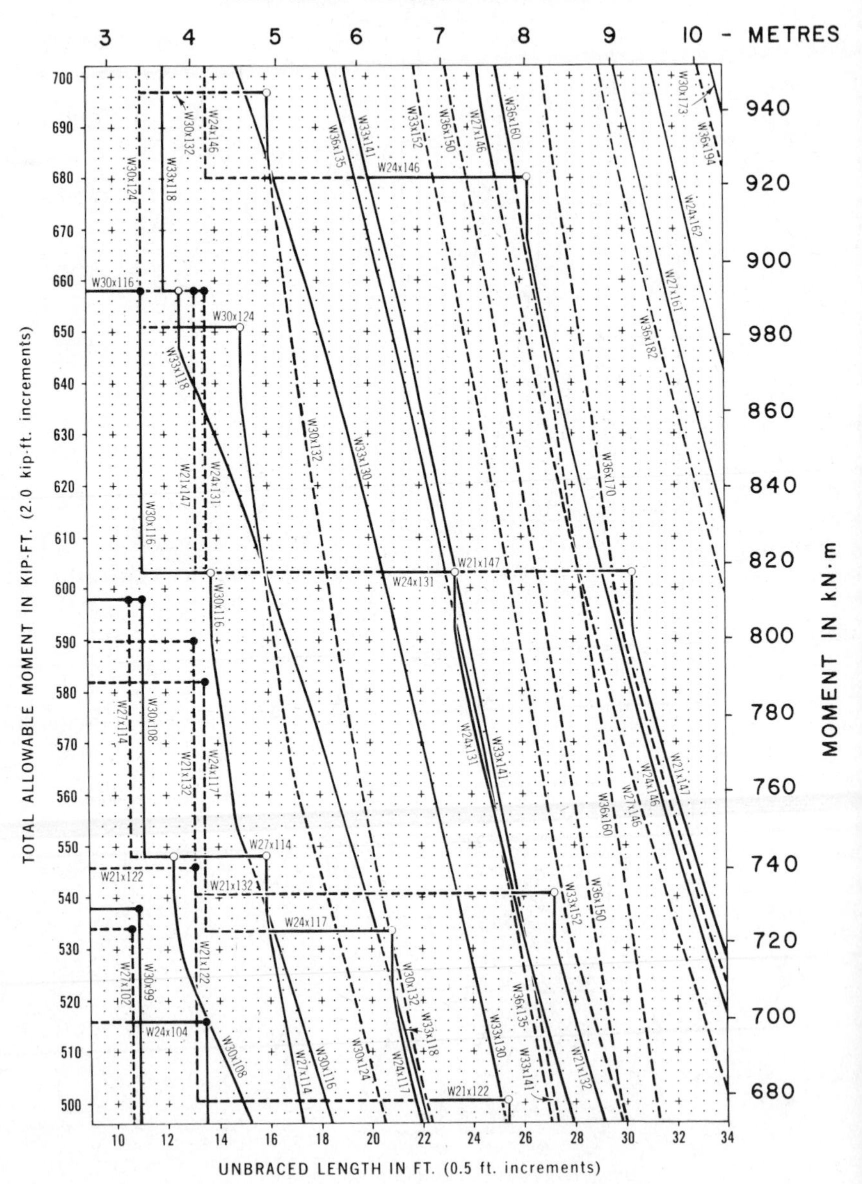

ALLOWABLE MOMENTS IN BEAMS
3 4 5 6 7 8 9 10 - METRES
TOTAL ALLOWABLE MOMENT IN KIP-FT. (2.0 kip-ft. increments)
700 690 680 670 660 650 640 630 620 610 600 590 580 570 560 550 540 530 520 510 500
MOMENT IN kN·m
940 920 900 980 860 840 820 800 780 760 740 720 700 680
UNBRACED LENGTH IN FT. (0.5 ft. increments)
10 12 14 16 18 20 22 24 26 28 30 32 34
W30x124
W33x118
W30x132
W24x146
W36x135
W33x141
W33x152
W36x150
W27x146
W36x160
W30x173
W36x194
W24x162
W27x161
W36x182
W36x170
W30x116
W21x147
W24x131
W30x132
W33x130
W24x131
W27x114
W30x108
W21x132
W24x117
W21x122
W36x160
W27x146
W24x146
W21x147
W33x152
W36x150
W30x99
W27x102
W24x104
W30x108
W27x114
W30x116
W30x124
W24x117
W33x118
W36x135
W33x130
W33x141
W21x132
W21x122

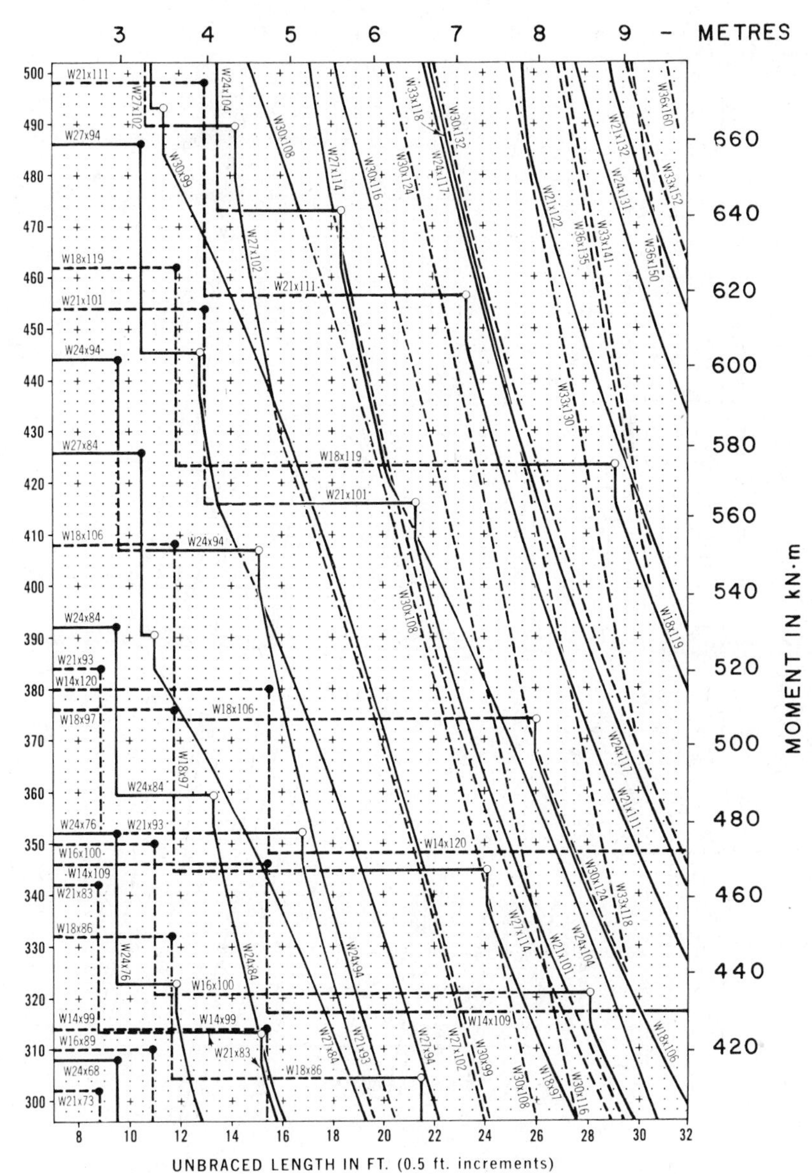
ALLOWABLE MOMENTS IN BEAMS
METRES
TOTAL ALLOWABLE MOMENT IN KIP-FT. (2.0 kip-ft. increments)
MOMENT IN kN·m
UNBRACED LENGTH IN FT. (0.5 ft. increments)

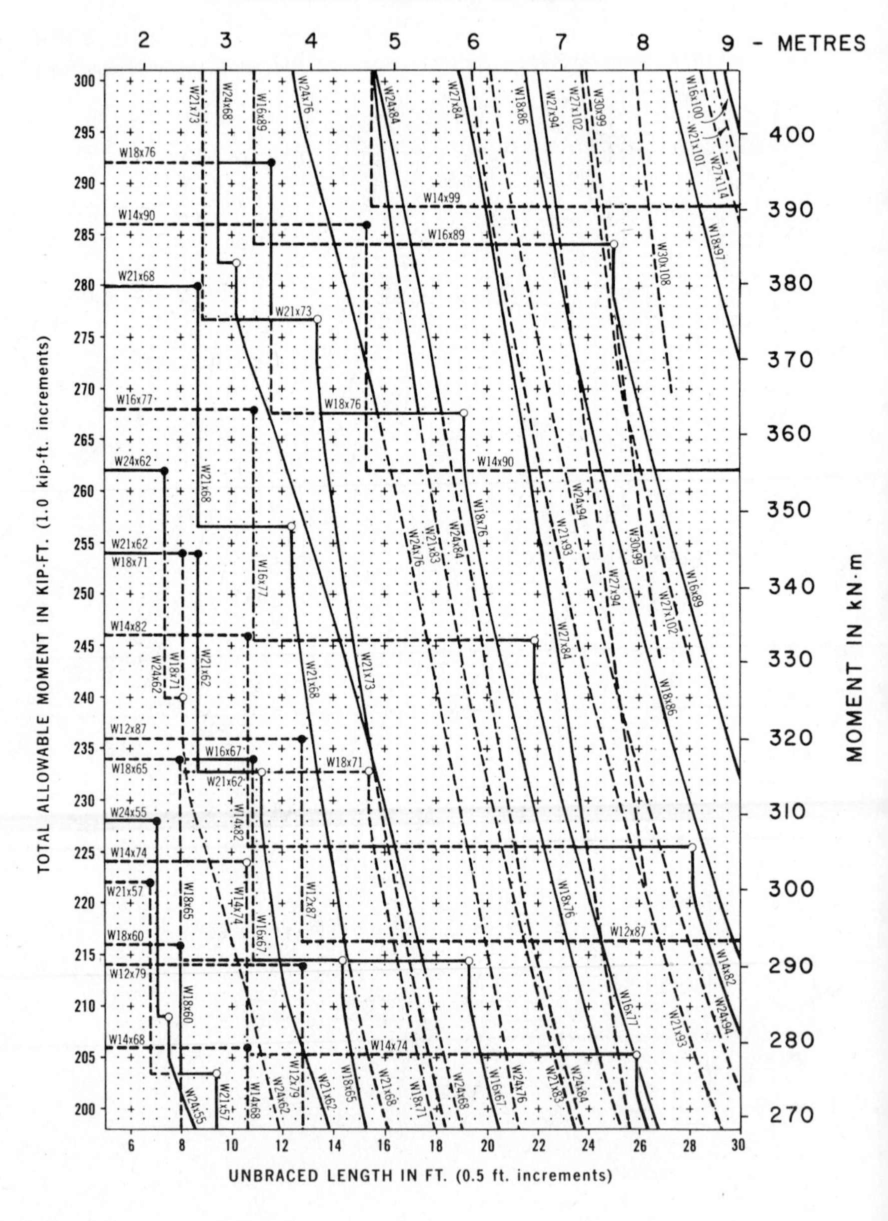
ALLOWABLE MOMENTS IN BEAMS
2 3 4 5 6 7 8 9 - METRES
TOTAL ALLOWABLE MOMENT IN KIP-FT. (1.0 kip-ft. increments)
MOMENT IN kN·m
UNBRACED LENGTH IN FT. (0.5 ft. increments)
W18x76
W14x90
W21x68
W16x77
W24x62
W21x62
W18x71
W14x82
W12x87
W16x67
W18x65
W24x55
W14x74
W21x57
W18x60
W12x79
W14x68
W14x99
W16x89
W21x73
W14x74
W12x87
W24x84
W27x84
W18x86
W27x94
W27x102
W30x99
W16x100
W21x101
W27x114
W30x108
W18x97
W16x89
W24x76

ALLOWABLE MOMENTS IN BEAMS

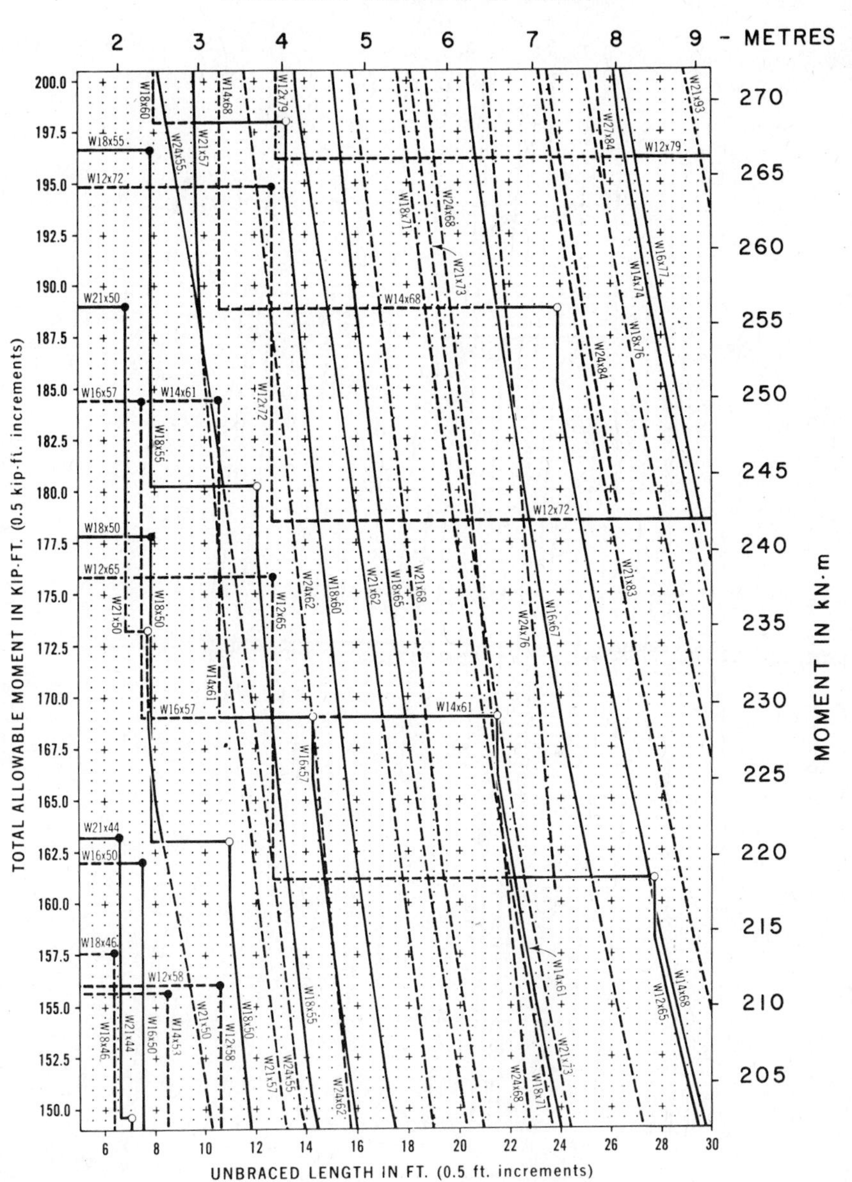

ALLOWABLE MOMENTS IN BEAMS

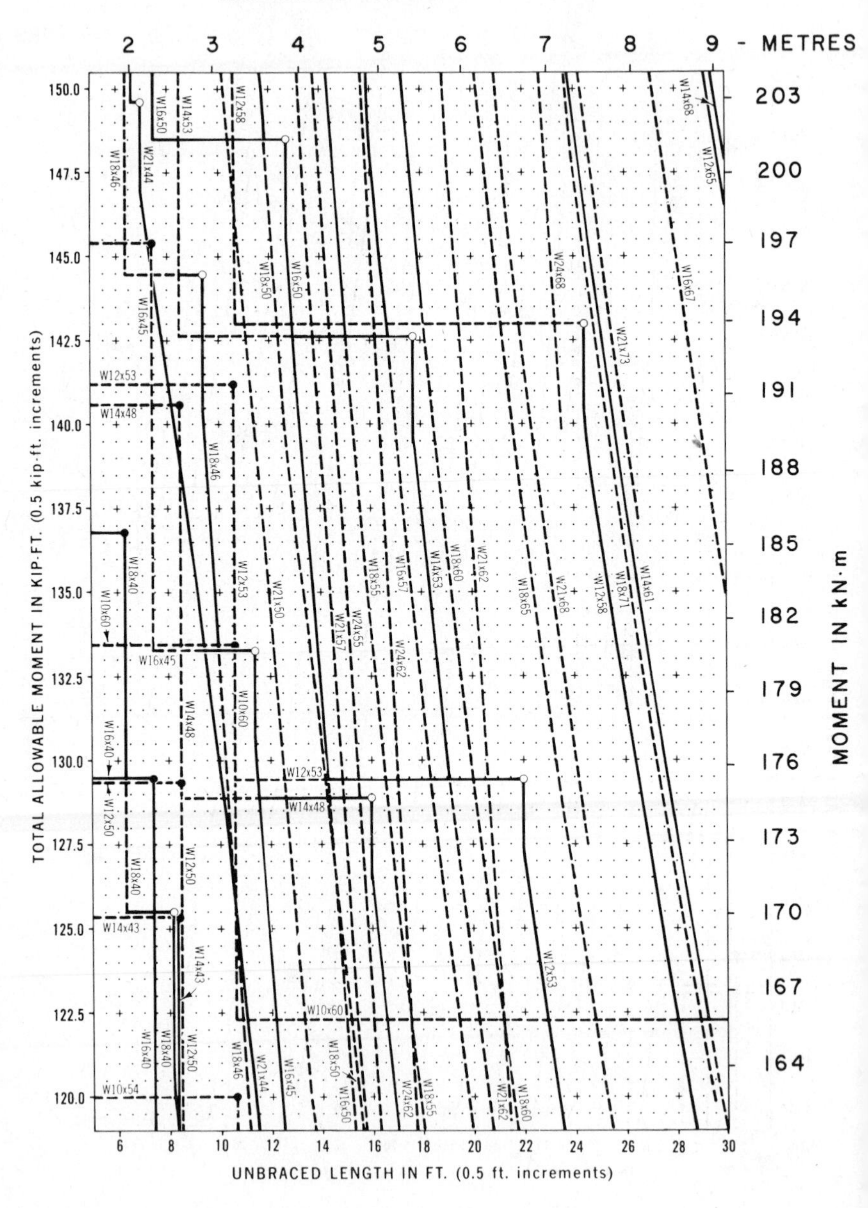

ALLOWABLE MOMENTS IN BEAMS

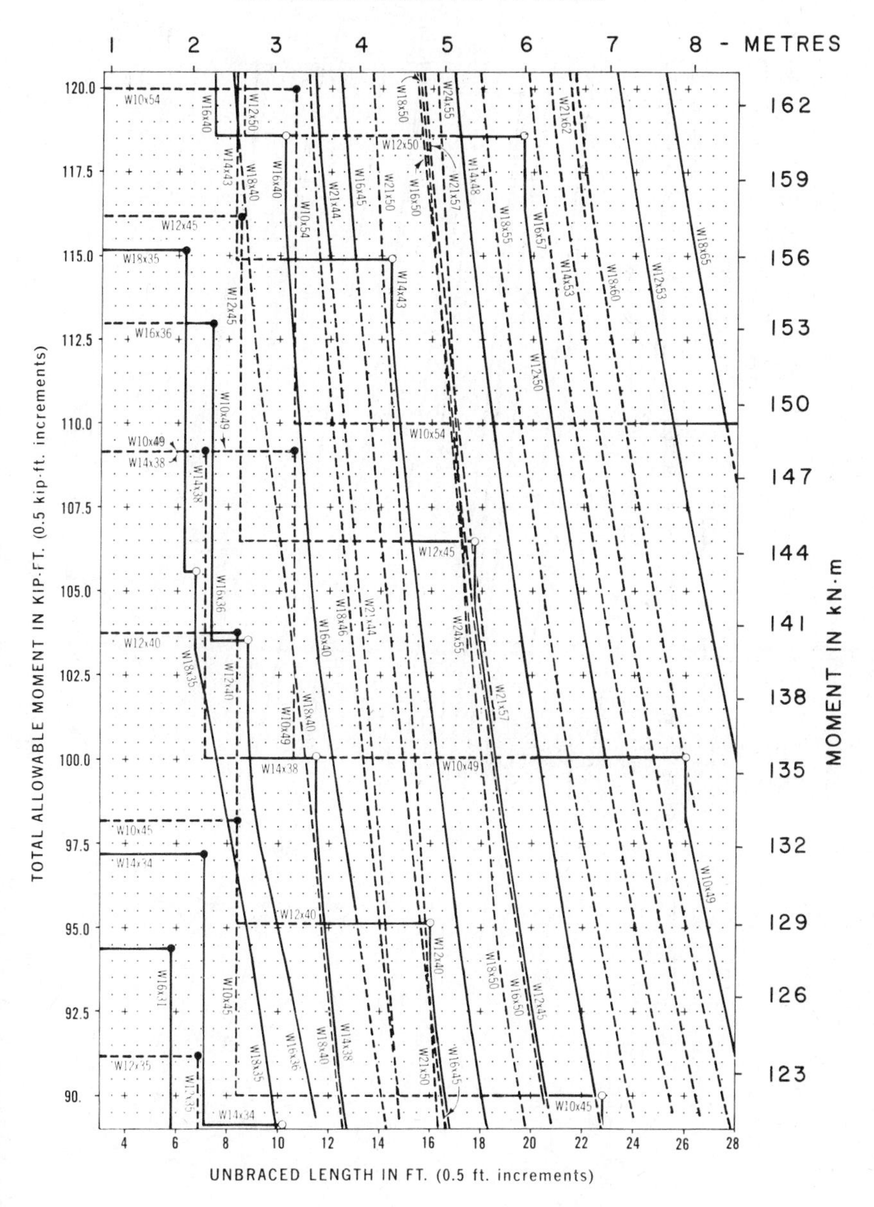

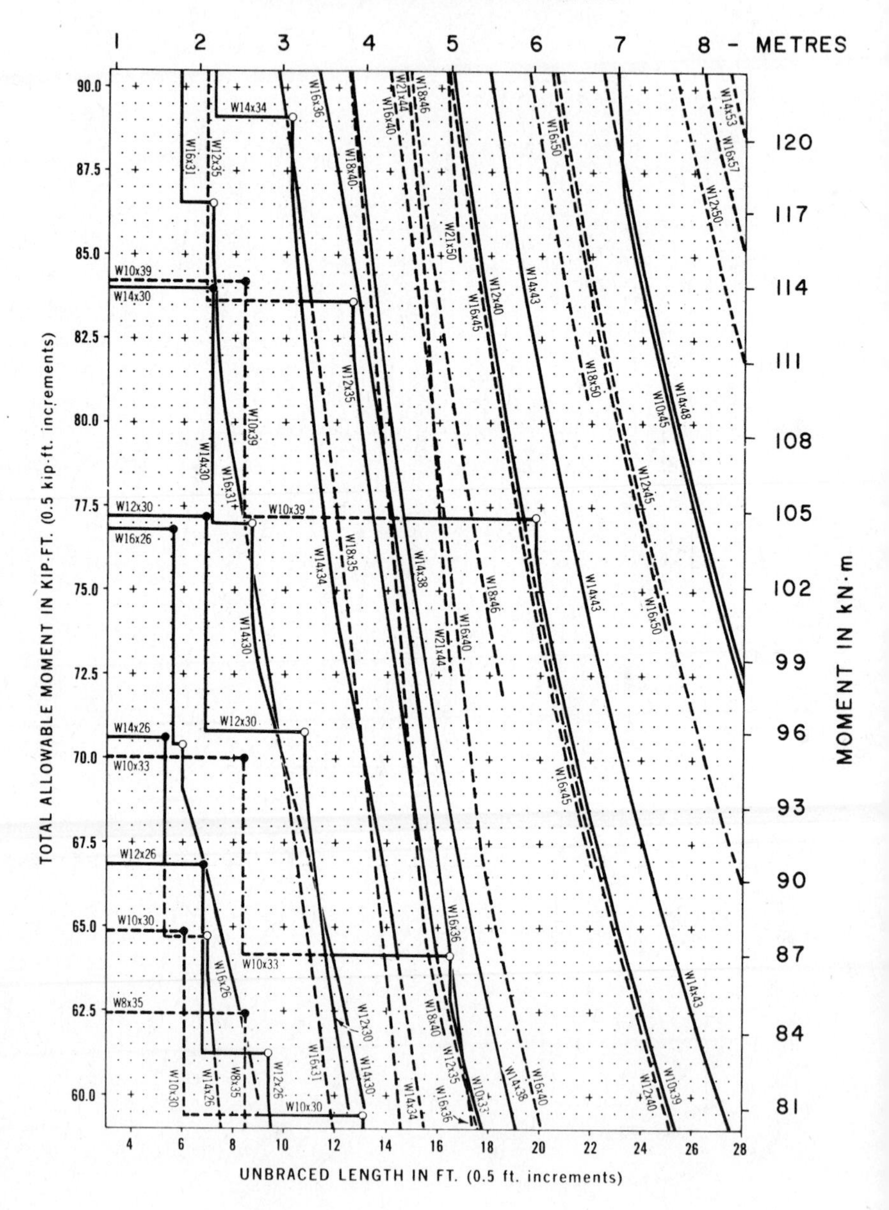
ALLOWABLE MOMENTS IN BEAMS
1 2 3 4 5 6 7 8 - METRES
TOTAL ALLOWABLE MOMENT IN KIP-FT. (0.5 kip-ft. increments)
MOMENT IN kN·m
UNBRACED LENGTH IN FT. (0.5 ft. increments)
90.0
87.5
85.0
82.5
80.0
77.5
75.0
72.5
70.0
67.5
65.0
62.5
60.0
120
117
114
111
108
105
102
99
96
93
90
87
84
81
4 6 8 10 12 14 16 18 20 22 24 26 28
W14x34
W16x31
W12x35
W10x39
W14x30
W12x30
W16x26
W14x26
W10x33
W12x26
W10x30
W8x35
W16x36
W18x40
W21x44
W16x40
W18x46
W21x50
W16x45
W12x40
W14x43
W16x50
W18x50
W10x45
W14x48
W12x45
W14x53
W16x57
W12x50
W18x35
W14x38
W12x30
W16x31
W14x30
W12x35
W10x33
W14x34
W16x36
W12x40
W10x39

ALLOWABLE MOMENTS IN BEAMS

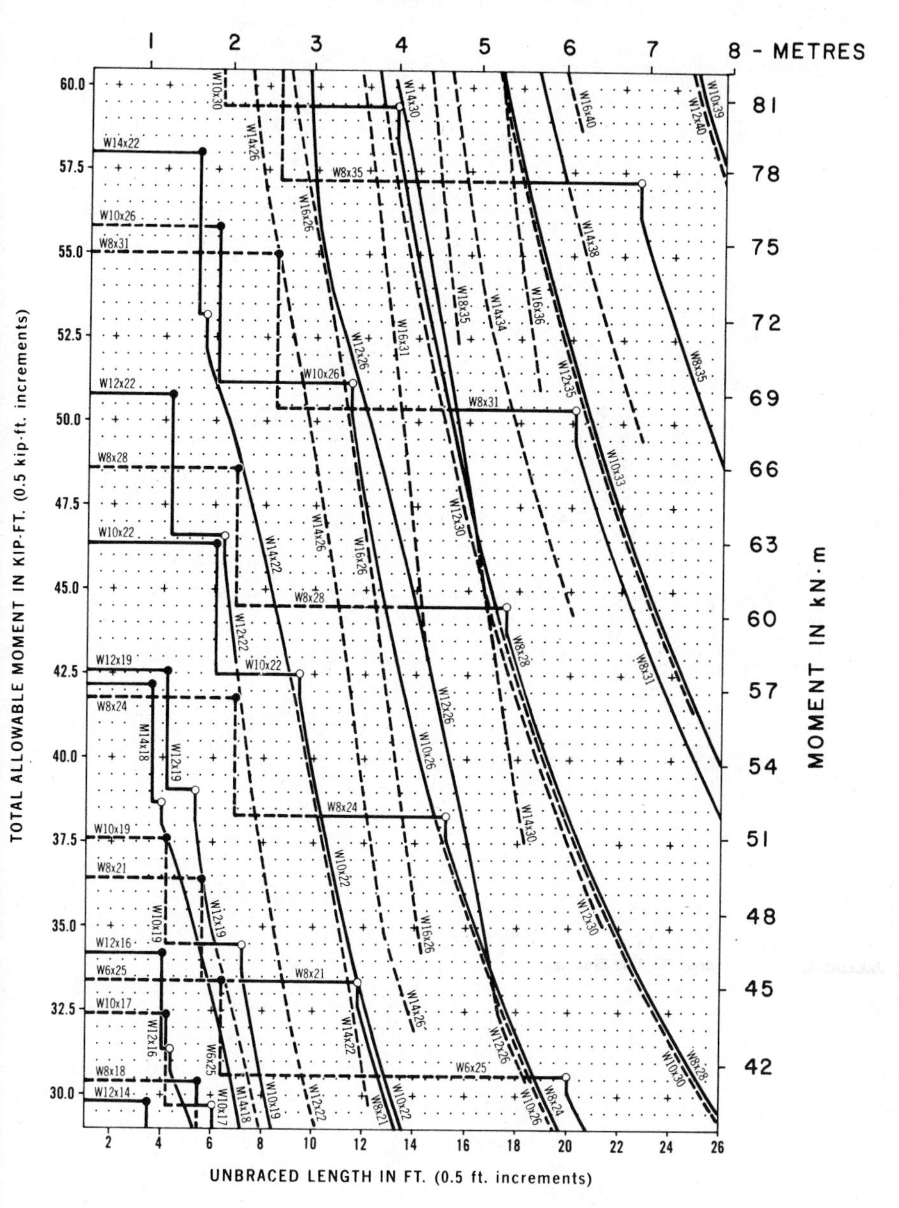

ALLOWABLE MOMENTS IN BEAMS

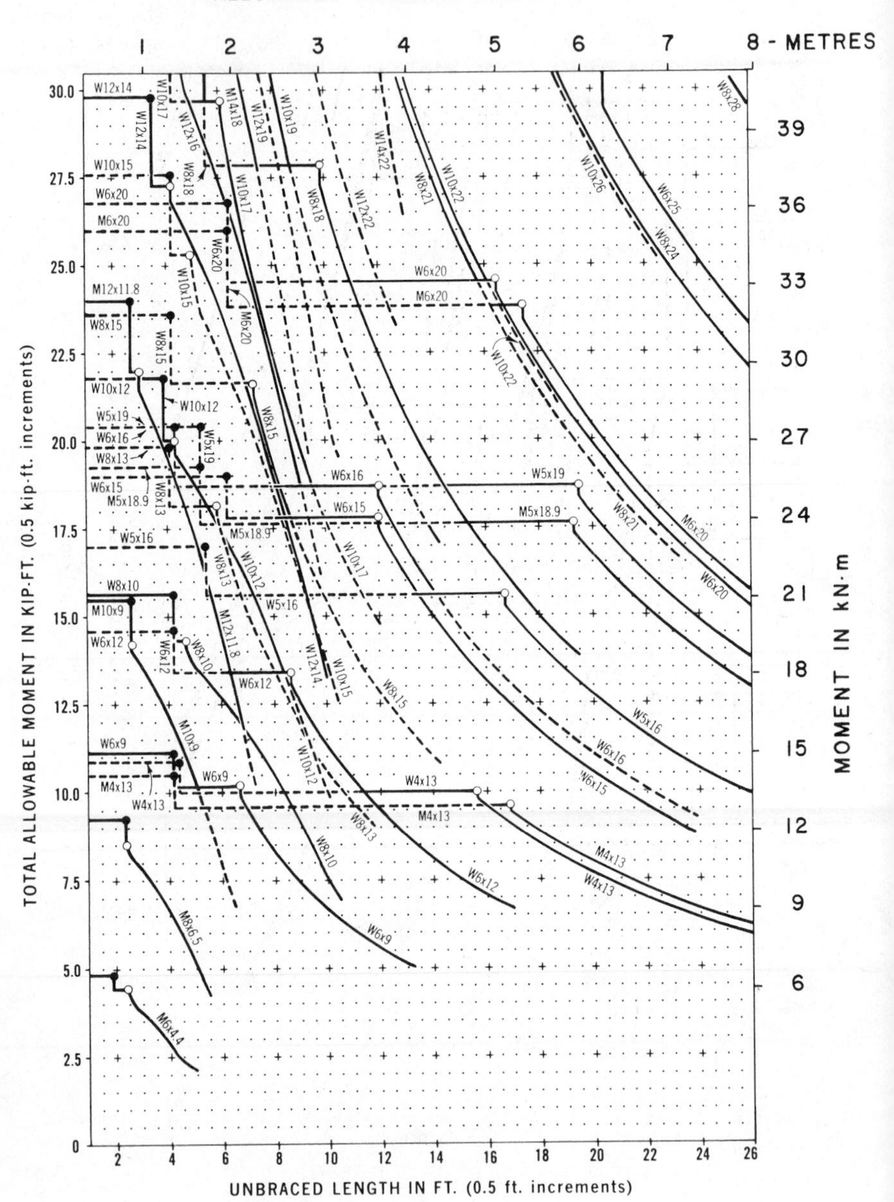

APPENDIX D

LOAD–SPAN VALUES FOR BEAMS

The following tables give values for the total allowable uniformly distributed load for simple span beams. Table values are based on a maximum bending stress of 24 ksi [165 MPa]; this assumes a yield stress of 36 ksi [250 MPa], moment due to loading in the plane of the *y*-axis of the section, and adequate conditions of lateral support. Smaller shapes, super-size shapes from the 9th edition of the AISC Manual, and shapes more commonly used as columns are omitted. Beams are listed in order of the value of their section modulus with respect to the *x*-axis (S_x).

Table values may be used for beams when the distance between points of lateral support does not exceed the L_c value for the shape. These values are given in the table. If the distance between points of lateral support exceeds L_c, the charts in Appendix C should be used instead of these tables.

Table values are not given for spans in excess of 27 times the beam depth, which is the approximate feasible limit based on reasonable limitation of deflection. Deflections in excess of 1/360 of the span will result for table loads to the right of the heavy vertical lines. Actual values of deflections in inches at the center of the span may be obtained by dividing the deflection factor given for the span by the depth of the beam in inches.

Examples of the use of these tables for common design situations appear in Sec. 5.5.

TABLE D.1 Load-Span Values for Beams[a]

Shape	Span (ft)	8	10	12	14	16	18	20	22	24	26	28	30
	Deflection factor[b] / L_c^c (ft)	1.59	2.48	3.58	4.87	6.36	8.05	9.93	12.0	14.3	16.8	19.5	22.3
M 8 × 6.5	2.4	9.24	7.39	6.16	5.28	4.62	4.11						
M 10 × 9	2.6	15.5	12.4	10.3	8.87	7.76	6.90	6.21	5.64				
W 8 × 10	4.2	15.6	12.5	10.4	8.92	7.81	6.94						
W 8 × 13	4.2	19.8	15.9	13.2	11.3	9.91	8.81						
W 10 × 12	3.9	21.8	17.4	14.5	12.5	10.9	9.69	8.72	7.93				
W 8 × 15	4.2	23.6	18.9	15.7	13.5	11.8	10.5						
M 12 × 11.8	2.7	24.0	19.2	16.0	13.7	12.0	10.7	9.60	8.73	8.00	7.38	6.86	
W 10 × 15	4.2	27.6	22.1	18.4	15.8	13.8	12.3	11.0	10.0				
W 12 × 14	3.5	29.8	23.8	19.9	17.0	14.9	13.2	11.9	10.8	9.93	9.17	8.51	
W 8 × 18	5.5	30.4	24.3	20.3	17.4	15.2	13.5						
W 10 × 17	4.2	32.4	25.9	21.6	18.5	16.2	14.4	13.0	11.8				
W 12 × 16	4.1	34.2	27.4	22.8	19.5	17.1	15.2	13.7	12.4	11.4	10.5	9.77	
W 8 × 21	5.6	36.4	29.1	24.3	20.8	18.2	16.2						
W 10 × 19	4.2	37.6	30.1	25.1	21.5	18.8	16.7	15.0	13.7				
W 8 × 24	6.9	41.8	33.4	27.9	23.9	20.9	18.6						
M 14 × 18	3.6	42.2	33.8	28.1	24.1	21.1	18.7	16.9	15.3	14.1	13.0	12.0	11.2
W 12 × 19	4.2	42.6	34.1	28.4	24.3	21.3	18.9	17.0	15.5	14.2	13.1	12.2	
W 10 × 22	6.1	46.4	37.1	30.9	26.5	23.2	20.6	18.5	16.9				
W 8 × 28	6.9	48.6	38.9	32.4	27.8	24.3	21.6						

TABLE D.1 (*Continued*)

Shape	Span (ft)	12	14	16	18	20	22	24	26	28	30	32	34
	Deflection factor[b] / L_c^c (ft)	3.58	4.87	6.36	8.05	9.93	12.0	14.3	16.8	19.5	22.3	25.4	28.7
W 12 × 22	4.3	33.9	29.0	25.4	22.6	20.3	18.5	16.9	15.6	14.5			
W 10 × 26	6.1	37.2	31.9	27.9	24.8	22.3	20.3						
W 14 × 22	5.3	38.7	33.1	29.0	25.8	23.2	21.1	19.3	17.8	16.6	15.5	14.5	
W 10 × 30	6.1	43.2	37.0	32.4	28.8	25.9	23.6						
W 12 × 26	6.9	44.5	38.2	33.4	29.7	26.7	24.3	22.3	20.5	19.1			
W 10 × 33	8.4	46.7	40.0	35.0	31.0	28.0	25.4						
W 14 × 26	5.3	47.1	40.3	35.3	31.4	28.2	25.7	23.5	21.7	20.2	18.8	17.6	
W 16 × 26	5.6	51.2	43.9	38.4	34.1	30.7	27.9	25.6	23.6	21.9	20.5	19.2	18.1
W 12 × 30	6.9	51.5	44.1	38.6	34.3	30.9	28.1	25.7	23.8	22.0			
W 14 × 30	7.1	56.0	48.0	42.0	37.3	33.6	30.5	28.0	25.8	24.0	22.4	21.0	
W 10 × 39	8.4	56.1	48.1	42.1	37.4	33.7	30.6						
W 12 × 35	6.9	60.8	52.1	45.6	40.5	36.5	33.2	30.4	28.1	26.0			
W 16 × 31	5.8	62.9	53.9	47.2	41.9	37.8	34.3	31.5	29.0	27.0	25.2	23.6	22.2
W 14 × 34	7.1	64.8	55.5	48.6	43.2	38.9	35.3	32.4	29.9	27.8	25.9	24.3	
W 10 × 45	8.5	65.5	56.1	49.1	43.6	39.3	35.7						

TABLE D.1 (*Continued*)

Shape	Span (ft)	16	18	20	22	24	26	28	30	32	34	36	38
	Deflection factor[b] / L_c^c (ft)	6.36	8.05	9.93	12.0	14.3	16.8	19.5	22.3	25.4	28.7	32.2	35.9
W 12 × 40	8.4	51.9	▮ 46.1	41.5	37.7	34.6	31.9	29.6					
W 14 × 38	7.1	54.6	48.5	▮ 43.7	39.7	36.4	33.6	31.2	29.1	27.3			
W 16 × 36	7.4	56.5	50.2	45.2	▮ 41.1	37.7	34.8	32.3	30.1	28.2	26.6	25.1	
W 18 × 35	6.3	57.8	51.4	46.2	42.0	38.5	▮ 35.6	33.0	30.8	28.9	27.2	25.7	24.3
W 12 × 45	8.5	58.1	▮ 51.6	46.5	42.2	38.7	35.7	33.2					
W 14 × 43	8.4	62.7	55.7	▮ 50.1	45.6	41.8	38.6	35.8	33.4	31.3			
W 12 × 50	8.5	64.7	57.5	51.7	47.0	43.1	39.8	37.0					
W 16 × 40	7.4	64.7	▮ 57.5	51.7	▮ 47.0	43.1	39.8	37.0	34.5	32.3	30.4	28.7	
W 18 × 40	6.3	68.4	60.8	54.7	49.7	45.6	▮ 42.1	39.1	36.5	34.2	32.2	30.4	28.8
W 14 × 48	8.5	70.3	62.5	▮ 56.2	51.1	46.9	43.3	40.2	37.5	35.1			
W 12 × 53	10.6	70.6	▮ 62.7	56.5	51.3	47.1	43.4	40.3					
W 16 × 45	7.4	72.7	64.6	58.2	▮ 52.9	48.5	44.7	41.5	38.8	36.3	34.2	32.3	
W 14 × 53	8.5	77.8	69.1	▮ 62.2	56.6	51.9	47.9	44.4	41.5	38.9			
W 18 × 46	6.4	78.8	70.0	63.0	57.3	52.5	▮ 48.5	45.0	42.0	39.4	37.1	35.0	33.2
W 16 × 50	7.5	81.0	72.0	64.8	▮ 58.9	54.0	49.8	46.3	43.2	40.5	38.1	36.0	

TABLE D.1 (*Continued*)

Shape	Span (ft)	16	18	20	22	24	27	30	33	36	39	42	45
	Deflection factor[b] / $L_c{}^c$ (ft)	6.36	8.05	9.93	12.0	14.3	18.1	22.3	27.0	32.2	37.8	43.8	50.3
W 21 × 44	6.6	81.6	72.5	65.3	59.3	54.4	48.3	43.5	39.6	36.3	33.5	31.1	29.0
W 18 × 50	7.9	88.9	79.0	71.1	64.6	59.3	52.7	47.4	43.1	39.5	36.5		
W 14 × 61	10.6	92.2	81.9	73.8	67.0	61.5	54.6	49.2	44.7				
W 16 × 57	7.5	92.2	81.9	73.8	67.0	61.5	54.6	49.2	44.7	41.0			
W 21 × 50	6.9	94.5	84.0	75.6	68.7	63.0	56.0	50.4	45.8	42.0	38.8	36.0	33.6
W 18 × 55	7.9	98.3	87.4	78.6	71.5	65.5	58.2	52.4	47.7	43.7	40.3		
W 18 × 60	8.0	108	96.0	86.4	78.5	72.0	64.0	57.6	52.4	48.0	44.3		
W 21 × 57	6.9	111	98.7	88.6	80.7	74.0	65.8	59.2	53.8	49.3	45.5	42.3	39.5
W 24 × 55	7.0	114	101	91.2	82.9	76.0	67.5	60.8	55.3	50.7	46.8	43.4	40.5
W 16 × 67	10.8	117	104	93.6	85.1	78.0	69.3	62.4	56.7	52.0			
W 18 × 65	8.0	117	104	93.6	85.1	78.0	69.3	62.4	56.7	52.0	48.0		
W 18 × 71	8.1	127	113	102	92.4	84.7	72.2	67.7	61.5	56.4	52.1		
W 21 × 62	8.7	127	113	102	92.4	84.7	72.2	67.7	61.5	56.4	52.1	48.4	45.1
W 24 × 62	7.4	131	116	105	95.3	87.3	77.6	69.9	63.5	58.2	53.7	49.9	46.6
W 16 × 77	10.9	134	119	107	97.4	89.3	79.4	71.5	65.0	59.5			
W 21 × 68	8.7	140	124	112	102	93.3	83.0	74.7	67.9	62.2	57.4	53.3	49.8
W 18 × 76	11.6	146	130	117	106	97.3	86.5	77.9	70.8	64.9	59.9		
W 21 × 73	8.8	151	134	121	110	101	89.5	80.5	73.2	67.1	61.9	57.5	53.7
W 24 × 68	9.5	154	137	123	112	103	91.2	82.1	74.7	68.4	63.2	58.7	54.7
W 18 × 86	11.7	166	147	133	121	111	98.4	88.5	80.5	73.8	68.1		
W 21 × 83	8.8	171	152	137	124	114	101	91.2	82.9	76.0	70.1	65.1	60.8

TABLE D.1 (*Continued*)

Shape	Span (ft)	24	27	30	33	36	39	42	45	48	52	56	60
	Deflection factor[b] / L_c^c (ft)	14.3	18.1	22.3	27.0	32.2	37.8	43.8	50.3	57.2	67.1	77.9	89.4
W 24 × 76	9.5	117	104	93.9	85.3	78.2	72.2	67.0	62.6	58.7			
W 21 × 93	8.9	128	114	102	93.1	85.3	78.8	73.1	68.3				
W 24 × 84	9.5	131	116	104	95.0	87.1	80.4	74.7	69.7	65.3			
W 27 × 84	10.5	142	126	114	103	94.7	87.4	81.1	75.7	71.0	65.5	60.8	
W 24 × 94	9.6	148	131	118	108	98.7	91.1	84.6	78.9	74.0			
W 21 × 101	13.0	151	134	121	110	101	93.1	86.5	80.7				
W 27 × 94	10.5	162	144	130	113	108	99.7	92.6	86.4	81.0	74.8	69.4	
W 24 × 104	13.5	172	153	138	125	115	106	98.3	91.7	86.0			
W 27 × 102	10.6	178	158	142	129	119	109	102	94.9	89.0	82.1	76.3	
W 30 × 99	10.9	179	159	143	130	120	110	102	95.6	89.7	82.8	76.9	71.7
W 24 × 117	13.5	194	172	155	141	129	119	111	103	97.0			
W 27 × 114	10.6	199	177	159	145	133	123	114	106	99.7	92.0	85.4	
W 30 × 108	11.1	199	177	159	145	133	123	114	106	99.7	92.0	85.4	79.7
W 30 × 116	11.1	219	195	175	159	146	135	125	117	110	101	94.0	87.7
W 30 × 124	11.1	237	210	189	172	158	146	135	126	118	109	101	94.7

TABLE D.1 (*Continued*)

Shape	Span (ft)	30	33	36	39	42	45	48	52	56	60	65	70
	Deflection factor[b]	22.3	27.0	32.2	37.8	43.8	50.3	57.2	67.1	77.9	89.4	105	122
	$L_c{}^c$ (ft)												
W 33 × 118	12.0	191	174	159	147	137	128	120	110	103	95.7	88.4	
W 30 × 132	11.1	203	184	169	156	145	135	127	117	109	101		
W 33 × 130	12.1	216	197	180	166	155	144	135	125	116	108	99.9	
W 27 × 146	14.7	219	199	183	169	156	146	137	126	117			
W 36 × 135	12.3	234	213	195	180	167	156	146	135	125	117	108	100
W 33 × 141	12.2	239	217	199	184	171	159	149	138	128	119	110	
W 33 × 152	12.2	260	236	216	200	185	173	162	150	139	130	120	
W 36 × 150	12.6	269	244	224	207	192	179	168	155	144	134	124	115
W 30 × 173	15.8	287	261	239	221	205	192	180	166	154	144		
W 36 × 160	12.7	289	263	241	222	206	193	181	167	155	144	133	124
W 36 × 170	12.7	309	281	258	238	221	206	193	178	166	155	143	132
W 30 × 191	15.9	319	290	268	245	228	213	199	184	171	159		
W 36 × 182	12.7	332	302	277	256	237	221	208	192	178	166	153	142
W 36 × 194	12.8	354	322	295	272	253	236	221	204	190	177	163	152
W 33 × 201	16.6	365	332	304	281	260	243	228	210	195	182	168	
W 36 × 210	12.9	383	349	319	295	274	256	240	221	205	192	177	164
W 33 × 221	16.7	404	367	336	310	288	269	252	233	216	202	186	
W 33 × 241	16.7	442	402	368	340	316	295	276	255	237	221	204	
W 36 × 230	17.4	446	406	372	343	319	298	279	257	239	223	206	191
W 36 × 245	17.4	477	434	398	367	341	318	298	275	256	239	220	204
W 36 × 260	17.5	508	462	423	391	363	339	318	293	272	254	234	218
W 36 × 280	17.5	549	499	458	422	392	366	343	317	294	275	253	235
W 36 × 300	17.6	592	538	493	455	423	395	370	341	317	296	273	254

[a] Total allowable uniformly distributed load in kips for simple span beams with yield stress of 36 ksi [250 MPa]. Loads to the right of the heavy vertical lines will cause deflections in excess of $\frac{1}{360}$ of the span.
[b] Maximum deflection in inches at the center of the span may be obtained by dividing the factor given by the depth of the beam in inches.
[c] Maximum permitted distance between points of lateral support. For greater distances use the charts in Appendix C to obtain the required beam size.

APPENDIX E

ASPECTS OF STRUCTURAL INVESTIGATION

This section contains a digest of concepts, relationships, and basic analytical formulas relating to elementary structural mechanics, as commonly developed in basic courses in applied mechanics, strength of materials, and elementary structural analysis. This is not intended as a text for study by persons with no previous experience in these subjects, but rather as a review of basic materials. For a fuller treatment the reader is referred to elementary texts, such as *Simplified Mechanics and Strength of Materials* by Harry Parker or *Elementary Structures for Architects and Builders* by R. E. Schaeffer.

E.1 FORCES

For structural work, *force* may be defined as an effort that tends to change the state of motion or the shape of an object. Visualization of the form of motion that must be resisted is a basis for perceiving the reactive effects required to prevent motion. Visualization of potential deformations of shape is a basis for perceiving the internal actions in a structure that prevent its collapse.

Forces that act on a structure (from gravity, wind, earthquakes, etc.) are called *loads*. External forces that are reactively developed to hold a structure in place—in opposition to the active loads—are called *reactions*. The loads and reactions constitute the total external force system acting on a structure, which must be balanced in a state of *equilibrium*, if the structure is to remain in place.

Forces, or loads, may be categorized in various ways, including the following:

Static or Dynamic. Forces constant over time (such as gravity) are *static*. Forces that vary with time (such as wind) are essentially *dynamic*, although their effects on a structure may be of a general static nature, depending on the dynamic properties of the structure.

Dead or Live. *Dead load* is permanent load; basically consisting of the weight of the structure itself, plus the weight of any other items permanently supported by the structure (mostly the rest of the construction). *Live loads* include other loads that are not constant over time, although building codes use the term "live load" to mean nonpermanent gravity loads on roof or floor surfaces.

Vector Properties. A single static force may be represented by a *vector*, graphically imaged as an arrow (Fig. E.1). The nature of the vector (force) may be fully defined by establishing of three properties: *magnitude, direction*, and *sense*. These may be graphically indicated as shown in Fig. E.1. They may also be described; such as "a vertical, downward force of 15 lb," with 15 lb being the magnitude, vertical the direction, and downward the sense.

System Arrangement. For equilibrium, forces must exist in sets; the minimum set being two collinear, equal forces of opposite sense. The geometric arrangement of sets of forces may be categorized by consideration of three qualifications that establish whether the forces are coplanar, parallel, or concurrent. On the basis of separate considerations of these

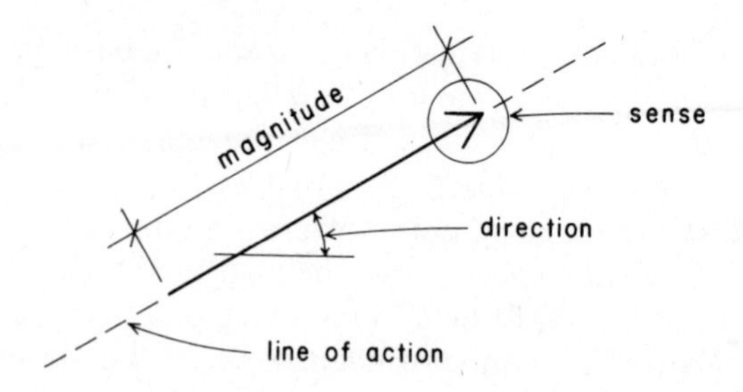

FIGURE E.1 Properties of a force.

conditions, the possible arrangements for force systems are those listed in Table E.1 and illustrated in Fig. E.2. Note that system No. 5 in the table is actually the same as system No. 1.

Sets of forces may be combined (called *composition*) to produce a net effect, called the *resultant*. In an opposite action, a single force may be visualized as having *component* actions, and be replaced by a set of its components; this process is called *resolution*, or the resolving of a force into its components (see Fig. E.3). A common form of resolution is to break single forces down into their orthogonal components (x, y, and z axes).

TABLE E.1 Classification of Force Systems

System Variation	Qualifications		
	Coplanar	Parallel	Concurrent
1	Yes	Yes	Yes
2	Yes	Yes	No
3	Yes	No	Yes
4	Yes	No	No
5	No[a]	Yes	Yes
6	No	Yes	No
7	No	No	Yes
8	No	No	No

[a] Not possible if forces are parallel and concurrent.

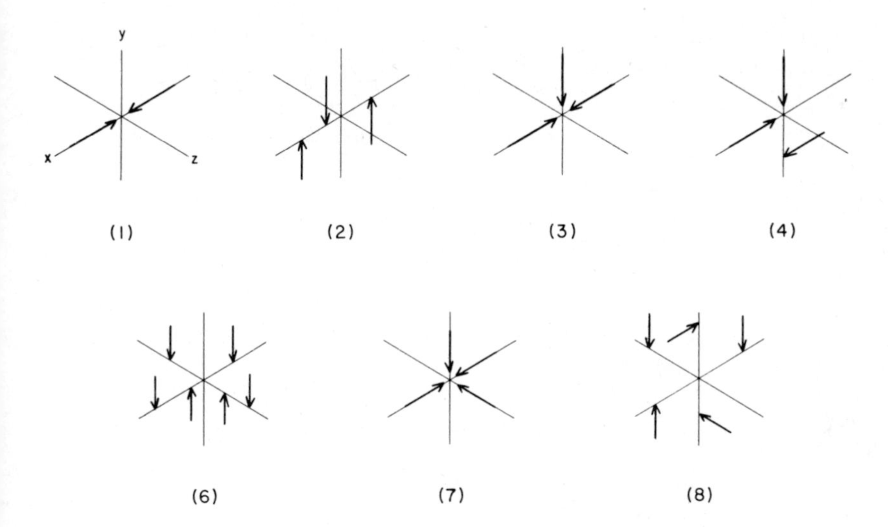

FIGURE E.2 Classification of force systems—orthogonal reference axes (see Table E.1).

Investigation of structures frequently involves the use of the conditions of equilibrium to determine reaction forces or some of the internal forces in a structure. A convenient form for expression of equilibrium makes use of the orthogonal system. Thus for system No. 1 in Table E.1, equilibrium may be algebraically established by satisfying the condition: $\Sigma F = 0$ (the sum of the forces is zero). For the other systems, however, more conditions must be established. For system No. 3, equilibrium may be established by satisfying the two equations: $\Sigma F_x = 0$ and $\Sigma F_y = 0$. (The sum of the force components in the x direction is zero and the sum of the force components in the y direction is zero.)

For systems of forces that are not concurrent, equilibrium involves not only the sums of force components, but the consideration of *moments*. A moment is the product of a force times a distance (the *moment arm*) and tends to produce motion in the form of rotation. Where moments occur, an additional algebraic condition must be included: $\Sigma M = 0$ (see Fig. E.4).

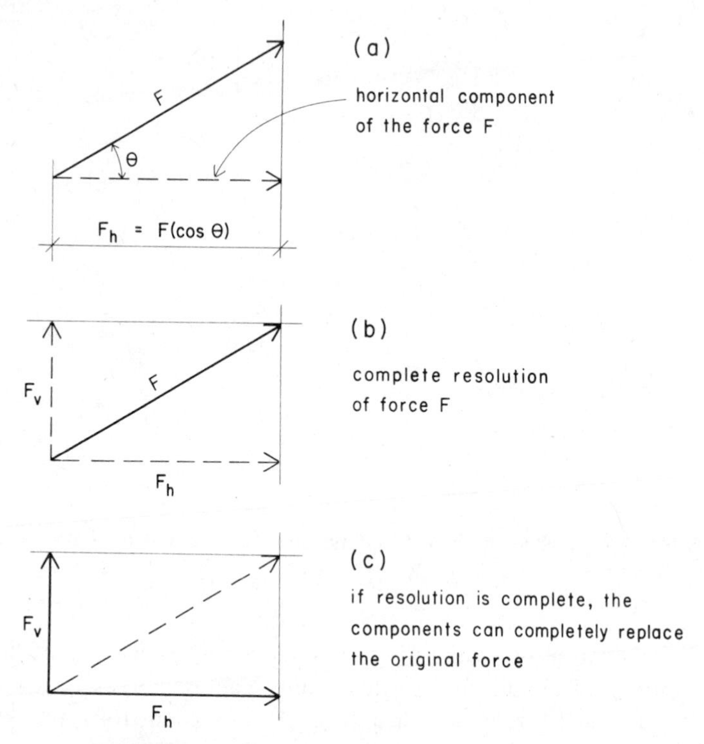

FIGURE E.3 Resolution of a force into components.

E.2 INTERNAL FORCES, STRESSES, AND STRAINS

Under the actions of the external forces (loads and reactions) a structure develops *internal forces* to resist collapse or unacceptable levels of deformation (sag, stretch, twist, etc.). The five types of internal force are *compression* (producing shortening), *tension* (producing lengthening), *shear* (producing slicing or lateral racking), *bending* (producing curvature of linear members), and *torsion* (producing twisting). All of these actions are produced within the material of a structure by intra-particle forces, called *stresses*.

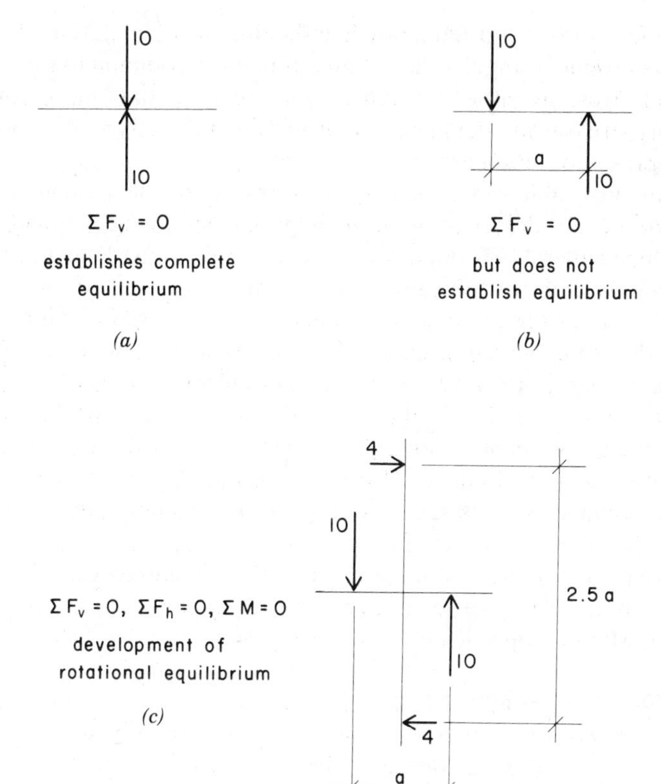

FIGURE E.4 Consideration of rotational equilibrium (sum of moments equals zero).

Stresses are basically of one of two types: *direct stress* or *shear stress*. Internal force actions and stresses are visualized by use of a *cut section,* produced by passing a cutting plane across a structure and removing the part on one side of the cut. The internal force actions or stresses are visualized by imagining the effects of the removed portion on the cut face of the remaining portion. Direct stress is visualized as stress perpendicular to the cut face; shear stress is in the plane of the cut face. Direct stress produces either shortening or lengthening in the material; shear stress produces angular distortions. It is more common to refer to direct stress as either tension or compression; thus more commonly stresses are defined as being of one of three types: tension, compression, or shear.

An inevitable consequence of stress is the development of *strain,* or unit deformation in the material. Strains accumulate to produce general deformations of the structure: sag (deflection) of beams, lateral drift of frames, stretch of cables, and so on.

A relationship of some significance for a material is the property described as *stiffness*. This may be expressed as the ratio of stress to strain. For direct stress, this ratio is expressed as the *modulus of elasticity* for the material. For an ideal, elastic material, the modulus of elasticity remains a constant through all levels of stress development. However, for most materials the modulus changes as stress is increased, which adds some complication to analyses for stress-strain behavior.

Analysis for internal forces and stresses involves use of various geometric properties of the cut section (called the *cross section*). Major properties used are the following:

The Area. Simple direct and shear stresses are considered to be uniformly distributed on the area of the cut section and are found as quantities expressed as force per unit area. Thus

$$f = \frac{P}{A}$$

in which f = unit stress (in lb/in.^{2}, etc.)
P = internal force (tension, compression, or shear in lb, etc.)
A = area of the cross section

The Centroid of the Area. This is visualized as the center of gravity of a planar element whose plan shape is that of the cut section. The net internal force must act at this point if the simple stress condition exists.

The Moment of Inertia of the Area. This is the second moment of the area about an axis in the plane of the section. The most significant reference axes are those passing through the centroid, two of which are the mutually perpendicular *principal axes,* called the *major axis* and the *minor axis*. For bending stress, an equation commonly used is

$$f = \frac{My}{I}$$

in which f = magnitude of bending stress at a distance y from the axis of reference (a centroidal axis),
I = moment of inertia about the same reference axis.

The maximum bending stress occurs at the farthest distance from the reference axis (the edge of the section), called the c distance. Thus the critical bending condition is typically computed as

$$f = \frac{Mc}{I}$$

Since I/c is a single value for a given section, a simpler value is often substituted for I/c (called the section modulus) and the maximum bending stress is computed as

$$f = \frac{M}{S}, \qquad \text{where } S = I/c$$

In the *working stress method* for investigation and design, limiting safe conditions are established for stresses, called the *allowable stresses*. Loads are investigated for the actual usage conditions of the structure (called the *service loads*) and stresses are investigated to assure that they do not exceed allowable values under the service loads.

In the *strength method* for investigation and design (also called the *ultimate strength method* or the *factored load method*), behavior of the structure is considered at its predicted failure, using the stresses and strains assumed or observed from tests to occur at the failure of the structure. Control for safety is assured by using a failure level load that is greater than the service load.

Materials typically have different responses and different magnitudes of resistance to different stresses. For example, concrete is strong in compression, but weak in tension and shear; wood is strong in tension or compression in the direction of its grain, but is weak in shear or in cross-grain stress resistance, especially tension.

Use of investigation for internal forces, stresses, and strains is illustrated in the discussions and example problems in various sections of this book. Some additional material regarding special problems of structural investigation is presented in the remaining sections of this appendix.

E.3 BEAMS

Major usage is made in building structures of elements that are variations on the form of structure called a *beam*. A beam is any linear element with a primary loading that is perpendicular to its longitudinal axis and which is required to span or cantilever to support the load (see Fig. E.5). Common elements that function as beams include roof and floor decks, joists, girders, lintels, headers, and spanning trusses. Walls under the action of lateral forces from wind or earth pressure may also have beam-type actions.

Figure E.5 shows the form of the most common beam (called a *simple beam*), with a single span, uniformly distributed load, and end supports that resist only direct force (not rotation). Aspects of basic beam behavior illustrated in Fig. E.5 are as follows:

Load and Support Conditions. These are as described, with the usual standard representations as shown in Fig. E.5*a*.

Reactions. These are determined by solving for the equilib-

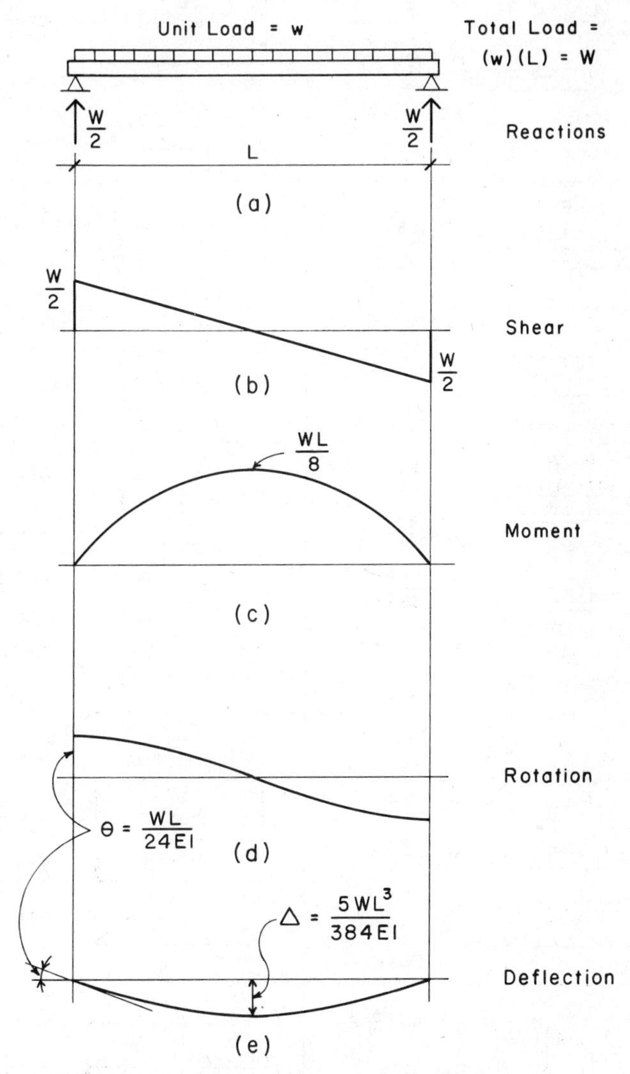

FIGURE E.5 Aspects of behavior of a simple beam.

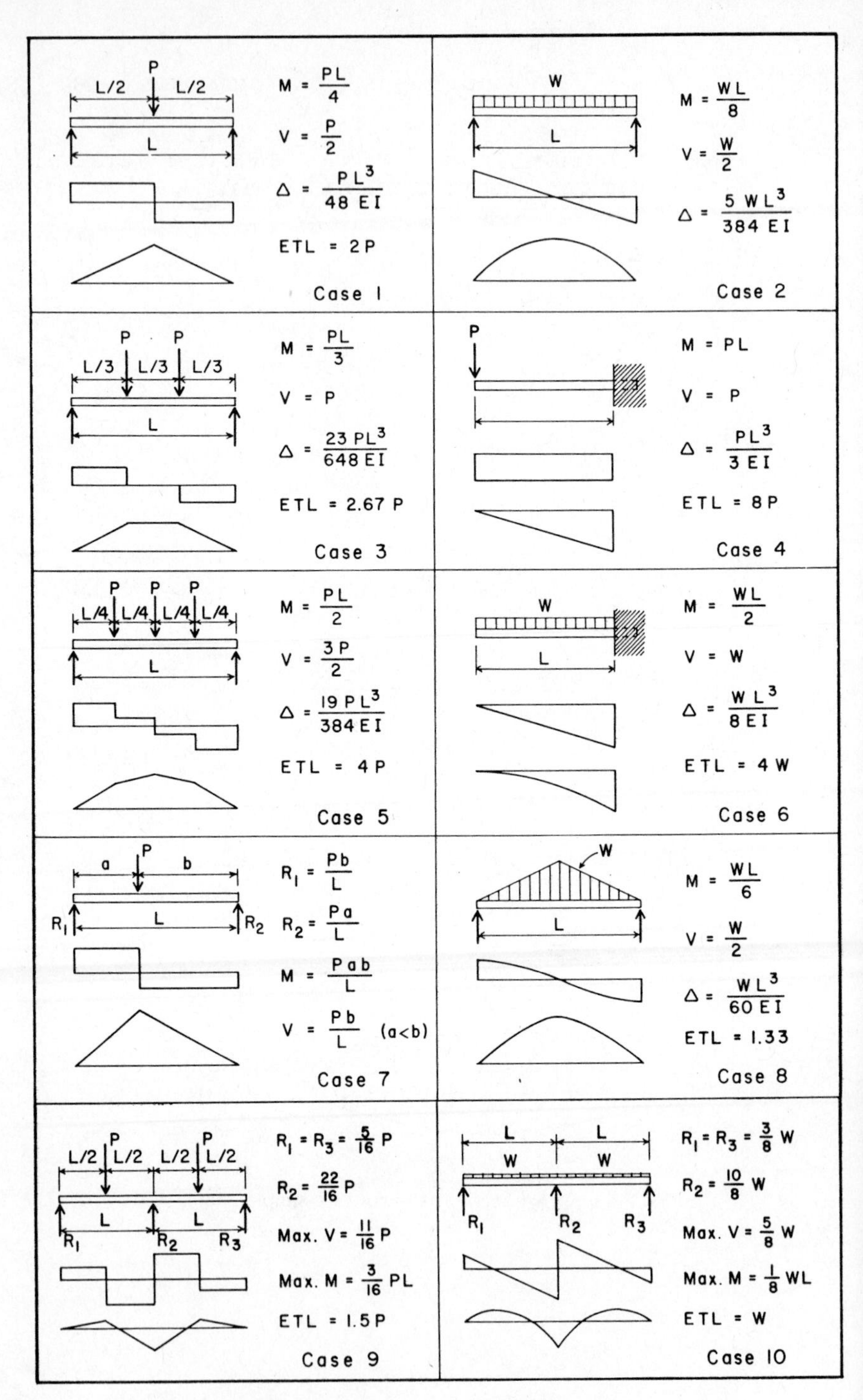

FIGURE E.6 Values for typical beam loadings.

rium of the parallel, planar force system (system No. 2 in Table E.1), consisting of the loads and reactions.

Internal Shear. This is determined by graphing the variation of the external forces from one end to the other on the beam (see Fig. E.5*b*).

Internal Bending Moment. This is determined by considering the moment of external forces on either side of selected points along the beam axis about the selected point.

Rotation. This is the angle of rotation of a vertical cut section, or of a tangent to the deflected axis of the beam at a cut section. For the simple beam, the rotation is zero at the center of the span and a maximum value at the supports.

Deflection. This is the distance of movement of the beam from its unbent position. Typically, the unloaded beam's longitudinal axis is visualized as straight; deflection is defined by a curved axis produced by the loads. For the simple beam the deflection is zero at the supports and a maximum value at the center of the span.

Values for some other common beam load and support conditions are given in Fig. E.6.

E.4 INVESTIGATION OF STRUCTURES

Investigation of proposed structures, whether done by hand computations, with computer assistance, or by observation of physical models, is usually done for two primary purposes. The first of these is the development of the finished design of the structure. Certain needs must be determined in order to make selections of materials, forms, and details. The second purpose of investigation is to establish the safety of the structure with some assurance, which is usually done by processes built into the design operations (use of allowable stresses or factored loads).

The remaining sections in this appendix deal with some of the basic considerations for structural investigations, as related to the example computations illustrated in various sections of this book.

E.5 ANALYSIS OF COPLANAR, CONCURRENT FORCES

The forces that operate on individual joints in planar trusses constitute sets of coplanar, concurrent forces. The following discussion deals with the analysis of such systems, both algebraically and graphically, and introduces some of the procedures that will be used in the examples of truss analysis in this book.

In the preceding examples, forces have been identified as F_1, F_2, F_3, and so on. However, a different system of notation will be used in the work that follows. This method consists of placing a letter in each space that occurs between the forces or their lines of action, each force then being identified by the two letters that appear in the adjacent spaces. A set of five forces is shown in Fig. E.7*a*. The common intersection point is identified as *BCGFE* and the forces are *BC*, *CG*, *GF*, *FE*, and *EB*. Note particularly that the forces have been identified by reading around the joint in a continuous clockwise manner. This is a convention that will be used throughout this book, since it has some relevance to the methods of graphic analysis that will be explained later.

In Fig. E.7*b* a portion of a truss is shown. Reading around the joint *BCGFE* in a clockwise manner, the upper chord member between the two top-joints is read as member *CG*. Reading around the joint *CDIHG*, the same member is read as *GC*. Either designation may be used when referring to the member itself. However, if the effect of the force in the member on a joint is

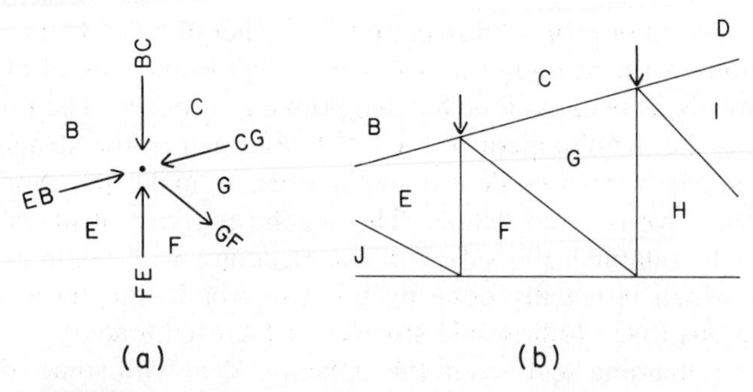

FIGURE E.7 Force notation for concurrent force systems.

being identified, it is important to use the proper sequence for the two-letter designation.

In Fig. E.8*a* a weight is shown hanging from two wires that are attached at separate points to the ceiling. The two sloping wires and the vertical wire that supports the weight directly meet at joint *CAB*. The "problem" in this case is to find the tension forces in the three wires. We refer to these forces that exist within the members of a structure as *internal forces*. In this example it is obvious that the force in the vertical wire will be the same as the magnitude of the weight: 50 lb. Thus the solution is reduced to the determination of the tension forces in the two sloping wires. This problem is presented in Fig. E.8*b*, where the force in the vertical wire is identified in terms of both direction and magnitude, while the other two forces are identified only in terms of their directions, which must be parallel to the wires. The senses of the forces in this example are obvious, although this will not always be true in such problems.

A graphic solution of this problem can be performed by using the available information to construct a force polygon consisting of the vectors for the three forces: *BC*, *CA*, and *AB*. The process for this construction is as follows:

1. The vector for AB is totally known and can be represented as shown by the vertical arrow with its head down and its length measured in some scale to be 50.
2. The vector for force *BC* is known as to direction and must pass through the point *b* on the force polygon, as shown in Fig. E.8*d*.
3. Similarly, we may establish that the vector for force *CA* will lie on the line shown in Fig. E.8*e*, passing through the point *a* on the polygon.
4. Since these are the only vectors in the polygon, the point *c* is located at the intersection of these two lines, and the completed polygon is as shown in Fig. E.8*f*, with the sense established by the continuous flow of the arrows. This "flow" is determined by reading the vectors in continuous clockwise sequence on the space diagram, starting with the

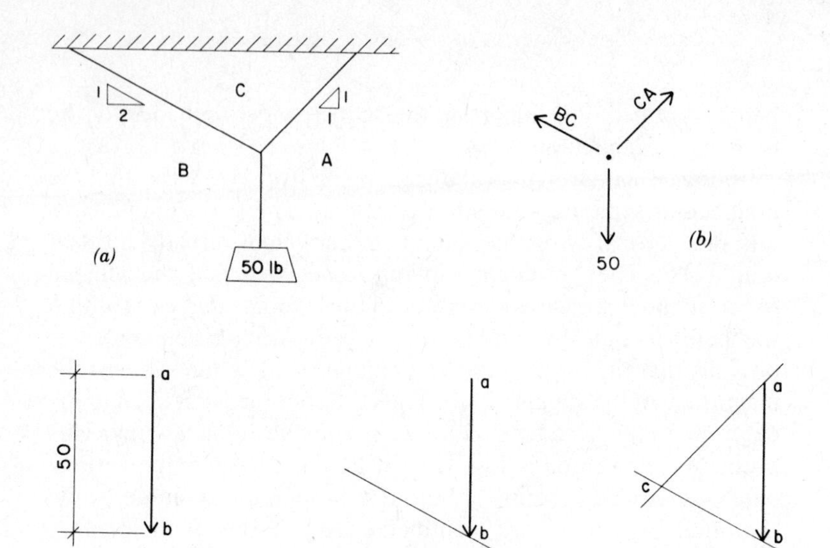

1
2
C
1
1
B
A
50 lb
(a)
BC
CA
50
(b)
a
50
b
(c)
a
b
(d)
a
c
b
(e)

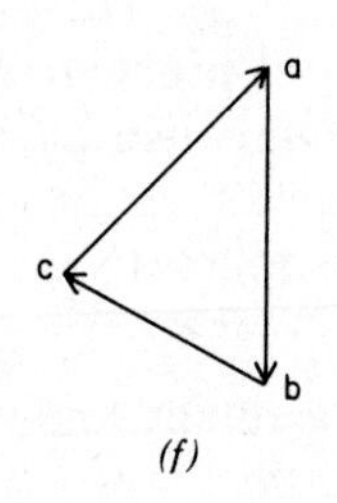

a
c
b
(f)

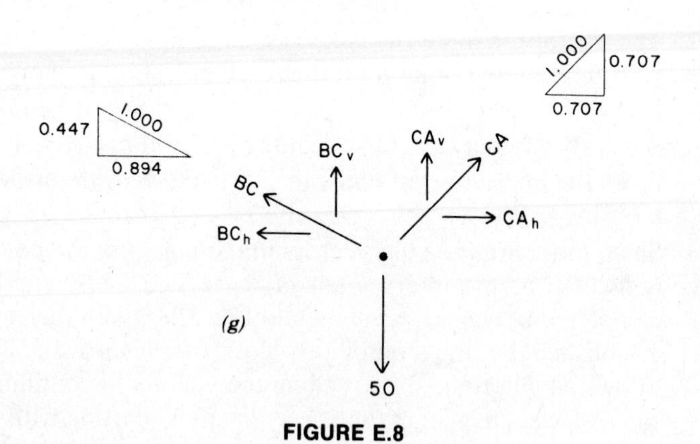

1.000
0.707
0.707
0.447
1.000
0.894
BCv
CAv
CA
BC
BCh
CAh
(g)
50

FIGURE E.8

vector of known sense. We thus read the direction of the arrows as flowing from *a* to *b* to *c* to *a*.

With the force polygon completed, we can determine the magnitudes for forces *BC* and *CA* by measuring their lengths on the polygon, using the same scale that was used to lay out force *AB*.

For an algebraic solution of the problem illustrated in Fig. E.8, we first resolve the forces into their horizontal and vertical components, as shown in Fig. E.8*g*. This increases the number of unknowns from two to four. However, we have two extra relationships that may be used in addition to the conditions for equilibrium, because the direction of forces *BC* and *CA* are known. As shown in Fig. E.8*a*, force *BC* is at an angle with a slope of 1 vertical to 2 horizontal. Using the rule that the hypotenuse of a right triangle is related to the sides such that the square of the hypotenuse is equal to the sum of the squares of the sides, we can determine that the length of the hypotenuse of the slope triangle is

$$l = \sqrt{(1)^2 + (2)^2} = \sqrt{5} = 2.236$$

We can now use the relationships of this triangle to express the relationships of the force *BC* to its components. Thus, referring to Fig. E.9,

$$\frac{BC_v}{BC} = \frac{1}{2.236}, \qquad BC_v = \frac{1}{2.236}(BC) = 0.447(BC)$$

$$\frac{BC_h}{BC} = \frac{2}{2.236}, \qquad BC_h = \frac{2}{2.236}(BC) = 0.894(BC)$$

These relationships are shown in Fig. E.8*g* by indicating the dimensions of the slope triangle with the hypotenuse having a value of 1. Similar calculations will produce the values shown for the force *CA*. We can now express the conditions required for equilibrium. (Sense up and right is considered positive.)

$$\Sigma F_v = 0 = -50 + BC_v + CA_v$$

$$0 = -50 + 0.447(BC) + 0.707(CA) \qquad (1)$$

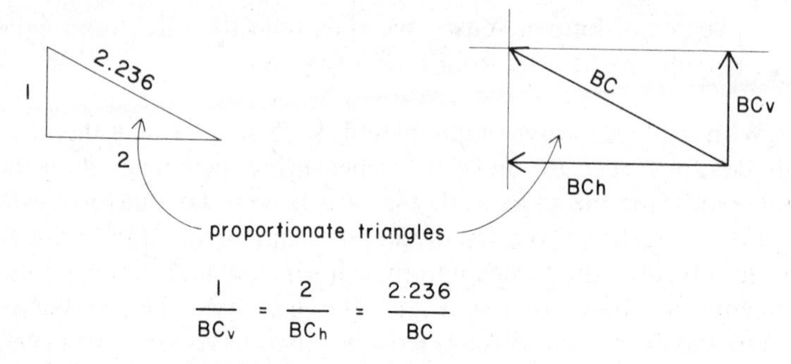

FIGURE E.9 Determination of force components.

and

$$\Sigma F_h = 0 = -BC_h + CA_h$$

$$0 = -0.894(BC) + 0.707(CA) \quad (2)$$

We can eliminate CA from these two equations by subtracting equation (2) from equation (1) as follows:

$$\text{Equation (1): } 0 = -50 + 0.447(BC) + 0.707(CA)$$

$$\text{Equation (2): } 0 = \quad + 0.894(BC) - 0.707(CA)$$

$$\text{Combining: } 0 = -50 + 1.341(BC)$$

Then

$$BC = \frac{50}{1.341} = 37.29 \text{ lb}$$

Using equation (2) yields

$$0 = -0.894(37.29) + 0.707(CA)$$

$$CA = \frac{0.894}{0.707}(37.29) = 47.15 \text{ lb}$$

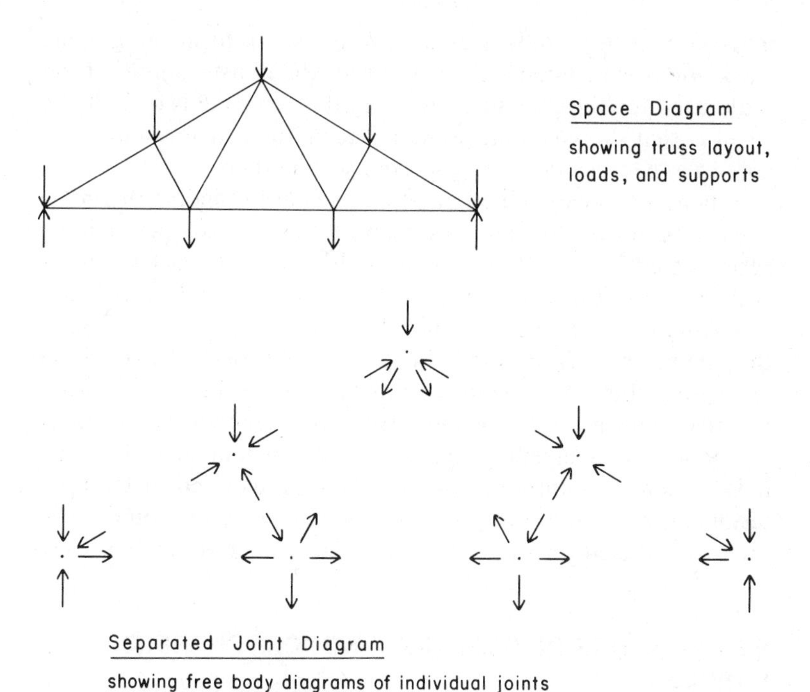

FIGURE E.10 Diagrams used to represent trusses and their actions.

The degree of accuracy of the answer obtained in an algebraic solution depends on the number of digits that are carried throughout the calculation. In this work we will usually round off numerical values to a three- or four-digit number, which is traditionally a level of accuracy sufficient for structural design calculations. Had we carried the numerical values in the preceding calculations to the level of accuracy established by the limits of an eight-digit pocket calculator, we would have obtained a value for the force in member *BC* of 37.2678 lb. Although this indicates that the fourth digit in our previous answer is slightly off, both answers will round to a value of 37.3, which is sufficient for our purposes. When answers obtained from algebraic solutions are compared to those obtained from graphic solutions, the level of correlation may be even less, unless great care is exercised in the graphic work and a very large scale is used for the constructions. If the scale used for the graphic solution in this example is actually as small as that shown on the printed page in Fig. E.8, it is unreasonable to expect accuracy beyond the second digit.

When the so-called method of joints is used, finding the internal forces in the members of a planar truss consists of solving a series of concurrent force systems. Figure E.10 shows a truss with the truss form, the loads, and the reactions displayed in a *space diagram*. Below the space diagram is a figure consisting of the free-body diagrams of the individual joints of the truss. These are arranged in the same manner as they are in the truss in order to show their interrelationships. However, each joint constitutes a complete concurrent planar force system that must have its independent equilibrium. "Solving" the problem consists of determining the equilibrium conditions for all of the joints. The procedures used for this solution are illustrated in Section 9.4.

E.6 ANALYSIS OF COPLANAR, NONCONCURRENT FORCES

Solution of equilibrium problems with nonconcurrent forces involves the application of the available algebraic summation equations. The following example illustrates the procedure for the solution of a simple parallel force system.

Figure E.11 shows a 20-kip force applied to a beam at a point between the beam's supports. The supports must generate the

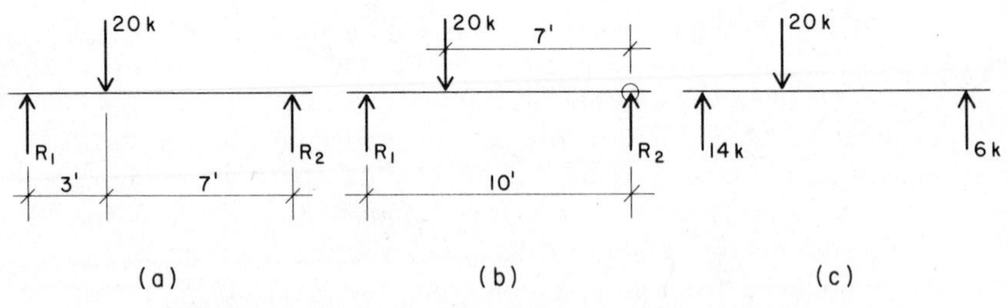

FIGURE E.11 Analysis of a simple parallel force system.

two vertical forces, R_1 and R_2, in order to oppose this load. (In this case we will ignore the weight of the beam itself, which will also add load to the supports, and we will consider only the effect of the distribution of the load added to the beam.) Since there are no horizontal forces, the complete equilibrium of this force system can be established by the satisfaction of two summation equations:

$$\Sigma F_v = 0$$

$$\Sigma M = 0$$

Considering the force summation first,

$$\Sigma F_v = 0 = -20 + (R_1 + R_2) \text{ (sense up considered positive)}$$

Thus

$$R_1 + R_2 = 20$$

This yields one equation involving the two unknown quantities. If we proceed to write a moment summation involving the same two quantities, we then have two equations that can be solved simultaneously to find the two unknowns. We can simplify the algebraic task somewhat if we use the technique of making the moment summation in a way that eliminates one of the unknowns. This is done simply by using a moment reference point that lies on the line of action of one of the unknown forces. If we choose a point on the action line of R_2, as shown in Fig. E.11*b*, the summation will be as follows:

$$\Sigma M = 0 = -20(7) + R_1(10) + R_2(0) \text{ (clockwise moment plus)}$$

Then

$$R_1(10) = 140$$

$$R_1 = 14 \text{ kips}$$

Using the relationship established from the previous force summation, we have

$$R_1 + R_2 = 20$$

$$14 + R_2 = 20$$

$$R_2 = 6 \text{ kips}$$

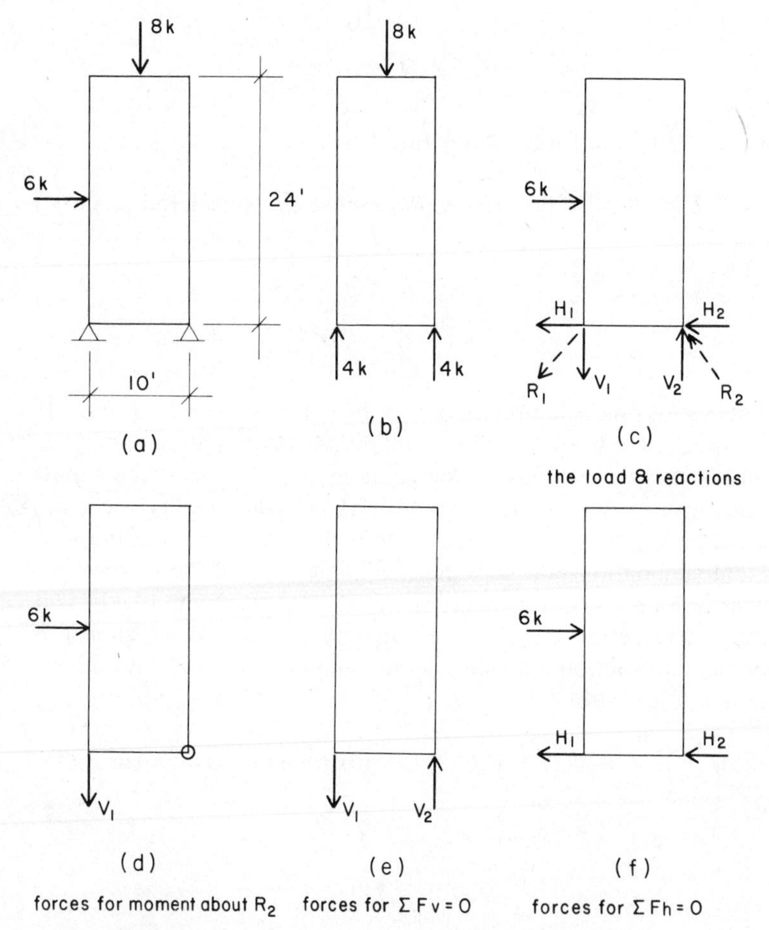

FIGURE E.12 Analysis of a general planar force system.

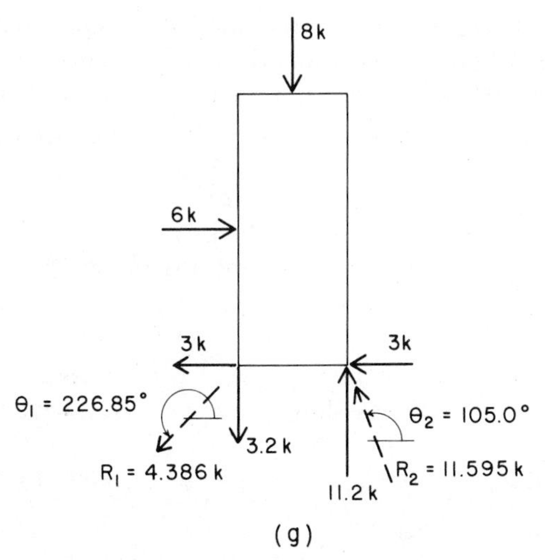

FIGURE E.12 *(Continued)*

The solution is then as shown in Fig. E.11*c*.

In the structure shown in Fig. E.12*a*, the forces consist of a vertical load, a horizontal load, and some unknown reactions at the supports. Since the forces are not all parallel, we may use all equilibrium conditions in the determination of the unknown reactions. Although it is not strictly necessary, we will use the technique of finding the reactions separately for the two loads and then adding the two results to find the true reactions for the combined load.

The vertical load and its reactions are shown in Fig. E.12*b*. In this case, with the symmetrically placed load, each reaction is simply one-half the total load.

For the horizontal load, the reactions will have the components shown in Fig. E.12*c*, with the vertical reaction components developing resistance to the moment effect of the load, and the horizontal reaction components combining to resist the actual horizontal force effect. The solution for the reaction forces may be accomplished by finding these four components; then, if desired, the actual reaction forces and their directions may be found from the components, as explained in Sec. E.5.

Let us first consider a moment summation, choosing the location of R_2 as the point of rotation. Since the action lines of V_2, H_1, and H_2 all pass through this point, their moments will be zero and the summation is reduced to dealing with the forces shown in Fig. E.12*d*. Thus

$$\Sigma M = 0 = +6(12) - V_1(10)$$ clockwise moment considered positive)

$$V_1 = \frac{6(12)}{10} = 7.2 \text{ kips}$$

We next consider the summation of vertical forces, which involves only V_1 and V_2, as shown in Fig. E.12*e*. Thus

$$\Sigma F_v = 0 = -V_1 + V_2$$ (sense up considered positive)

$$0 = -7.2 + V_2$$

$$V_2 = +7.2 \text{ kips}$$

For the summation of horizontal forces the forces involved are those shown in Fig. E.12*f*. Thus

$$\Sigma F_h = 0 = +6 - H_1 - H_2$$ (force toward right considered positive)

$$H_1 + H_2 = 6 \text{ kips}$$

This presents an essentially indeterminate situation that cannot be solved unless some additional relationships can be established. Some possible relationships are the following.

1. R_1 offers resistance to the horizontal force, but R_2 does not. This may be the result of the relative mass or stiffness of the supporting structure or the type of connection between the supports and the structure above. If a sliding, rocking, or rolling connection is used, some minor frictional resistance may be developed, but the support is essentially without significant capability for the development of horizontal resistance. In this case $H_1 = 6$ kips and $H_2 = 0$.
2. The reverse of the proceeding; R_2 offers resistance, but R_1 does not. $H_1 = 0$ and $H_2 = 6$ kips.

3. Details of the construction indicate an essentially symmetrical condition for the two supports. In this case it may be reasonable to assume that the two reactions are equal. Thus $H_1 = H_2 = 3$ kips.

For this example we will assume the symmetrical condition for the supports with the horizontal force being shared equally by the two supports. Adding the results of the separate analyses we obtain the results for the combined reactions as shown in Fig. E.12*g*. The reactions are shown both in terms of their components and in their resultant form as single forces. The magnitudes of the single force resultants are obtained as follows:

$$R_1 = \sqrt{(3)^2 + (3.2)^2} = \sqrt{19.24} = 4.386 \text{ kips}$$
$$R = \sqrt{(11.2)^2 + (3)^2} = \sqrt{134.44} = 11.595 \text{ kips}$$

The directions of these forces are obtained as follows:

$$\theta_1 = \arctan \frac{3.2}{3} = \arctan 1.0667 = 46.85°$$

$$\theta_2 = \arctan \frac{11.2}{3} = \arctan 3.7333 = 75.0°$$

Note that the angles for the reactions as shown in Fig. E.12*g* are measured as counterclockwise rotations from a right-side horizontal reference. Thus, as illustrated, the angles are actually

$$\theta_1 = 180 + 46.85 = 226.85°$$

and

$$\theta_2 = 180 - 75.0 = 105.0°$$

If this standard reference system is used, it is possible to indicate both the direction and sense of a force vector with the single value of the rotational angle. The technique is illustrated in Fig. E.13. In Fig. E.13*a* four forces are shown, all of which are rotated 45° from the horizontal. If we simply make the statement, "the force is at an angle of 45° from the horizontal," the situation may

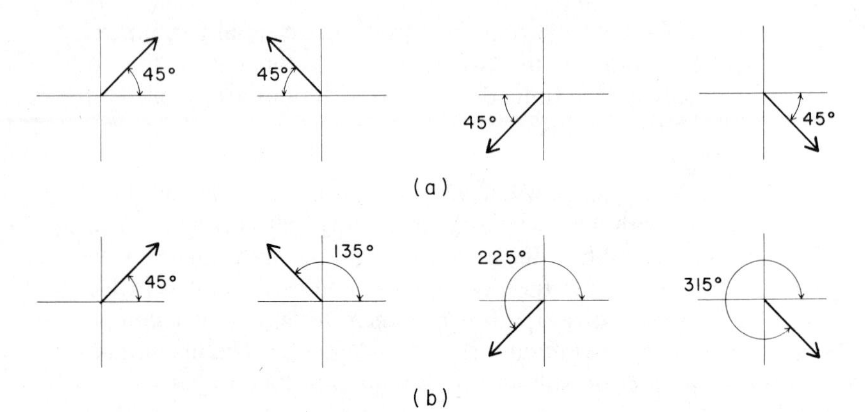

FIGURE E.13 Reference notation for angular direction: (a) horizontal reference (slope, inclination); (b) polar reference.

be any one of the four shown. If we use the reference system just described, however, we would describe the four situations as shown in Fig. E.13*b*, and they would be identified unequivocally.

E.7 INVESTIGATION OF BEAMS

For design work, investigation of beams is typically performed only to the degree necessary to determine the beam clearly and safely. However, to come to some level of understanding of what beams do, it is necessary to study the various aspects and stages of the complete investigation of a beam. The following discussion presents the basic elements of investigation of ordinary beams.

Reactions

For statically determinate beams the first step in the investigation of beam behavior is the determination of the effects of the supports on the beam—called the *reactions*. For the simplest case the reactions are a set of vertical forces that respond to the vertical loads on the beam, constituting with the loads a system of coplanar, parallel forces. This system yields to solution by con-

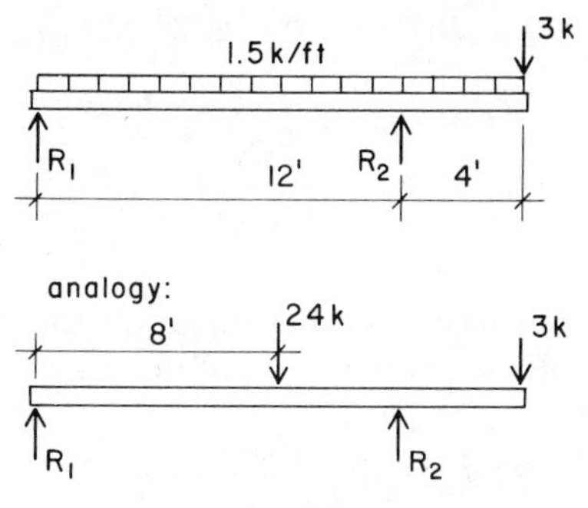

FIGURE E.14

sideration of static equilibrium if there are not more than two unknowns. Using the beam shown in Fig. E.14, we demonstrate the usual procedure for finding the reactions in the following example.

Example 1. Find the reaction forces at the supports, R_1 and R_2, for the beam in Fig. E.14.

Solution. The general mathematical technique is to write two equations involving the two unknowns, and then to solve them simultaneously. This procedure is simplest if one equation can be written that involves only one of the unknowns. We may thus consider a summation of moments of the forces about a point on the line of action of one of the reactions, as follows:

$$\Sigma M \text{ about } R_1 = (24 \times 8) \circlearrowright + (R_2 \times 12) \circlearrowleft + (3 \times 16) \circlearrowright = 0$$

from which

$$R_2 = \frac{192 + 48}{12} = 20 \text{ kips}$$

For a second step we may now write any equilibrium equation that includes the action of R_1 and it will be an equation with one unknown. The simplest choice for this is a summation of vertical forces; thus

$$F = 0 = R_1 \uparrow \; + 24 \downarrow \; + 20 \uparrow \; + 3 \downarrow$$

from which

$$R_1 = 7 \text{ kips}$$

For a check we can write another equation for R_1 to see if the system works, such as a summation of moments about R_2. Thus

$$\Sigma\, M \text{ about } R_2 = (7 \times 12) \circlearrowright + (24 \times 4) \circlearrowleft + (3 \times 4) \circlearrowright = 0$$

$$84 \circlearrowright + 96 \circlearrowleft + 12 \circlearrowright = 0$$

which verifies the answer for R_1.

This is the usual form of solution for a beam with two supports, loaded only with loads perpendicular to the beam.

Shear

Internal shear in a beam is the effort required of the beam to maintain the equilibrium of the external forces. Since the shear is itself a direct force, it is thus possible to use a simple summation of forces to establish the necessary equilibrium. Consider the beam shown in Fig. E.15, which is the same beam that was used for the solution for reactions in the preceding example. We therefore present it here with the given values for the reaction forces and proceed to consider the problem of determining values for the internal shear in the beam.

If we want to find the value of internal shear at some point in the beam—say at 3 ft from the left end—we cut a section through the beam at that point and remove the portion of the beam on one side—say the right side—and consider the remaining portion as a free body. This free body will be acted on by the loads and reactions that directly effect it plus the actions of the removed portion; the latter representing the internal forces at the section. To find the internal shear at the section, we now consider the equilibrium of vertical forces on the free body; thus

$$\Sigma\, F_v = 0 = 7 \uparrow \; + 4.5 \downarrow \; + V$$

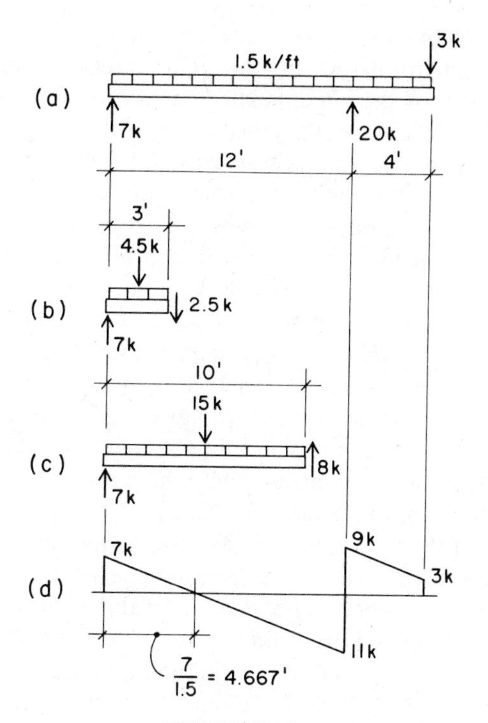

FIGURE E.15

from which

$$V = 2.5 \text{ kips} \downarrow \qquad \text{as shown in Fig. E.15}b$$

Let us consider the internal shear at a second section, say 10 ft from the left end. Using a similar summation yields

$$F_v = 0 = 7\uparrow + 15\downarrow + V$$

from which

$$V = 8 \text{ kips} \uparrow \qquad \text{as shown in Fig. E.15}c$$

If we continue this process, we may find the values for internal shear at as many points as we desire, until we have a general description of the variation of shear along the beam length. Our usual technique for displaying this general form is through use of a simple graph of the shear, called the *shear diagram*. For this beam the shear diagram takes the form shown in Fig. E.15*d*. Note that the shear has a sign as well as magnitude. The convention used to establish the sign for the shear graph shown in Fig. E.15*d* is to proceed as demonstrated with the free bodies in Fig. E.15*b* and *c* and to consider shear force down in sense as positive.

In effect, the shear diagram is simply a plot of the external vertical forces along the beam length. To produce the graph shown in Fig. E.15*d*, we simply start at the left end of the beam and plot the forces as we encounter them, proceeding to the other end of the beam. We start at zero, since the beam end is discontinuous and there is literally nothing beyond its ends. We should also end at zero at the opposite end, which is a means of checking the correctness of our work.

Note that the shear graph passes through the zero axis, indicating a switch in the sign of the internal shear, at some point between the two supports. The location of this point may be found by noting that the rate of decline of the graph is the "rate" or unit of the uniformly distributed load on the beam in this portion of its length: 1.5 kips/ft. The distance from the left end to the point of zero shear is thus the distance required for the drop from 7 kips at the rate of 1.5 kips/ft; thus

$$x = \frac{7}{1.5} = 4.667 \text{ ft}$$

Bending Moment

Internal moment in a beam is the effort required by the beam to maintain rotational equilibrium under the action of the external forces. A technique for finding internal moments is to use the same procedures that were demonstrated previously for the finding of internal shear. To show the procedure we will use the same beam that was used before—shown now in Fig. E.16 with the reactions and shear diagram complete.

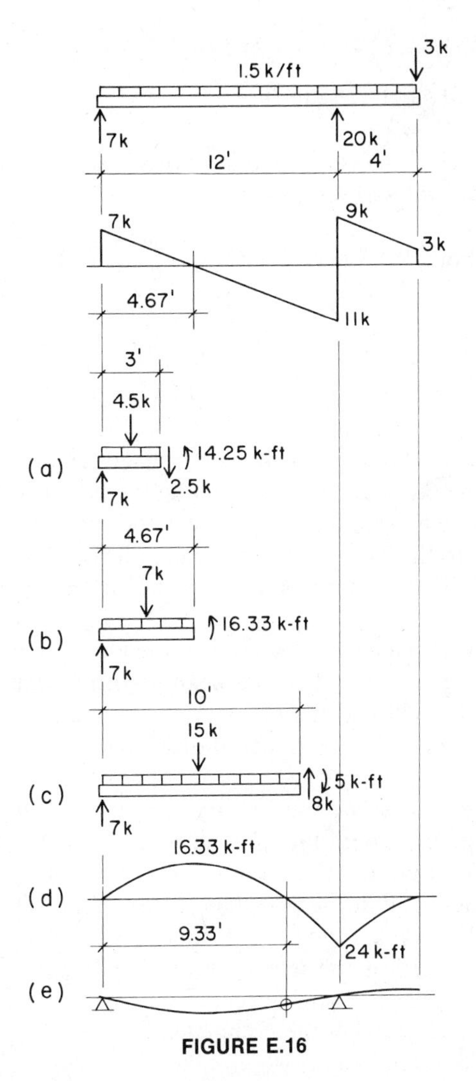

FIGURE E.16

At Fig. E.16*a* we show the free body for the portion of the beam cut 3 ft from the left end. Previously we found the internal shear at this section to be 2.5 kips with sense as shown in Fig. E.16*a*. We now consider the equilibrium of moments for this free body. If we sum moments at the cut section, we write

$$\Sigma M = (7 \times 3) \circlearrowright + (4.5 \times 1.5) \circlearrowleft + M_3 = 0$$

$$M_3 = 14.25 \text{ kip-ft} \circlearrowleft$$

To verify this answer, we may write a second equation for moment about the left end; thus

$$\Sigma M \text{ about } R_1 = (4.5 \times 1.5) \circlearrowright + (2.5 \times 3) \circlearrowright + M_3 = 0$$

or

$$6.75 \circlearrowright + 7.5 \circlearrowright + 14.25 \circlearrowleft = 0$$

The value of 14.25 kip-ft for M_3 is the magnitude of the internal moment at the cut section, representing the effect of the portion of beam that was removed. Note that the moment has a sign. With respect to the cut section, this sign indicates the existence of compression in the upper part of the beam and tension in the lower part.

Let us now consider a cut section at the location of zero shear: 4.667 ft from the left end of the beam. With a similar moment summation, we will find the moment at this section to be 16.333 kip-ft with a sign the same as that of the previous section (see Fig. E.16*b*).

Finally, we consider a cut section at 10 ft from the left end of the beam. The consideration of equilibrium in this case will produce an internal moment of 5 kip-ft with a sign of opposite rotation from that found at the previous two sections. This switch of moments indicates that the beam at this location has compression in the bottom portion and tension in the top (see Fig. E.16*c*).

If we continue to investigate sections along the beam length we may eventually establish the complete pattern of moment variation along the beam length, which takes the form of the graph in Fig. E.16*d*. This graph is called the *moment diagram*. Below the moment diagram is shown a sketch of the general form of the deflected beam (Fig. E.16*e*).

Mathematically it can be demonstrated that the load, shear, moment, rotation, and deflection for the beam are interrelated in sequence. Thus the shear is the first integration of the load, the moment is the second integration, the rotation is the third integration, and the deflection is the fourth integration. This process can also be reversed using derivatives. Based on these relationships, plus consideration of the investigation just performed and illustrated in Fig. E.16, we can make the following statements with regard to investigation of beams.

1. The internal shear at any point along a beam is the sum of the loads and reactions on one side of that point. Either side may be used.
2. The internal moment at any point along a beam is the sum of moments of the loads and reactions on one side of that point. Either side may be used.
3. The change in moment between any two points on the beam is equal to the area under the shear diagram between the two points.
4. Points of maximum value on the moment diagram correspond to points of zero value on the shear diagram.
5. The sign of the internal moment is related to the type of curvature of the deflected shape.
6. Points of zero moment along the beam indicate points of change of curvature—called inflection or contraflexure—of the deflected shape.

The reader should verify these statements with the information displayed in Fig. E.16. The shear and moment diagrams as displayed are produced by following a particular sign convention, as follows:

For the Shear Diagram. Begin from the left end, considering force up to be positive.

For the Moment Diagram. Consider internal moment that causes compression in the top of the beam positive and moment that causes tension in the top of the beam negative.

Regarding statement 3, if the sign conventions just described are used, it may be observed that areas of positive shear on the shear diagram relate to positive changes of moment, while areas of negative shear relate to negative changes of moment. This is the basis for statement 4.

Visualization of the deflected shape is a very useful device in the investigation of members subjected to bending. In most cases this visualization can be made on the basis of consideration of the loads, supports, and beam form before any other investigation is done. If so, it provides immediate clues to the character of the moment in the beam.

Examples of various situations of beam investigation are given in various parts of this book, especially in the work in Chapters 4, 5, and 12. Use of design aids of various kinds generally makes it unnecessary to perform all of the steps of beam investigation illustrated here. The common occurrence of typical loading and span conditions permits use of tabular data and simple formulas for critical values of the significant design factors.

E.8 PROPERTIES OF SECTIONS

The discussions in Sec. E.2 illustrated the use of various geometric properties of the forms described by cut sections. The following materials present some further discussions and derivations regarding these properties.

Mass

The mass of a physical shape may be expressed as the product of a unit density times the number of units of volume of the shape. By this means, mass becomes analogous with the concept of geometry of the physical shape. A shape may be visualized as a force in the form of the weight exerted by the mass. An area may be considered as a mass by conceiving it as having a unit density with the unit being some constant thickness. Thus instead of being an area in the pure mathematical sense, it consists of a shape cut from some thin material, such as cardboard. It is helpful to think of areas in this way in the problems that follow.

Moment of a Physical Shape

Considered as a mass, and thus as a force (weight), a physical shape may be seen to have the capability of exerting direct force and also of exerting moment. The latter occurs when the shape exerts its mass force other than along a line through its mass center. Referring to the ball on a string, shown in Fig. E.17*a*:

When the string is vertical, that is, it is aligned with the line of action of the weight of the ball, the effect of the ball is to exert only a direct pull on the string.

When the string is other than vertical, there is a moment, or a rotational effect, caused by the component of weight that is perpendicular to the string.

Centroid

The concept of centroid (or mass center, or center of gravity) may be visualized in terms of the lack of rotational effect. The mass center is the point about which the sum of the moments of the incremental units of the mass is zero. If defined only geometrically, this point is called the *centroid*. Referring to Fig. E.17*b*, the shape defined as an area has a moment about the axis O–O expressed as

$$\Sigma M_{O-O} = \Sigma wxda = (wA)\bar{x}$$

where ΣM_{O-O} = total moment of the area about O–O,
w = unit density of the area (thickness),
wda = unit weight of the area,
wA = total weight of the area,
$\bar{x}$ = distance from the axis O–O to the center of gravity or centroid of the area.

If only the geometric property of the location of the centroid is desired, the unit density w may be dropped from the expression and the centroidal distance can be expressed as

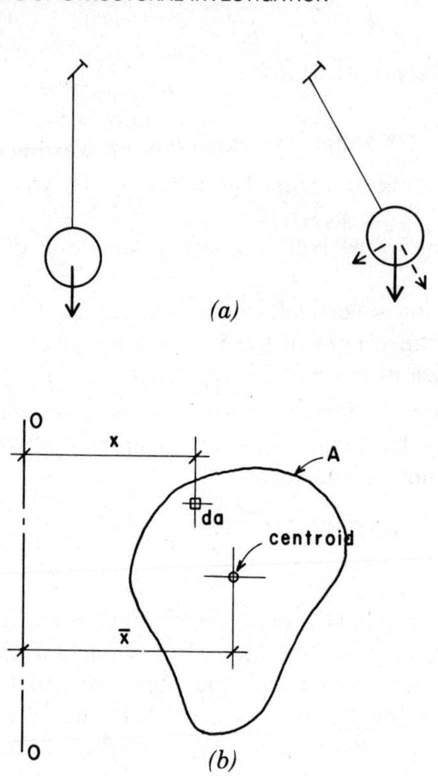

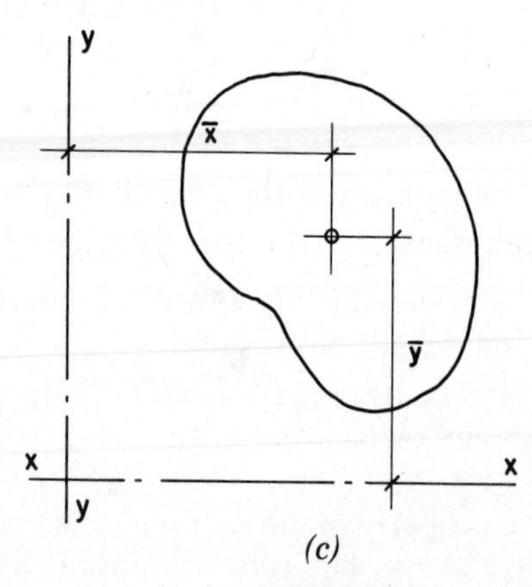

FIGURE E.17

$$\bar{x} = \frac{\Sigma M_{O-O}}{A}$$

For an area the precise location of the centroid can be established by determining centroidal distances from two reference axes that are not parallel. Using the customary x and y axis references, as shown in Fig. E.17c, the precise location of the centroid for the area shown can be expressed as

$$\bar{x} = \frac{\Sigma M_{y-y}}{A} \quad \text{and} \quad \bar{y} = \frac{\Sigma M_{x-x}}{A}$$

The following examples illustrate the process for determination of the centroid for an area.

Example 1. The shape in Fig. E.18a represents the area of a planar figure of uniform density. Find the location of the centroid of the area.

Solution. Consider the area as consisting of two units as shown; the areas and centroids of which are determinable. Then

$$\text{total } A = A_1 + A_2 = 5 + 4 = 9 \text{ in.}^2$$

$$\Sigma M_y = (A_1 \times 0.5) + (A_2 \times 1.5) = 2.5 + 6 = 8.5 \text{ in.}^3$$

$$\Sigma M_x = (A_1 \times 3.5) + (A_2 \times 0.5) = 17.5 + 2 = 19.5 \text{ in.}^3$$

$$\bar{x} = \frac{\Sigma M_y}{A} = \frac{8.5}{9} = 0.944 \text{ in.}$$

$$\bar{y} = \frac{\Sigma M_x}{A} = \frac{19.5}{9} = 2.167 \text{ in.}$$

Example 2. Find the location of the centroid of the area shown in Fig. E.18b.

Solution. In this case we consider the area to be comprised of two units: the area without the hole, and the hole. The uncut area

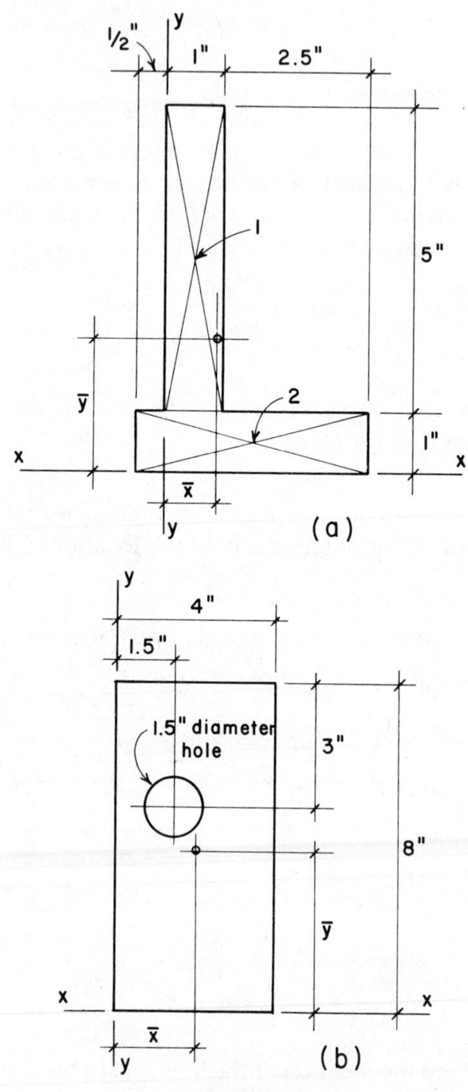

FIGURE E.18

is considered as a positive area and the hole as a negative area. Thus

$$\text{total } A = A_1 + A_2 = (4 \times 8) - \frac{\pi(1.5)^2}{4}$$

$$= 32 - 1.77 = 30.23 \text{ in.}^2$$

$$\Sigma M_y = (A_1 \times 2) - (A_2 \times 1.5) = 64 - 2.66 = 61.34 \text{ in.}^3$$

$$\Sigma M_x = (A_1 \times 4) - (A_2 \times 5) = 128 - 8.85 = 119.15 \text{ in.}^3$$

$$\bar{x} = \frac{\Sigma M_y}{A} = \frac{61.34}{30.23} = 2.02 \text{ in.}$$

$$\bar{y} = \frac{\Sigma M_x}{A} = \frac{119.15}{30.23} = 3.94 \text{ in.}$$

Moment of Inertia

In certain derivations of relationships involving the behavior of structures a mathematical expression occurs that is related to the geometry of a physical shape. If the shape is a planar area, the expression is of the form

$$\int x^2 \, da$$

where x is the distance of an incremental area da from some fixed reference axis in the plane of the area.

This expression is similar to the one for the moment of an area, except for the second power of x. For this reason it is sometimes called the second moment of an area. The more common name given to it in structural engineering is the *moment of inertia*. The symbol commonly used for this property is I.

For the area shown in Fig. E.19*a*:

$I_x = \int y^2 \, da$ = moment of inertia of the area with respect to the x-axis

$I_y = \int x^2 \, da$ = moment of inertia of the area with respect to the y-axis

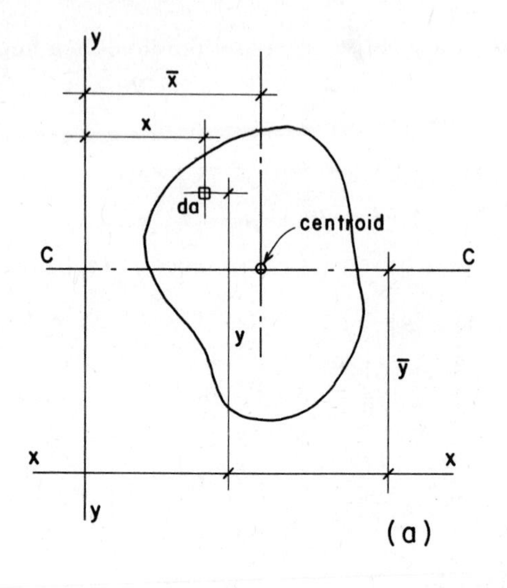
y
$\overline{x}$
x
da
centroid
C
C
y
$\overline{y}$
x
x
y
(a)

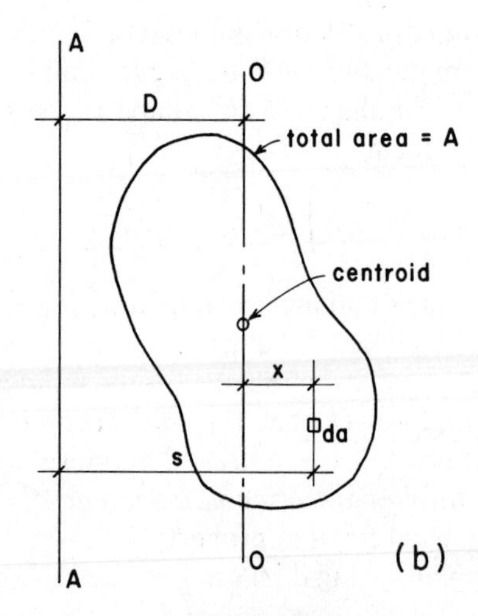
A
O
D
total area = A
centroid
x
da
s
O
A
(b)

FIGURE E.19

$\bar{I}_x = \int (y - \bar{y})^2 \, da$ = moment of inertia of the area with respect to a centroidal axis parallel to the x-axis

For structural investigations, the most used properties of this kind are the values of moment of inertia about centroidal axes. Centroidal I values for common geometric shapes are given in Fig. E.20.

Parallel Axis Theorem

A relationship of use in determining the moment of inertia of composite shapes is the parallel axis theorem. The derivation of this relationship is as follows. Referring to Fig. E.19*b*, we have

$$I_{A\text{–}A} = \int s^2 \, da = \int (D + x)^2 \, da = \int (D^2 + 2Dx + x^2) \, da$$
$$= \int D^2 \, da + \int 2Dx \, da + \int x^2 \, da$$

We note:

$\int D^2 \, da = D^2 \int da = A \, D^2$

$\int 2Dx \, da = 0$ since $\int x \, da$ = moment of area about its centroid, which by definition is zero

$\int x^2 \, da = \bar{I}_{O\text{–}O}$ = moment of inertia of the area about the centroidal axis parallel to A–A

and therefore

$$I_{A\text{–}A} = \bar{I}_{O\text{–}O} + AD^2$$

or, stated differently,

$$\bar{I}_{O\text{–}O} = I_{A\text{–}A} - AD^2$$

The main use of the parallel axis theorem is in the transferring of moments of inertia from one axis to another. This is used in finding the moment of inertia of composite shapes for which moments of inertia of component units of the area are determinable. The following example demonstrates this process.

Example 3. Find the moment of inertia of the area shown in Fig. E.21*a* with respect to the centroidal axis that is parallel to the base of the shape.

Solution. The axis through the base is designated *A–A* in the figure and the designated centroidal axis is *C–C*. As the location

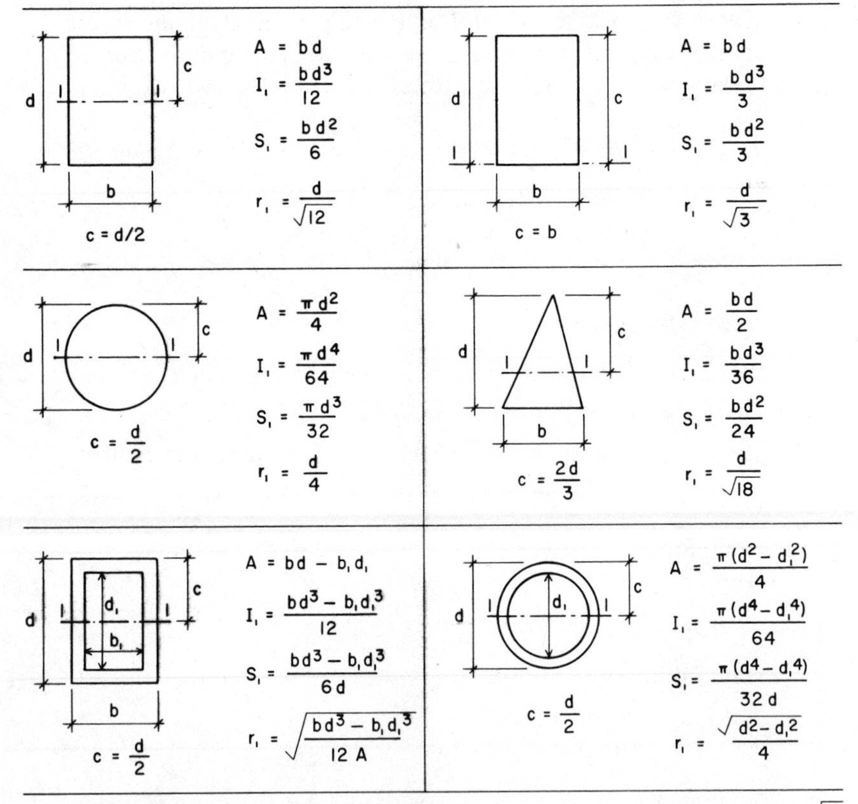

FIGURE E.20 Properties of geometric shapes.

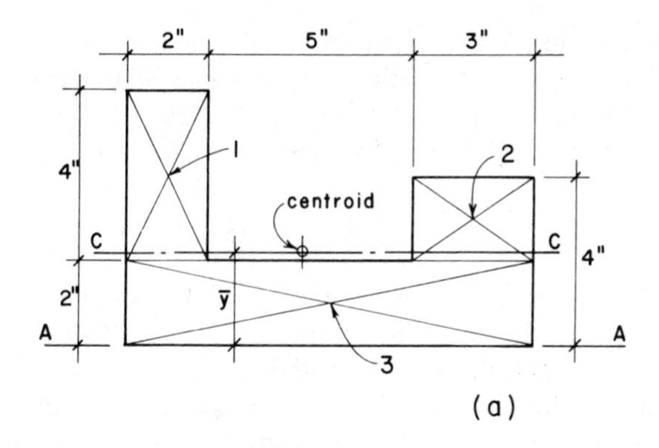

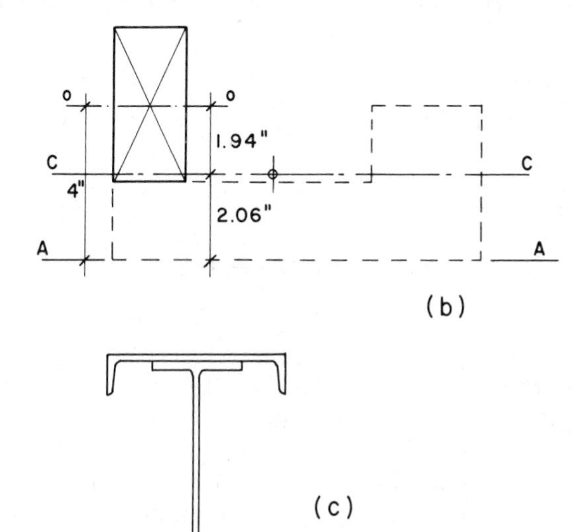

FIGURE E.21

of the centroid is not given, the first step is to find its location—in this case only with respect to the *A–A* axis. The process for this is demonstrated in Examples 1 and 2. Thus

$$\bar{y} = \frac{(A_1 \times 4) + (A_2 \times 3) + (A_3 \times 1)}{A_1 + A_2 + A_3} = 2.06 \text{ in.}$$

With the location of the centroid known, the moment of inertia of the individual parts about the $C–C$ axis can be determined with the parallel axis theorem. Referring to Fig. E.21*b*, the moment of inertia of part 1 is found as follows:

$$I_{C\text{-}C} = I_{O\text{-}O} + AD^2$$

in which $I_{C\text{-}C}$ is the moment of inertia of the component area of part 1; $I_{O\text{-}O}$ is the moment of inertia of part 1 about its own centroidal axis parallel to $C–C$; and AD^2 is the area of part 1 times the square of the distance from its centroid to the centroid of the whole shape. For the simple rectangular shape of part 1, the moment of inertia about its own centroid is found as (see Fig. E.20)

$$I_{O\text{-}O} = \frac{bd^3}{12} = \frac{2(4)^3}{12} = 10.67 \text{ in.}^4$$

Then for part 1,

$$I_{C\text{-}C} = 10.67 + 8(1.94)^2 = 10.67 + 30.11$$
$$= 40.78 \text{ in.}^4$$

and using similar procedures for the other parts, the moment of inertia of the whole figure about the axis $C–C$ is found to be 77.22 in.^{4}.

A procedure similar to that shown in the example can also be used for composite sections such as that shown in Fig. E.21*c*, consisting of a combination of two standard structural steel rolled shapes. Although the component units are more complex in this case, the areas, the locations of the centroids, and the centroidal moments of inertia are all given for such shapes in tables of properties in the AISC Manual.

Radius of Gyration

This property of a cross section is related to the design of compression members. Just as the section modulus is a measure of the resistance of a beam section to bending, the radius of gyration (which is also related to the size and shape of the cross section) is an index of the stiffness of a structural section when used as a column or other compression member. The radius of gyration is found from the formula

$$r = \sqrt{\frac{I}{A}}$$

and is expressed in inches, since the moment of inertia is in inches to the fourth power and the cross-sectional area is in square inches.

If a section is symmetrical about both major axes, the moment of inertia, and consequently the radius of gyration, is the same for each axis. But most common sections, particularly steel columns, are not symmetrical about the two major axes, and *in the design of columns the least moment of inertia, and therefore the least radius of gyration, is the one used in computations*. By *least* we mean the smallest in magnitude. Note in Table A.5 that the least radius of gyration of angle sections occurs about the *Z–Z axes*.

Example. Verify the tabulated values for radii of gyration for a W 12 × 58, as given in Table A.1.

Solution. The table shows the area of this section to be 17.0 in.^{2} [10968 mm^2] and I with respect to the X–X axis to be 475 in.^{4} [197.7 × 10^6 mm^4]. Then

$$r_x = \sqrt{\frac{I}{A}} = \sqrt{\frac{475}{17}} = \sqrt{27.94} = 5.29 \text{ in. } [134.3 \text{ mm}]$$

The table value for I with respect to the Y–Y axis is 107 in.^{4} [44.53 × 10^6 mm^4]. Therefore

$$r_y = \sqrt{\frac{I}{A}} = \sqrt{\frac{107}{17}} = \sqrt{6.29} = 2.51 \text{ in. } [63.7 \text{ mm}]$$

Compare these with the values given in Table A.1. It may be noted that there is a minor discrepancy in the value for r_x, having to do with the manner of rounding off for the last digit.

E.9 BENDING STRESS

When a linear structural member is subjected to a bending moment that lies in a plane parallel to the longitudinal axis of the member, the effect is called *bending*. Such a situation is shown in Fig. E.22*a* and we observe the following regarding its actions.

1. The bending tends to cause a curling up or curving of the member.
2. The curving indicates that the material on one side of the member is being lengthened due to tension, while the material on the opposite side is being shortened due to compression.
3. Due to the reversal of stress from side to side, there will be some point of transition at which the stress is zero on the member's cross section.

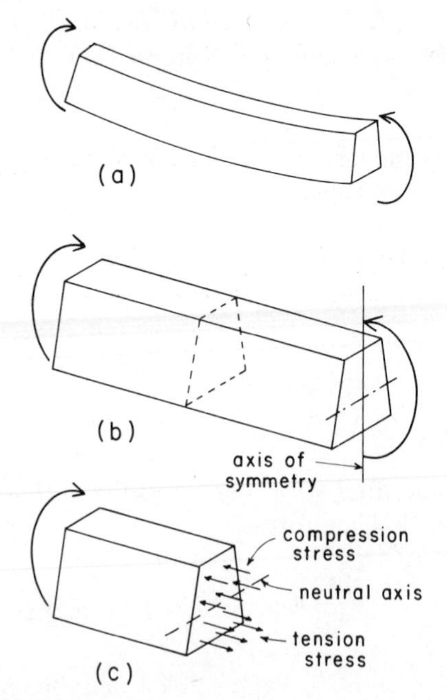

FIGURE E.22 Development of bending stress.

4. Internal bending resistance at a cross section is developed by an internal force couple, produced by the resultants of the tension and compression stresses.

If the member being bent has a symmetrical cross section and the moment exists in the plane containing the axis of symmetry of the section (see Fig. E.22*b*), we may make the following observations:

1. The neutral stress points referred to in item 3 above will lie on an axis through the centroid of the section and perpendicular to the axis containing the bending moment. This axis at which no stress occurs is called the *neutral axis*.
2. If the material of the member is homogeneous, isotropic, and elastic (basically meaning of a single elastic material), planar cross sections of the member at right angles to its longitudinal axis will retain their planar form during bending. As the beam curves, these plane sections will rotate about their neutral axes.
3. The radius of curvature of the member at any point along its length (assuming an initial straight condition) will be inversely proportional to the bending moment at that point.

The observations just made can be verified by tests, and because of them, we can make the following derivations in terms of the stresses and strains in the member. We first observe that due to the plane section observation, the strain at any point is proportional to its distance from the neutral axis. Accepting this, if the material is elastic—meaning that stress is proportional to strain—and the stress is below the material's failure limit, the stress at any point is proportional to the distance of the point from the neutral axis. Thus (see Fig. E.23)

$$\frac{\varepsilon_y}{y} = K_1 \quad \text{and} \quad \frac{f_y}{y} = K_2$$

The total internal resisting moment at a section may be determined by a summation of the increments of moment resistance

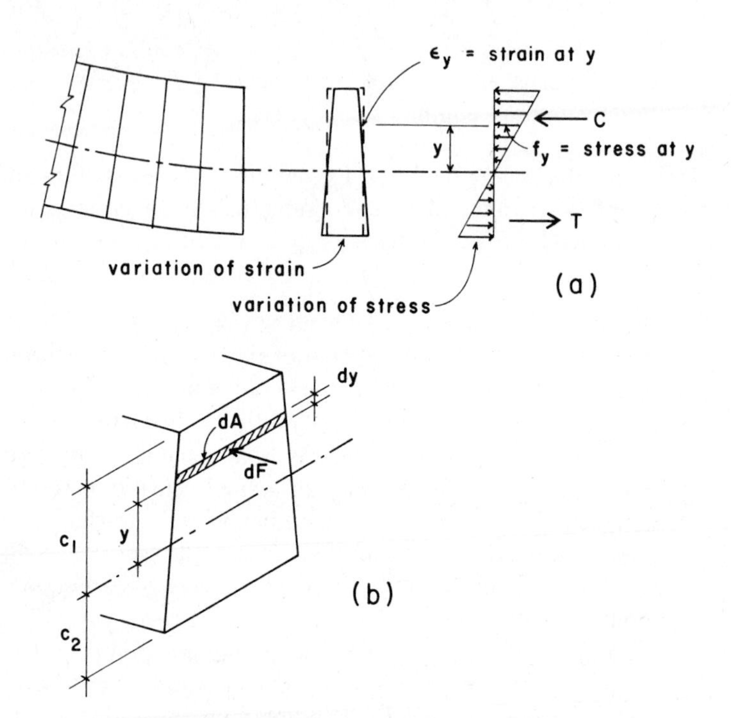

FIGURE E.23 Bending stress on a cross section.

consisting of the effects of stresses on unit areas of the cross section. If we use a unit area that consists of a slice dy thick and y distance from the neutral axis (see Fig. E.23*b*), we can assume the stress to be a constant on the unit area. Thus the increment of internal force developed on the unit area is the product of the stress times the unit area, or

$$dF = f_y \, dA$$

and the unit moment developed by this unit force is

$$dM = dF \, y = (f_y \, dA)y$$

Using the calculus, the total resisting moment of the cross section can be expressed as

$$M = \int_{c_2}^{c_1} dM = \int_{c_2}^{c_1} (f_y \, dA)y = \int_{c_2}^{c_1} f_y y \, dA$$

Since f_y/y is a constant, we can make the following transformation:

$$M = \int_{c_2}^{c_1} \frac{f_y}{y} y^2 \, dA = \frac{f_y}{y} \int_{c_2}^{c_1} y^2 \, dA$$

Astute students of geometry and calculus will recognize that the expression following the integral sign in this formula is the basic form for indication of the second moment of the area (or moment of inertia; see Sec. E.8) of the section about its neutral centroidal axis. We designate this basic geometric property as I and thus express the formula more simply as

$$M = \frac{f_y}{y} I$$

or, transforming it into an expression for stress, we have

$$f_y = \frac{My}{I}$$

which is the general formula for bending stress.

Referring to Fig. E.23, we note that the maximum stress in compression will be obtained when $y = c_1$ and the maximum in tension when $y = c_2$. Further, if the section is one with symmetry about the neutral axis, c_1 will be equal to c_2. With these considerations, a simpler expression for maximum bending stress that is used commonly takes the form

$$f = \frac{Mc}{I}$$

An additional simplification can be made by use of a single term for the combination of I and c. This term is called the section modulus, defined as

$$S = \frac{I}{c}$$

Substituting this in the stress formula reduces it to

$$f = \frac{M}{S}$$

For beam design purposes, the formula is transposed to

$$S = \frac{M}{f}$$

When designing a beam, once the required resisting moment is determined and the material to be used is established (yielding a value for the limit of stress), a single required property for the beam—S—can be found for the selection of the beam.

The following examples illustrate simple problems using the formulas for bending stress.

Example 1. A beam has a T-shaped section as shown in Fig. E.24. A bending moment occurs in the axis of symmetry of the section, producing compression in the upper portion and tension

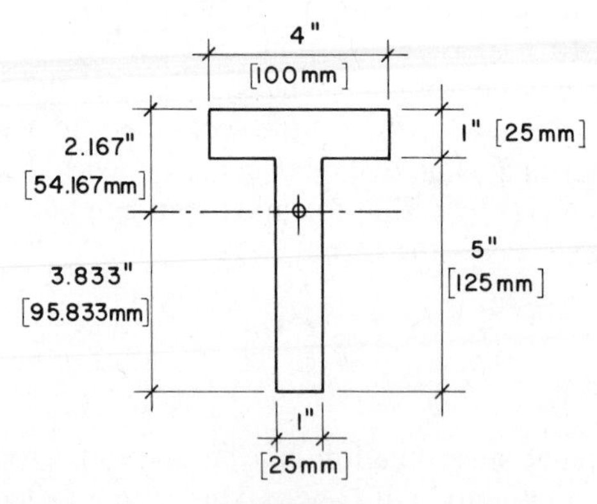

FIGURE E.24

in the lower portion of the section. If the bending moment is 10 kip-ft [14 kN-m], and the moment of inertia of the section about the neutral axis is 30.71 in.^{4} [12.02×10^6 mm^4], find the values for maximum tension and compression stress due to bending.

Solution. For compression at the top of the section we use the distance from the neutral axis of 2.167 in. and the stress formula in the form

$$f_y = \frac{My}{I} = \frac{[10(12)]2.167}{30.71} = 8.47 \text{ ksi } [63.1 \text{ MPa}]$$

Note that the moment in kip-ft must be changed to inch units, as the other values use inches.

For tension stress at the bottom we use the distance of 3.833 in.; thus

$$f_y = \frac{My}{I} = \frac{[10(12)]3.833}{30.71} = 14.98 \text{ ksi } [111.6 \text{ MPa}]$$

Example 2. For the beam section in Fig. E.24, what is the value for S (section modulus) that indicates the moment resistance of the section based on maximum bending stress?

Solution. If the section is limited by the maximum stress, we must use the greatest distance from the neutral axis to any edge, or 3.833 in. for the section given. Thus

$$S = \frac{I}{c} = \frac{30.71}{3.833} = 8.01 \text{ in.}^3 \; [125.4 \times 10^3 \text{ mm}^3]$$

E.10 SHEAR STRESS IN BEAMS

Development of shear stress in a beam tends to produce lateral deformation. This may be visualized by considering the beam to consist of a layer of loose boards. Under the beam loading on a simple span, the boards tend to slide over each other, taking the

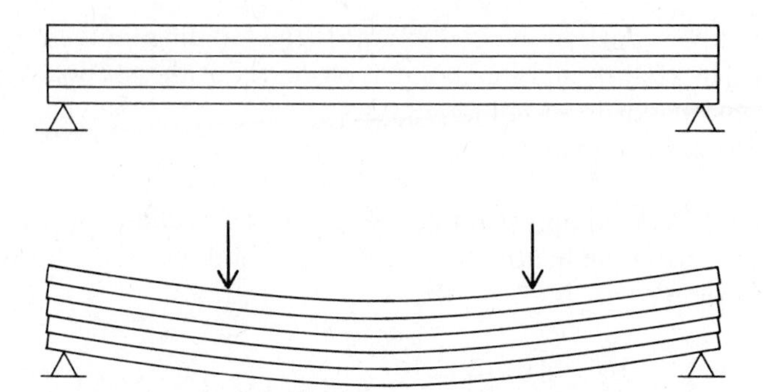

FIGURE E.25 Development of shear in beams.

form shown in Fig. E.25. This type of deformation also tends to occur in a solid beam, but is resisted by the development of horizontal shearing stresses.

Shear stresses in beams are not distributed evenly over cross sections of the beam as was assumed for the case of simple direct shear (see Sec. E.2). From observations of tested beams and derivations considering the equilibrium of beam segments under combined shear and moment actions, the following expression has been obtained for shear stress in a beam:

$$f_v = \frac{VQ}{Ib}$$

where V = shear force at the beam section,

Q = moment about the neutral axis of the area of the section between the point of stress and the edge of the section,

I = moment of inertia of the section with respect to the neutral (centroidal) axis,

b = width of the section at the point of stress.

It may be observed from this formula that the maximum value for Q will be obtained at the neutral axis of the section, and that the stress will be zero at the edges of the section farthest from the neutral axis. The form of shear stress distribution for various geometric shapes of beam sections is shown in Fig. E.26.

The following examples illustrate the use of this stress relationship.

Example 1. A beam section with depth of 8 in. and width of 4 in. [200 and 100 mm] sustain a shear force of 4 kips [18 kN]. Find the maximum shear stress.

Solution. For the rectangular section the moment of inertia about the centroidal axis is

$$I = \frac{bd^3}{12} = \frac{4(8)^3}{12} = 170.7 \text{ in.}^4 \ [67 \times 10^6 \text{ mm}^4]$$

The static moment (Q) is the product of the area a' and its centroidal distance from the axis of the section ($\bar{y}$) as shown in Fig. E.27*b*. We thus compute Q as

$$Q = a'\bar{y} = [4(4)]2 = 32 \text{ in.}^3 \ [500 \times 10^3 \text{ mm}^3]$$

The maximum shear stress at the neutral axis is thus

$$f_v = \frac{VQ}{Ib} = \frac{4000(32)}{170.7(4)} = 187.5 \text{ psi} \ [1.34 \text{ MPa}]$$

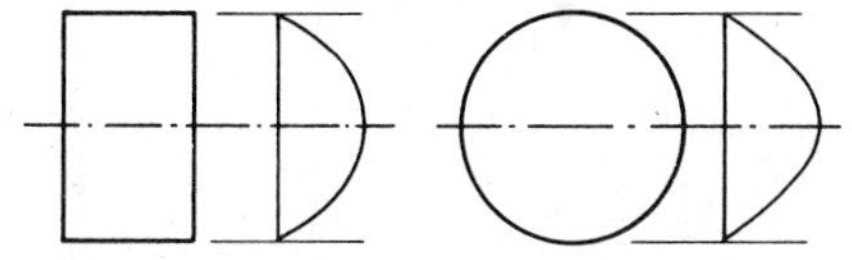

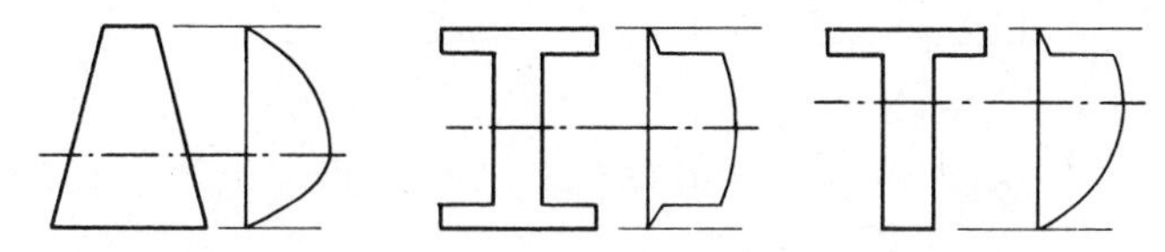

FIGURE E.26 Variation of shear stress on beam cross sections.

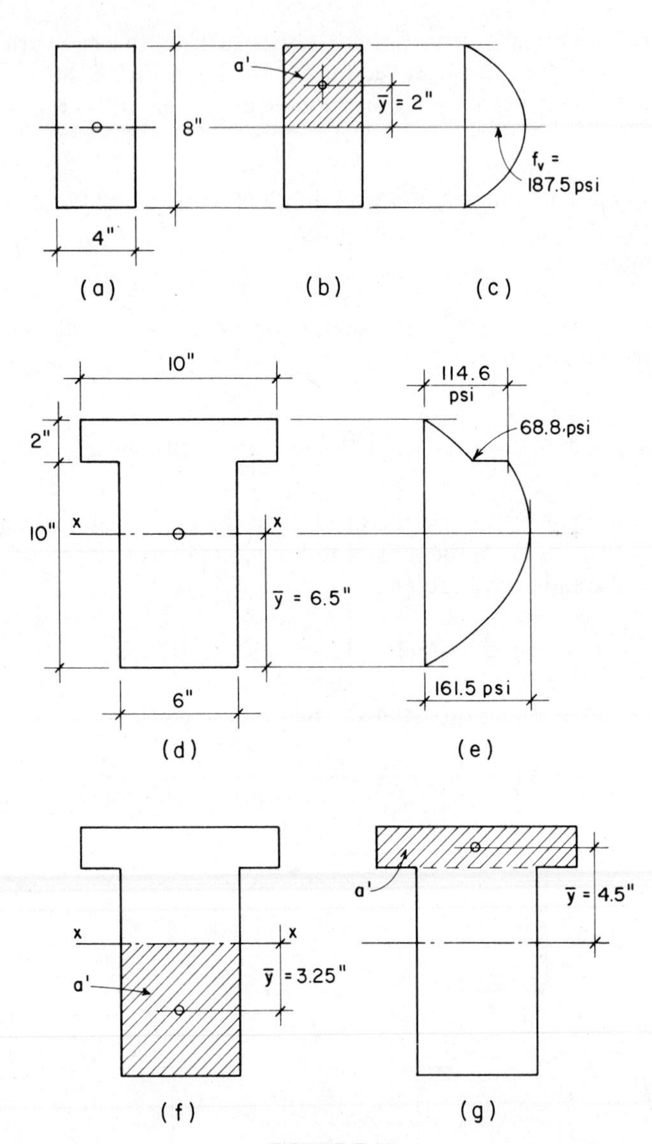

a'
$\overline{y}$ = 2"
8"
4"
f_v =
187.5 psi
(a)
(b)
(c)
10"
114.6
psi
68.8 psi
2"
10"
x
x
$\overline{y}$ = 6.5"
6"
161.5 psi
(d)
(e)
x
x
a'
$\overline{y}$ = 3.25"
a'
$\overline{y}$ = 4.5"
(f)
(g)

FIGURE E.27

TABLE E.2 Computation of Properties for the Section in Example 2

Part	Area (in.^{2})	y from Bottom	Ay_1	I_0	Ay_2^2	I_x (in.^{4})
1	6(12) = 72	6	432	$\frac{6(12)^3}{12} = 864$	$72(0.5)^2 = 18$	882
2	2(2)(2) = 8	11	88	$2(2)(2)^3 = 2.7$	$8(4.5)^2 = 162$	164.7
Σ	80 in.^{2} [50 × 10^3 mm^2]		520 in.^{3} [8.125 × 10^6 mm^3]			1046.7 in.^{4} [4.088 × 10^8 mm^4]

$$y_x = \frac{520}{80} = 6.5 \text{ in. [162.5 mm] (see Fig. E.27}d\text{).}$$

Example 2. A beam with the T-section shown in Fig. E.27*d* is subjected to a shear of 8 kips [36 kN]. Find the maximum shear stress and the shear stress at the location of the juncture of the web and the flange of the T.

Solution. For this section the location of the centroid and the determination of the moment of inertia about the centroidal axis must be accomplished using processes explained in Sec. E.8. This work is summarized in Table E.2. For determination of the maximum shear stress at the neutral axis (centroidal axis x–x, as shown in Fig. E.27*d*) we find Q using the bottom portion of the web, as shown in Fig. E.27*f*. Thus

$$Q = a'\bar{y} = [6.5(6)]3.25 = 126.75 \text{ in.}^3 \ [1.98 \times 10^6 \text{ mm}^3]$$

and the maximum stress at the neutral axis is thus

$$f_v = \frac{VQ}{Ib} = \frac{8000(126.75)}{1046.7(6)} = 161.5 \text{ psi} \ [1160 \text{ kPa}]$$

For the stress at the juncture of the web and flange we use the area shown in Fig. E.27*g* for Q; thus

$$Q = [2(10)]4.5 = 90 \text{ in.}^3 \ [1.41 \times 10^6 \text{ mm}^3]$$

and the two shear stresses at this location, as displayed in Fig. E.27*e*, are

$$f_{v1} = \frac{8000(90)}{1046.7(6)} = 114.6 \text{ psi [828 kPa]}$$

$$f_{v2} = \frac{8000(90)}{1046.7(10)} = 68.8 \text{ psi [497 kPa]}$$

In most design situations it is not necessary to use the complex form of the general expression for beam shear. In wood structures the beam sections are almost always of simple rectangular shape. For this shape we can make the following simplification.

$$I = \frac{bd^3}{12}, \qquad Q = (b)\frac{d}{2}\frac{d}{4} = \frac{bd^2}{8}$$

$$f_v = \frac{VQ}{Ib} = \frac{V(bd^2/8)}{(bd^3/12)b} = \frac{3}{2}\frac{V}{bd}$$

For steel beams—which are mostly I-shaped cross sections—the shear is taken almost entirely by the web. (See shear stress distribution for the I-shape in Fig. E.26.) Since the stress distribution in the web is so close to uniform, it is considered adequate to use a simplified computation of the form

$$f_v = \frac{V}{dt_w}$$

in which d is the overall beam depth and t_w is the thickness of the beam web.

There are situations in which the general form of the shear stress formula must be used. Most of these involve the use of complex geometries for beam cross sections which commonly occur in beams of prestressed concrete and the built-up, compound sections of steel or timber + plywood.

APPENDIX F

STUDY AIDS

The materials in this section are given to provide the reader with some means to measure his or her comprehension and skill development with regard to the book presentations. It is recommended that upon completion of an individual chapter, the materials given here for that chapter be used to find out what has been accomplished. Answers to the questions are given at the end of this section.

Words and Terms

Using the text of each indicated chapter together with the Index, review the meanings of the following words and terms.

Chapter 1

AISC Manual
Cold-forming
Formed steel

Hot-rolling
Miscellaneous metals
Structural steel

Chapter 2

Cold-formed products
Ductility
Fabricated components
Installed cost
Nominal size (depth, diameter, etc.)
Plastic stress-strain behavior
Rolled shapes: W, M, S, C, MC, L, T
Yield: stress, strength, point

Chapter 4

Deck–beam–girder system
Framing plan
Spacing of framing

Chapter 5

Allowable stress design
Compact section
Crippling of beam web
Equivalent tabular load
Lateral unsupported length
Least-weight member
Plastic design
Superimposed load

Chapter 6

Bending factors for columns (B_x, B_y)
Biaxial bracing conditions
Built-up sections

Effective column length
Slenderness
Slenderness ratio
Strut

Chapter 7

Bent
Braced frame
Diaphragm
Eccentric bracing
Knee-brace
Moment-resistive connection
Rigid frame
X-brace

Chapter 8

Composite construction (AISC term)
Composite deck
Composite structural element (general)
Flitched beam
Manufactured system

Chapter 9

Lateral bracing for truss
Chord
Panel (unit, point)
Purlin
Web member

Chapter 10

Bolted Connections:
Bearing-type connection
Double shear

Edge distance
Effective net area
Framed beam connection
Friction-type connection
Gage (of angle)
Gusset plate
High-strength bolt
Net cross-section area (general)
Pitch
Single shear
Tearing
Unfinished bolt
Upset end

Welded Connections:
Base metal
Penetration of welds: complete, partial
Types of welds: butt, tee, lap, groove, fillet, plug, slot
Weld: throat, root, reinforcement, boxing of
Welding work done in shop; field

Chapter 11

Building code: model, legally enforced, regulations
Dead load
Design wind pressure
Drift
Lateral load
Live load
Live-load reduction
Overturning
Uplift
Wind speed

Chapter 13

Ductility
Plastic: behavior, hinge, moment, range, section modulus
Strain hardening

Appendix E

Allowable stress
Beam
Bending
Centroid
Component of a force
Composition
Compression
Cut section
Deflection
Equilibrium
Force
Force, types: static, dynamic
Internal force (five types)
Load
Load, types: dead, live
Modulus of elasticity
Moment
Moment arm
Moment of inertia
Principal axes: major, minor
Reaction
Resolution
Resultant
Service load
Shear

Simple beam
Stiffness
Strain
Strength method
Stress
Stress, type: direct, shear
Tension
Torsion
Vector
Working stress method

GENERAL QUESTIONS

(*Note:* Answers follow the last question.)

Chapter 1

1. What form of construction is described by the following terms?
  (a) structural steel.
  (b) Formed steel.
  (c) Miscellaneous metals.

Chapter 2

1. Why is the yield point generally of more concern than the ultimate strength of the steel for most structural applications?

2. Why is the specified depth of a wide-flange shape referred to as a nominal dimension?

3. Cold-formed steel products are usually formed from what basic rolled product?

Chapter 3

1. How does the constant value of modulus of elasticity for all grades of steel limit the potentiality for use of higher grades of steel for structural applications?

2. What is the usual method for developing the fire resistance of steel structures?

3. Why is it not possible to predict the finished cost of a steel structure simply on the basis of the unit cost of the steel as raw material?

4. Steel is usually considered to be vulnerable in exposed situations. What are the primary concerns for this?

Chapter 4

1. Spacing of beams in a deck–beam–girder system must relate to the design and layout of what other elements of the structure?

2. How do panel point locations on a truss relate to layout of framing supported by the truss?

Chapter 5

1. What single property of a beam cross section is most predictive of bending strength?

2. What is significant about the properties identified as L_c and L_u for a rolled W shape?

3. For rolled shapes of A36 steel used as beams, what makes it possible to say that all beams with the same depth will deflect the same at their limiting safe loads on a given span?

4. What is the significance of a compact section?

5. What is the most common means for dealing with torsional effects on wide-flange beams?

6. What basis structural action is represented by web crippling in steel beams?
7. Steel decks serve primarily to support gravity loads. What other structural functions are also frequently required of the decks?

Chapter 6

1. For evaluation of simple axial compression capacity, what are the significant properties of a column cross section?
2. What generally exists to make it necessary to check buckling conditions on both axes of a W-shape steel column?

Chapter 7

1. What essential interaction is required between members of a rigid frame?
2. Why are columns in rigid frames usually larger in cross section than those in simple post and beam frames?
3. With regard to effects on beams and columns in the frame, what is the difference between X-bracing and K-bracing?
4. What form of bracing for a steel frame usually produces the most deformation-resistive structure?

Chapter 8

1. What is generally special about a manufactured system?
2. Why is steel often a component in most composite construction?

Chapter 9

1. What is the essential geometric character of a truss?
2. What truss members usually do the most work in developing the resistance of the truss to: **(a)** shear? **(b)** bending?

3. What signifies a condition of equilibrium in a force polygon for a set of planar, concurrent forces?
4. What usually produces stress reversal in a truss member?
5. What type of loading usually produces combined axial force and bending in a truss member?

Chapter 10

1. When so-called "high-strength" steel bolts are used for a connection between steel members, what basic action develops the initial load resistance in the joint?
2. Tearing in a bolted connection is resisted by what combination of stress developments?
3. Other than spacing and edge distances, what basic dimension limits the number of bolts that can be used in a framed beam connection?
4. Why is it generally not desired to have supporting steel beams of the same depth as the beams they support?
5. What is the significance of the throat dimension in a fillet weld?
6. What structural advantage is gained by boxing of welds in a joint?

Chapter 11

1. What is the reason for tabulating dead and live loads separately for structural design?
2. What is the usual basis for live-load reduction?
3. Why are the lateral (horizontal) effects, rather than the vertical effects, of wind and earthquakes generally more critical for building design?
4. Why are wind surface pressures greater on tall buildings?

Chapter 13

1. What basic property of the material is required for the development of a plastic hinge or plastic moment?

2. Development of the plastic section modulus involves the assumption of what stress condition for a beam cross section?

ANSWERS TO THE GENERAL QUESTIONS

Chapter 1

1. **(a)** Major structural frames, assembled from rolled products.
   (b) Elements consisting of sheet steel that has been folded, corrugated, stamped, and so on.
   (c) Generally, the nonstructural steel elements of the building construction.

Chapter 2

1. Because of the magnitude of deformations that occur in the plastic (ductile) range, and the resulting permanent deformations, the yield stress is the usual practical limit for useful strength of steel.

2. The actual depth dimensions vary from the nominal depth in most shapes.

3. Steel sheet.

Chapter 3

1. Many structural actions are limited by some form of buckling or by deflections, both of which are affected by stiffness (as measured by the modulus of elasticity), not by the strength of the material. Increasing allowable stresses (for higher grades of steel) makes buckling and deflection even more critical.

2. By protecting the steel with fire-resistive, thermal insulation.

3. Installed cost includes many other components, including labor, processing, and transportation.

4. Fire and corrosion (mainly rust).

Chapter 4

1. Span of the deck, spacing of columns, and panel points of the truss if the girder is a truss.

2. Framing should coincide with panel points (chord joints) of the truss to avoid bending in chords.

Chapter 5

1. The section modulus.

2. They represent staged points of capacity reduction due to length between points of lateral bracing.

3. If safe load is limited by bending stress, and the maximum allowable bending stress is the same for all the beams, the strain producing deflection will also be the same.

4. The general properties of slenderness of the web and flanges of the section permit the development of the maximum bending stress for the member.

5. Providing bracing that is adequate to prevent the torsional effect.

6. Buckling of the thin web due to compression in the plane of the web.

7. Bracing of beams against lateral buckling and serving as a horizontal diaphragm for resistance of lateral loads.

Chapter 6

1. Area and radius of gyration.

2. Different unbraced lengths, and/or different K-factors on the two axes.

Chapter 7

1. Transfer of bending moments through the joints.

2. They must develop bending as well as axial compression.

3. X-bracing develops only truss actions with the beams and columns, producing axial tension or compression forces in them. K-bracing produces truss actions, but also induces bending and shear in the columns and beams.

4. Vertical diaphragms (shear walls).

Chapter 8

1. It is a predesigned, package system, with individual components developed in coordinated sets.

2. Because of its high strength and stiffness, it can absorb force in very high proportion to its total mass.

Chapter 9

1. Triangular arrangement of its members.

2. **(a)** The web, or interior, members.
(b) The chords.

3. The polygon closes, with the head of the last force vector touching the tail of the first.

4. A different combination of loads on the truss.

5. Load not applied at panel points (truss joints).

Chapter 10

1. Friction between the connected parts.

2. Tension and shear.

3. Depth of the beam.

4. Both flanges of the supported beams must be cut back to achieve standard framed connections, considerably reducing the shear-resisting web of the supported beam.

5. It defines the critical cross section of the weld for the development of shear stress, which is the form of resistance used to establish the weld strength.

6. Increased resistance to tearing of the welds due to twisting actions on the joint.

Chapter 11

1. In order to deal with them as separate loadings for some computations (e.g., live-load deflection) and for addition with wind, seismic, and other loadings to determine the critical combinations of loading.

2. Statistical likelihood that the total load area will be totally loaded decreases in proportion to the amount of the load area.

3. The structure is routinely designed for the vertical orientation of gravity loads.

4. The higher above ground that the surface exposed to wind is, the less becomes the effects of ground surface drag or surrounding sheltering by trees, hills, or other buildings.

Chapter 13

1. Significant ductility.

2. Development of yield stress on the entire cross section.

ANSWERS TO EXERCISE PROBLEMS

Chapter 5

5.2.A. M 14 × 18 is lightest shape, W 10 × 19 is lightest W

5.2.B. W 12 × 22

5.2.C. M 14 × 18

5.2.D. W 24 × 55

5.2.E. W 12 × 26 (with U.S. data); W 14 × 22 (with SI data)

5.2.F. W 16 × 31

5.2.G. M 14 × 18

5.2.H. W 12 × 22 or W 14 × 22

5.2.I. W 14 × 22

5.2.J. W 16 × 26

5.3.A. 168.3 kips

5.3.B. 51.1 kips

5.3.C. 37.1 kips

5.4.A. 0.80 in.

5.4.B. 0.69 in.

5.4.C. 0.83 in.

5.5.A. (a) M 12 × 11.8, (b) W 8 × 15 is lightest in table; actually a W 5 × 19 will work

5.5.B. (a) W 16 × 26, (b) W 10 × 45

5.5.C. (a) W 21 × 50, (b) W 16 × 57

5.5.D. (a) W 12 × 19, (b) W 10 × 26

5.5.E. (a) W 18 × 35, (b) W 14 × 48

5.5.F. (a) W 24 × 76, (b) W 21 × 101

5.7.A. Beam is OK; actually W 18 × 76 is lightest choice

5.7.B. W 30 × 99 is lightest but deflects too much; required shape is W 30 × 108

5.10.A. R = 65.2 kips

5.10.B. P = 93.9 kips, stiffeners not required

5.11.A. B = 15 in., t = 1 in.

5.11.B. B = 17 in., t = $1\frac{1}{8}$ in.

5.12.A. 28K7

5.12.B. 30K7

5.12.C. (a) 22K4, (b) 18K7

5.12.D. (a) 20K3, (b) 14K6

5.13.A. WR20

5.13.B. WR18

5.13.C. IR22 or WR22

Chapter 6

6.3.A. 95.05

6.3.B. 119.2

6.5.A. 430 kips [1912 kN]

6.5.B. 278 kips [1237 kN]

6.5.C. 3612 kips [16065 kN]

6.5.D. 375 kips [1669 kN]

6.6.A. W 8 × 31

6.6.B. W 12 × 58

6.6.C. W 12 × 79

6.7.A. 4 in.

6.7.B. 5 in.

6.7.C. 6 in.

6.7.D. 8 in.

6.8.A. By formulas in Sec. 6.4, C_c = 111.6 and FS = 1.91; then F_a = 14.5 ksi and allowable load is 73.6 kips; table says 74 kips

6.8.B. TS 4 × 4 × $\frac{1}{4}$

6.9.A. From Table 6.6, 78 kips

6.9.B. Lightest choices: $4 \times 3 \times \frac{5}{16}$ or $3\frac{1}{2} \times 2\frac{1}{2} \times \frac{3}{8}$

6.10.A. W 12 × 58

6.10.B. W 14 × 120 (maybe W 14 × 109)

6.12.A. Possibility: 15 in. × 16 in. × $1\frac{1}{4}$ in.

6.12.B. Possibility: 12 in. × 14 in. × $1\frac{1}{8}$ in.

Chapter 8

8.3.A. 21.2 kips

Chapter 10

10.5.A. 6 bolts in two rows, $\frac{5}{8}$-in. thickness for 8-in. plates, 1-in. thickness for 10-in. plate

10.5.B. 7 bolts required but use 9 in three rows for symmetry, $\frac{5}{8}$-in. thickness for outer plates, $1\frac{1}{8}$-in. thickness for middle plate

10.7.A. Could use eight $\frac{3}{4}$-in. bolts, six $\frac{7}{8}$-in. bolts, five 1-in. bolts; most critical with $\frac{3}{4}$-in. bolts, but still OK for shear, bearing, and tearing

10.7.B. Could use seven $\frac{3}{4}$-in. bolts, five $\frac{7}{8}$-in. bolts, four 1-in. bolts; OK with $\frac{3}{4}$-in. bolts

10.7.C. Could use four $\frac{3}{4}$-in. bolts, three $\frac{7}{8}$-in. bolts; OK with either

10.13.A. Rounding off, use L_1 = 11 in., L_2 = 5 in.

10.13.B. Minimum: $4\frac{1}{2}$-in. long welds on each side

Chapter 13

13.9.A. W 16 × 31

13.9.B. W 14 × 26

INDEX

D

E

F

G

H

I

X

Y